FOOD PACKAGING

Innovations and Shelf-Life

Editor

Rui M.S. Cruz

MeditBio and Instituto Superior de Engenharia
Universidade do Algarve
Faro, Portugal

CRC Press
Taylor & Francis Group
Boca Raton London New York

CRC Press is an imprint of the
Taylor & Francis Group, an **informa** business

A SCIENCE PUBLISHERS BOOK

CRC Press
Taylor & Francis Group
6000 Broken Sound Parkway NW, Suite 300
Boca Raton, FL 33487-2742

First issued in paperback 2021

© 2020 by Taylor & Francis Group, LLC
CRC Press is an imprint of Taylor & Francis Group, an Informa business

No claim to original U.S. Government works

Version Date: 20191001

ISBN-13: 978-0-367-08574-2 (hbk)
ISBN-13: 978-1-03-208623-1 (pbk)

Visit the Taylor & Francis Web site at
http://www.taylorandfrancis.com

and the CRC Press Web site at
http://www.crcpress.com

Preface

FOOD PACKAGING: Innovations and Shelf-life reviews current developments in food packaging and their influence on food quality, preservation and shelf-life extension and provides sustainable and environmentally friendly packaging solutions.

The book is divided into 10 chapters, combining the effort of 35 experts from around the world. Topics such as active and intelligent packaging, development on new packaging systems from bio-based materials, micro and nano materials, packaging design, packaging production and characterization and interaction between packaging and food as well as regulatory aspects are discussed.

The book is an ideal reference source for government, industry, and academia professionals, food engineers and researchers from R&D laboratories working in the area of food science and technology and also for professionals from institutions related to food packaging.

I would like to acknowledge all authors for their invaluable contribution, essential to accomplish this project.

Finally, I would like to dedicate this work to my dear friend Professor Paula Cabral whose wisdom and friendship will perpetuate forever in my memory.

Rui M.S. Cruz

Contents

Protective Packaging for Light-sensitive Foods

María Gabriela Passaretti,[1,2] *Mario Daniel Ninago,*[3,4]
Marcelo Armando Villar[1,2] *and Olivia Valeria López*[1,5,*]

1. Introduction

Consumer expectations and interests in nutritional aspects and visual appearance, when purchasing food, are important factors that should be taken into account to improve the food industry (Gadioli et al., 2013). Currently, food packaging is strongly linked to material transparency, which must not hide the product but enable its visibility through the packaging (Eldesouky et al., 2015). Nevertheless, many food natural components as well as those added during their preparation are susceptible to light, causing the occurrence of photo-degradative and oxidative reactions (Duncan and Chang, 2012). Foods are exposed to several sources of light during their entire chain of production and commercialization. Some common light sources and their locations are (a) sunlight outdoors (storefronts and windows), (b) incandescent lamps (coolers and storage facilities) and (c) fluorescent lamps (processing areas, display cases and preparation zones). Foods are also exposed to other sources of light, such as germicidal lamps used in walk-in coolers and storage area rooms to reduce bacterial and mould counts as well as black lights used to detect the presence of insects, rodent excreta and other kinds of contamination. When light strikes a food packaging, it can be reflected off the material surface, absorbed by the packaging material, scattered and absorbed by the food and transmitted through the food. Particularly, light that is

[1] Planta Piloto de Ingeniería Química, PLAPIQUI (UNS-CONICET), Buenos Aires, Argentina.
[2] Departamento de Ingeniería Química, Universidad Nacional del Sur (UNS), Buenos Aires, Argentina.
[3] Facultad de Ciencias Aplicadas a la Industria, Universidad Nacional de Cuyo (FCAI-UN Cuyo), Mendoza, Argentina.
[4] Consejo Nacional de Investigaciones Científicas y Técnicas (CONICET), Buenos Aires, Argentina.
[5] Departamento de Química, Universidad Nacional del Sur (UNS), Buenos Aires, Argentina.
* Corresponding author: olivialopez@plapiqui.edu.ar

absorbed, mainly leads to functionality loss of food compounds (Duncan and Chang, 2012). In most solid foods, light only penetrates the outer layer and so deterioration occurs only at a superficial level. Meanwhile, in liquid foods, light penetration can be greater and their deterioration is more extensive.

A substance that absorbs light following a photochemical reaction is called photosensitizer (Hussmann Corporation, 2017). Molecules present in food formulations, which are more susceptible to photochemical reactions, are lipids, amino acids and nucleotides that make up fats, pigments, proteins and vitamins (Limbo et al., 2007). Among vitamins, the light-sensitive vitamins are A, B_{12}, D, K and E, as well as folic acid, pyridoxine and riboflavin and regarding pigments, anthocyanins, carotenoids, chlorophylls, myoglobin and haemoglobin are light-sensitive. Amino acids, tryptophan, phenylalanine, tyrosine, and histidine, that are present in foods, degrade during light exposure, while food fats, unsaturated fatty acids and phospholipids usually suffer oxidation due to light. In Fig. 1 are shown the most common photosensitizers present in food products.

Foods' light sensitivity depends on many factors, including the strength of the light source and type of light that it emits, besides light-source-food distance, exposure time, optical properties of packaging materials, oxygen concentration inside the packaging and storage temperature (Chiste et al., 2010; Duncan and Hannah, 2012; Mestdagh et al., 2005).

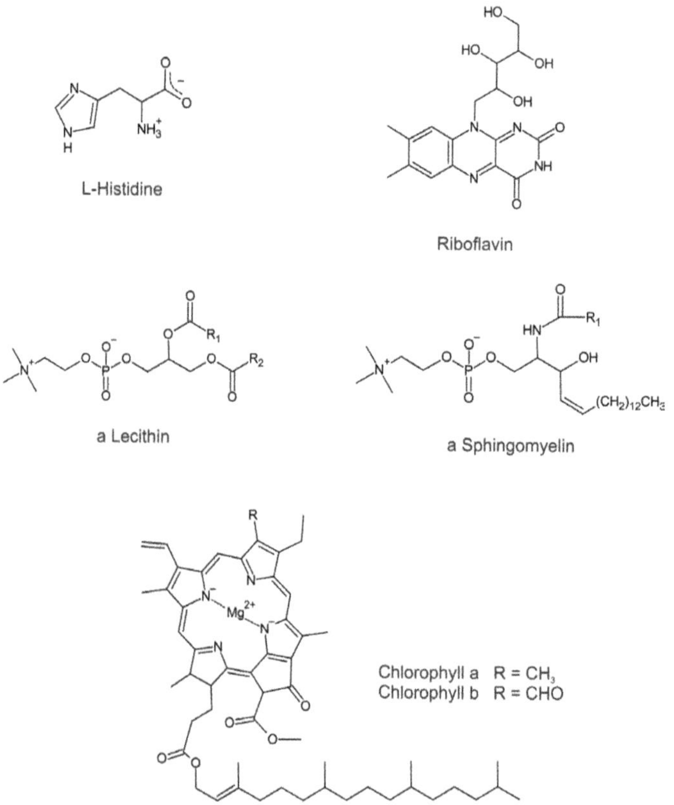

Fig. 1. Most common photosensitizers present in food products.

Changes in foods induced by photochemical reactions usually start when light is absorbed by a product component that will directly undergo chemical reaction or one food component that generates a product, which undergoes a reaction catalyzed by light. These undesirable reactions in foods cause nutritional loss and produce off-flavor, toxic by-products and colored compounds during product storage and marketing, making them less acceptable or unacceptable to consumers (Barrett et al., 2010).

Susceptibility of food compounds to light exposure is mainly attributed to the presence of electrons of conjugated bonds in their chemical structure, which require very little energy to trigger a reaction. Since all light emitted is a form of energy, the higher the input of energy the more rapid the potential degradation (Hussmann Corporation, 2017). Photosensitizers are mostly activated in the ultraviolet (UV) and the blue/green region of the visible spectrum and are shorter with more energetic wavelengths. Oxidation reactions can be carried out by either diradical triplet oxygen or non-radical singlet oxygen, which can be formed from triplet oxygen during photosensitized reactions (Min and Boff, 2002). Photochemical reactions are impossible to eliminate because foods are exposed to various light types and intensities throughout their supply chain processes: post-harvesting, transport, primary and/ or secondary processing, packaging and distribution. Once a food photosensitizer molecule absorbs light, the cascade of reactions cannot be avoided. Thus, it is relevant to obtain knowledge and comprehension about these photochemical reactions to give well-founded practical advice to the industry on how to minimize photo-oxidation through packaging and retail (Wold et al., 2009). In this way, the use of a packaging material that protects food from light exposure is one of the most effective means of preventing light-induced chemical spoilage.

Natural pigments, such as anthocyanins, carotenoids, betalains and chlorophylls have low light-stability and cause food discoloration (Cortez et al., 2017); particularly anthocyanins, which are stable under acidic conditions, but under normal processing and storage conditions, they transform to colorless compounds and subsequently to insoluble brown pigments (Arslan, 2015). In the case of carotenoids, the same extensive conjugated double-bond system, that makes these natural pigments powerful antioxidants, also makes them susceptible to oxidation (Boon et al., 2010). In accordance with Schwartz et al. (2017), due to the many double bonds in carotenoids structure, their oxidation may result in several products from isomerization to extensive molecule cleavage, decrease or loss in the provitamin activity. Besides, light has been shown to degrade betanin, a type of betalains, by around 15 percent in six days at 15°C (Kaimainen, 2014).

Photochemical reactions could liberate a wide range of volatile compounds, which are responsible for off-flavors and off-odors (Ball, 2006). For example, off-flavor in milk is the result of fats' oxidation to aldehydes and degradation of sulfur-containing amino acids (Brothersen et al., 2016). These reactions are facilitated when the product is exposed to near-UV and visible light from any source, such as sunshine, fluorescent light, or light-emitting diodes.

In some cases, food compounds' degradation leads the formation of toxins, which are harmful to health (Rather et al., 2017). In accordance with Lu and Zhao (2017), photo-oxidation of phytochemicals can cause a series of problems, such as sensory quality change and even the generation of hazardous and noxious substances. For example, hydroperoxides produced by oxidative reactions are important intermediates that may cause off-flavor and even toxic compounds.

Moreover, some chemical reactions induced by light cause loss of essential nutrients, such as vitamins and food browning (Rahman, 2007). In accordance with this, one nutrient abundant in milk that is highly light-sensitive is riboflavin (vitamin B_2), which is destroyed when exposed to light and produces by-products responsible for rancid milk. This degradation leads to the formation of free radicals, superoxide and singlet molecular oxygen, which degrade other nutrients, such as vitamin C (Francis and Dickton, 2015).

Nowadays, bioactive substances are being intensively studied due to their human health benefits (Barba et al., 2017). Since bioactives typically occur in small quantities in foods, the intake of a specific bioactive could be less than the dose that can exert a specific health effect (effective dose). To overcome this, the food industry is developing new products containing higher concentrations of selected bioactives. These bioactive-enriched foods (BEF) are also known as functional food products, which have a potentially positive effect on health. However, photo-degradation or oxidative reactions of bioactive compounds can reduce nutritional and health advantages of these functional foods (Duncan and Chang, 2012).

The extent of these undesirable changes in food products depends on many factors, including food composition and light source. Not all types of natural or artificial lights are equally absorbed or equally destructive. The effects of light on selected foods are described below.

Raw meat exposed to oxygen typically has an attractive red color. When exposed to light in the UV wavelengths for prolonged period, the fats become rancid (Nerin et al., 2006). While not a health issue per se, a perceived loss of quality due to deteriorating color, odor and taste does constitute a serious threat to the product's marketability. Processed or cured meats, such as luncheon and items like ham, are affected more rapidly than fresh meats (Moller and Skibsted, 2006). Nitrites, found in these products, give them their characteristic pink color, but when exposed to visible light, they combine with oxygen and slowly produce a brownish-grey color on the meat surface. Like changes in fresh meat caused by light, this deterioration of quality in processed meats will also impact the retailer sale. This undesirable color is called light fading and can be prevented by employing vacuum packaging, oxygen impermeable films or opaque packaging materials (Suman and Joseph, 2013).

Milk is considered an indispensable food in a balanced diet and is high-light sensitive. Thus, packaging and handling of milk are of great relevance in its nutritional quality. High light intensity to which milk is exposed in retail stores can induce significant photo-degradation of its components, changing its flavor and losing vitamins A, B_{12} and C, and also riboflavin and pyridoxine. Besides, amino acids and proteins, photo-degradation as well as lipid oxidation in milk produce an undesirable flavor (Hansen and Skibsted, 2000). Light-induced changes in milk depend on the light intensity, the type of holding-container, milk composition, agitation, and several other factors. Loss in milk quality produced by light is faster in clear glass and polycarbonate (PC) or polyethylene (PE) containers. An alternative to provide a desirable protection of milk against light is opaque fiberboard packaging. Another option is the employment of packaging containers based on dyed plastics with protective light capacity (Mestdagh et al., 2005).

With respect to dairy products, such as cheese, butter, yogurt, ice cream and cream, among others, they also suffer oxidative degradation when light exposed. Proteins' and fats' photo-degradation are responsible for changes in flavor, odor and color of dairy products with these effects being more important in soft cheese (Dalsgaard et al., 2010). Current trend to consume natural and fresh dairy products without preservatives, in addition to the use of transparent packages, is a challenge to the packaging industry to provide options that, at the same time, protect the product and be attractive to consumers. In fruits and vegetables, vitamins A, C, and E along with thiamine and riboflavin are among the components most sensitive to light. For example, vitamin A, exposed to daylight, resulted, according to Allwood and Martin (2000), in as much as 80 percent loss, affecting nutritional quality of foods that contain this nutrient. Regarding green fruits and vegetables, chlorophyll photo-degradation is responsible for less desirable colors, such as various shades of browns and greys, thereby influencing acceptability by consumers.

Degradation of fats, oils and foods with high content of them is accelerated by exposure to sunlight and/or fluorescent light. Thus, undesirable organoleptic changes in flavor, odor and color are detected after exposure of food to light. Light appears to accelerate the autoxidation of fats and oils, resulting in flavor and odor changes (Caponio et al., 2005). Fats and oils have different sensitivities to light, depending on their composition, different amounts and types of sensitizers and the protective effect of other constituents. Snack foods, like potato chips, are an example of a food rich in oils that are susceptible to photo-degradation, developing off-odors and off-flavors when exposed to light. This is why snack foods are packed in opaque bags that retain their quality longer than those packaged in clear ones (Petersen, 1999).

Sun-struck flavor and off-odor detected in beer and red wines is the result of beverages' exposure to light. This is the main reason why both brews are bottled in dark containers (Huvaere et al., 2005). For beer, this light-induced flavor is caused when hops as the constituent react with breakdown products of sulfur-containing amino acids that diminish the beer's flavor and aroma. Therefore, most beers are packaged in brown or green bottles or aluminum cans in order to reduce photo-degradation. In the case of wines, light often causes color change that reduces consumer acceptance. It is important to highlight that light sensitivity of wines depends on the type of wine and the color of the bottle in which it is packaged (Grant-Preece, 2015).

Several things that can be done to reduce photo-degradation of foods are (a) reducing the exposure of sensitive foods to light; (b) packaging foods in selectively absorbent or opaque packaging materials; (c) reducing the oxygen concentration to very low levels; (d) decreasing the level of light in display cases; and (e) choosing lights that have low photochemical activity. By following some of these recommendations, the product's shelf-life can be improved and quality be maintained for longer periods. Nowadays, photoreactions are mainly prevented by protecting food from light by using adequate packaging. Thus, the selection of packaging material is a relevant aspect to modify how the light strikes on the food item. In accordance with this, Andersen and Skibsted (2010) stressed that light-transmission properties of packaging materials determine which wavelengths are able to pass through the material and how much the intensity of light is. However, some problems related to foods' photo-degradation still remain and it is important to find more effective solutions, like developing new materials with enhanced light-barrier capability.

2. Degradation of Plastic Packaging

Plastic packaging is one of the most important contributors for protecting food from spoiling. There are many types of plastic packaging that have different functional properties, such as being safe for food, flexible, transparent or opaque and chemical and heat resistant (Eldesouky, 2015). For enhancing the performance and appearance of food packaging, as well as minimizing polymer degradation during their processing and shelf-life, a wide range of additives are commercially available (López Rubio et al., 2004).

Polymer degradation routes are generally attributed to macromolecular radicals formation due to the action of some external factors (temperature, mechanical stress and radiation, among others) and to the subsequent reactions of such radicals with polymer macromolecules (La Mantia et al., 2017). Thus, heat, humidity, solar UV light, ozone, impurities, mechanical load, chemicals and microorganisms, individually or in combination, can lead to degradation of polymeric materials. Most polymers exposed to environmental factors will undergo significant changes on their service life and properties and can only be prevented or slowed down by the addition of UV stabilizers and antioxidants. The result will be a steady decline in their (mechanical) properties caused mainly by modification of average molecular weight, molecular weight distribution and composition of the polymer (Singh and Sharma, 2008).

2.1 Types of Polymer Degradation

Several external factors facilitate polymers degradation and, consequently, this process can be classified into photo-oxidative, thermal-oxidative, ozone-induced, mechano-chemical and hydrolytic degradation as well as biodegradation.

2.1.1 Photo-oxidative

Among the wavelengths that reach the earth's surface, UV radiation is one of the most destructive that affects polymeric materials, producing scission, oxidation, crosslinking and side-group abstraction (Zhang et al., 2015). Degradation, caused by light, is called photo-degradation. Thus, sunlight photon energy is the most important factor that induces degradation and most of the polymers are susceptible to UV and visible light (Bhuvaneswari, 2018). UV radiation spectrum (10–400 nm) of sunlight plays a key role in plastic degradation and particularly UV-B rays (280–315 nm) are responsible in initiating photo-oxidative degradation of most common polymers, such as low-density polyethylene (LDPE), high-density polyethylene (HDPE), polypropylene (PP) and aliphatic polyamides (nylons) (Niaounakis, 2017). The combination of UV radiation and oxygen results in breaking of polymer chains, producing radicals and reducing the molecular weight (Yousif and Haddad, 2013). Typical steps that are involved in polymers photo-degradation are shown in Fig. 2.

2.1.2 Thermal-oxidative

Thermal degradation can be defined as the process that the polymer undergoes in inert atmosphere, because of heat action. In general, thermal stresses result in the formation of decomposition products, depending mainly on polymer composition and

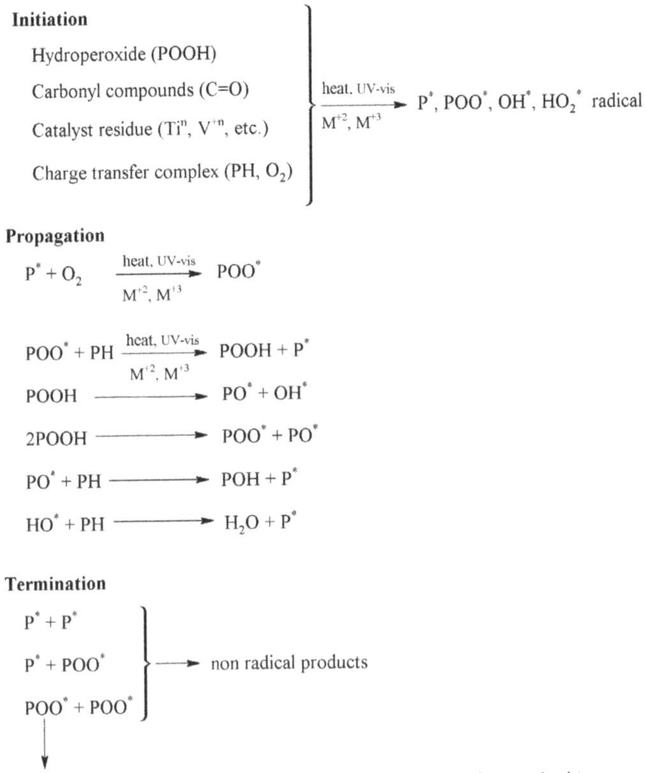

Initiation

Hydroperoxide (POOH)

Carbonyl compounds (C=O)

Catalyst residue (Ti^n, V^{+n}, etc.)

Charge transfer complex (PH, O$_2$)

$$\xrightarrow[M^{+2},\ M^{+3}]{\text{heat, UV-vis}} P^\bullet,\ POO^\bullet,\ OH^\bullet,\ HO_2{}^\bullet \text{ radical}$$

Propagation

$$P^\bullet + O_2 \xrightarrow[M^{+2},\ M^{+3}]{\text{heat, UV-vis}} POO^\bullet$$

$$POO^\bullet + PH \xrightarrow[M^{+2},\ M^{+3}]{\text{heat, UV-vis}} POOH + P^\bullet$$

$$POOH \longrightarrow PO^\bullet + OH^\bullet$$

$$2POOH \longrightarrow POO^\bullet + PO^\bullet$$

$$PO^\bullet + PH \longrightarrow POH + P^\bullet$$

$$HO^\bullet + PH \longrightarrow H_2O + P^\bullet$$

Termination

$$P^\bullet + P^\bullet$$
$$P^\bullet + POO^\bullet$$
$$POO^\bullet + POO^\bullet$$
$$\longrightarrow \text{non radical products}$$

etc. (Where P$^\bullet$ is polymer radical, M$^+$ is a metal ion, and PH is a polymer molecule)

Fig. 2. Scheme of general oxidation and photo-oxidation in polymers.

temperature (Utracki et al., 2014). Meanwhile, in the presence of oxygen, thermal-oxidative degradation occurs and if also mechanical stress is involved during the thermal treatment, degradation is referred to as thermo-mechanical (La Mantia et al., 2017). Generally, the resistance to degradation will depend on the chemical composition of polymers. For example, polymers such as PP, polyvinylchloride (PVC) and polybutadiene (PBD) are very susceptible to thermal degradation, and their degradation can only be reduced when they are formulated with UV stabilizers and antioxidants; whereas polymers such as polyether sulfones (PES), polysulfone (PSU), polyetherketone (PEEK) and polysiloxanes (silicones) possess excellent resistance to thermal and oxidative degradation due to the strong bonds in the long-chain backbones and in the side-groups (Ray and Cooney, 2018). The general mechanism of thermal-oxidative degradation of polymers is presented in Fig. 3.

For common thermoplastic polymers, at the initiation step, free radicals are formed during the course of breaking down of the main backbone chain into smaller molecules. Thermal-oxidative degradation is usually initiated when polymer chains form radicals, either by hydrogen abstraction or by homolytic scission of a carbon-carbon bond. This could occur randomly either at the weak sites of the macromolecule or at the chain end, forming highly reactive peroxide intermediates. Propagation of thermal degradation involves a number of reactions. The first step is the reaction

Initiation Step

R-H $\rightarrow$ R· + H·

Propagation Step

R· + O_2 $\rightarrow$ ROO·

ROO· + RH $\rightarrow$ ROOH

ROOH $\rightarrow$ RO· + ·OH

RO· + RH $\rightarrow$ R· + ROH

·OH + RH $\rightarrow$ R· + H_2O

Termination Step

R· + R· $\rightarrow$ R-R

2 ROO· $\rightarrow$ ROOR + O_2

R + ROO· $\rightarrow$ ROOR   Termination by recombination of radicals or disproportionation / hydrogen abstraction

R· + RO· $\rightarrow$ ROR

HO· + ROO· $\rightarrow$ ROH + O_2

R_n· + R_m· $\rightarrow$ R_{n-2}-CH=CH_2 + R_m Termination by chain scission

2RCOO· $\rightarrow$ RC=O + ROH + O_2

Fig. 3. Scheme of thermal-oxidative degradation mechanisms.

of a free radical with an oxygen molecule to form a peroxyl radical, which then abstracts a hydrogen atom from another polymer chain to form a hydroperoxide. The hydroperoxide then splits into two new free radicals, which abstract labile hydrogen from other polymer chains. Since each initiating radical can produce two new free radicals, the process can accelerate, depending on how easy it is to remove hydrogen from other polymer chains and how quickly the free radicals undergo termination via recombination and disproportionation. In addition, during the propagation step, different types of decomposition reactions may simultaneously take place resulting in cross-linking, main-chain unsaturation and intra-molecular or intermolecular hydrogen transfer along with further degradation of the main chain via random or chain-end pathway. During this process, adjacent degraded chains may interact to form cross-links and volatile products. The termination step is achieved by recombination of two radicals or by disproportionation/hydrogen abstraction. These reactions result in an increase of molecular weight and cross-linking density (Tyler, 2004). On the other hand, termination by chain scission results in the decrease of molecular weight, leading to polymer softening and reducing its mechanical properties (Singh and Sharma, 2008). Eventual termination steps will depend on polymer type and conditions. For example, polyolefins with short alkyl side groups, like PP and polybutylene (PB), and unsaturated polymers, like natural rubber (polyisoprene, PIP) undergo predominantly chain scission, whereas PE and rubbers with somewhat less active double bonds, like PBD and polychloroprene (PCP), suffer embrittlement due to crosslinking during aging (Grand and Wilkie, 2000).

2.1.3 Ozone-induced

Ozone is a high reactive form of oxygen that is present in the atmosphere, but it is found in small quantities. However, even in very small concentrations, ozone significantly increases the rate of polymeric materials aging (Agboola et al., 2017). Polyolefin ozonolysis leads to a change in structural and physical and dynamic parameters of polymer matrixes, such as the degree of crystallinity, density of amorphous phase and mobility of macromolecules chains (Cataldo, 2001). Both ozone and sunlight rapidly attack unprotected polymers which can significantly reduce plastic service life. Particularly, polymers with high unsaturation will suffer ozone degradation because double bonds readily react with ozone, causing chain scission. However, ozone also reacts with saturated polymers but at a comparatively slower rate. The general mechanism of ozone degradation is shown in Fig. 4.

Ozone attacks the unsaturation in polymers and this reaction generally occurs in three main steps. The first step is the cycloaddition of ozone to olefin double bonds to form an ozoneolefin adduct, referred to as the *primary ozonide*, which is an unstable species because it contains two very weak O-O bonds. The second step in the ozonolysis mechanism is the decomposition of primary ozonide to carbonyl compounds and a carbonyl oxide. Carbonyl oxide is considered to be the key intermediate in the $C = C$ bond ozonolysis mechanism. The third stage is the fate of carbonyl oxide, which depends on its source as well as on its environment condition. Carbonyl oxide flips over with the nucleophilic oxyanion, attacking the carbon atom of the carbonyl group (Cataldo, 2001).

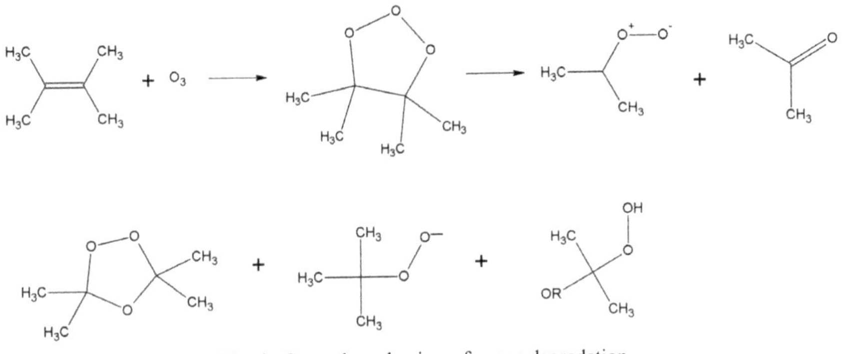

Fig. 4. General mechanism of ozone degradation.

2.1.4 Mechano-chemical

Mechano-chemical degradation of polymers takes place under mechanical stress, where molecular chains break down as a result of a mechanical force or under shear. This type of degradation is common in the machining process, such as grinding and ball milling (Agboola et al., 2017; Mitchell, 2003). When excessive stress is applied, the molecular chain breaks and produces a pair of free radicals, which can take part in subsequent reactions. Mechanically-generated radicals are believed to result from the cleavage of main backbone segments of polymer chains in the stressed amorphous regions connecting crystallites. Figure 5 shows the general mechano-chemical degradation mechanism (Singh and Sharma, 2008).

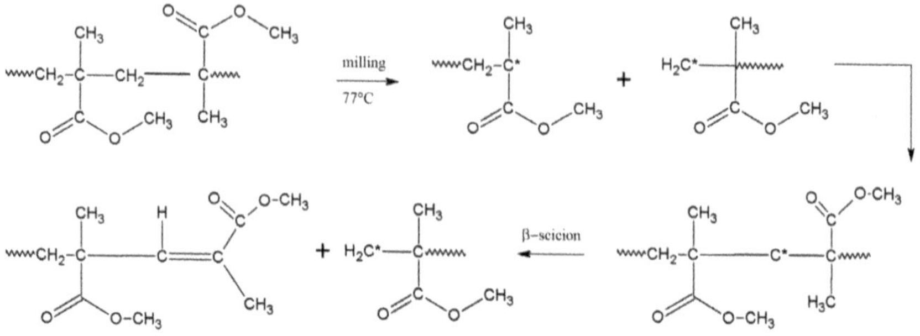

Fig. 5. Mechano-chemical degradation mechanism.

In addition, polymers can be degraded by flow-induced scission when used as inkjet printing material. In this sense, Wheeler et al. (2014) reported the use of ink formulations containing polystyrene (PS) and (polymethyl methacrylate) (PMMA). Main polymer degradation is attributed to the continuous mechanical recycling of polymeric ink through the pump, resulting in a molecular weight reduction. On the other hand, mechanical processing of polymers under nitrogen atmosphere does not change the plasticity or molecular length of the polymeric chains, but in oxygen presence, degradation occurs immediately and rapidly. In this aspect, under mechanical shear, macromolecules break into radicals and react with oxygen in the atmosphere, leading to permanent chain breakage. Besides, when polymers are subjected to very high vibrations, such as ultrasounds, they can suffer mechano-chemical degradation. Localized shear gradient produces tearing of molecules, leading to bond scissions and decreasing the chain length (Singh and Sharma, 2008); otherwise, when polymers in an extruder are in a molten state and under strong shearing forces, it has been found that mechanical degradation reduces the average molecular weight, impacting their final mechanical properties (Agboola et al., 2017).

2.1.5 Hydrolytic

The action of water on polymers is another source of degradation. Degradation by hydrolysis consists in breaking certain chemical bonds, such as ester, ether and amide, by attack of water molecules (Ojeda, 2013). Water-insoluble polymers, containing pendent anhydride or ester groups, may be solubilized if these functional groups are hydrolyzed to form ionized acids. Hydrolysis of polymer backbone is most desirable since it produces low molecular weight by-products that result from chain cleavage (Massey et al., 2007; Sevim and Pan, 2018). Synthetic polyesters are degraded mainly by simple hydrolysis in which the mechanism includes different steps. The first stage of the degradation process involves non-enzymatic, random hydrolytic ester cleavage and its duration is determined by the initial molecular weight of the polymer as well as its chemical structure. PE also undergoes random degradation through migration of a hydrogen atom from one carbon to another, thus generating two fragments. Massey et al. (2007) reported a degradation scheme of LDPE by water action (Fig. 6). The first step of the degradation is the loss of a proton due to the polarity of water molecules. The rupture of a carbon hydrogen bond occurs by electrostatic attraction of the proton, resulting in an alcohol group formation by association of the macroradical and an hydroxyl group.

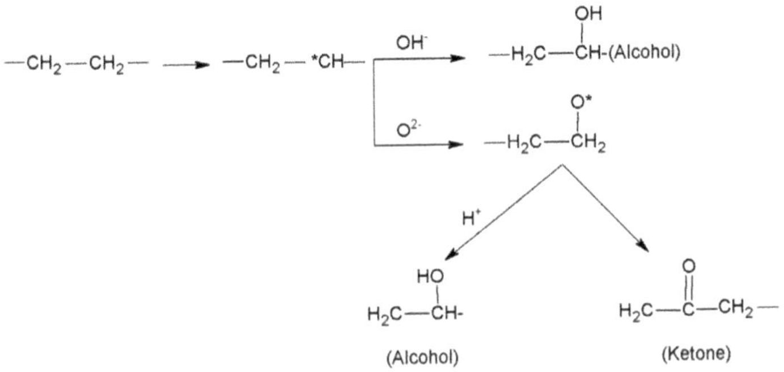

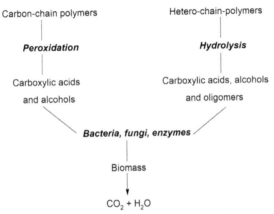

$$-CH_2-CH_2- \longrightarrow -CH_2-{}^*CH- \xrightarrow[]{OH^-} -H_2C-\overset{\displaystyle OH}{\underset{|}{C}}H\text{-(Alcohol)}$$

$$\xrightarrow[]{O^{2-}} -H_2C-\overset{\displaystyle O^*}{\underset{|}{C}}H_2$$

$$\xrightarrow[H^+]{} \quad \overset{\displaystyle HO}{\underset{|}{H_2C-CH-}} \qquad H_2C-\overset{\displaystyle O}{\overset{\displaystyle \|}{C}}-CH_2-$$

(Alcohol)   (Ketone)

Fig. 6. Hydrolytic degradation of synthetic polyolefins.

2.1.6 Biodegradation

Biodegradation is a process in which naturally-occurring microorganisms, such as bacteria, fungi, or algae act on the polymer material breaking down completely into non-plastic and non-toxic constituents (Sudhakar et al., 2007). Microorganisms identify the polymer as an organic and energy source, useful for growth. Thus, polymer chemically reacts under the influence of either cellular or extracellular enzymes whereby chains are fragmented (Nair et al., 2017). Considering this fact, biodegradation is responsible for changes in polymer surface properties, loss of mechanical strength, or backbone chain breakage and subsequent reduction in the average molecular weight of polymeric chains. In addition, physical properties, such as crystallinity, orientation and morphological properties, such as surface area, affect the rate of degradation (Singh and Sharma, 2008).

In the case of synthetic polymers, abiotic hydrolysis and/or peroxidation normally precede biotic attack (Fig. 7), (Scott, 1999). However, the time-scale over which hydrocarbon-based polymers biodegrade may vary by many orders of magnitude. In addition, since polyolefins are hydrophobic, with higher average molecular weight, they are not easily degraded by abiotic or biotic factors. These molecules are unable to enter microbial cells to be digested by intracellular enzymes due to their size and are inaccessible to the action of extracellular enzymes produced by microorganisms, due

Carbon-chain polymers   Hetero-chain-polymers

Peroxidation   **Hydrolysis**

Carboxylic acids   Carboxylic acids, alcohols
and alcohols   and oligomers

Bacteria, fungi, enzymes

Biomass

$CO_2 + H_2O$

Fig. 7. Biodegradation mechanisms of synthetic polymers.

to their excellent barrier properties (Agboola et al., 2017). Therefore, to promote their degradation, blends of polyolefins with natural macromolecules or pro-oxidants have been studied. Blending allows enhancing the biodegradation process since the rate of degradation of natural polymers is usually of an order of magnitude higher than that of synthetic polymers (Ammala et al., 2011; Ojeda et al., 2011; Gomes et al., 2014).

Biodegradable polymers are degraded to carbon dioxide and water by bacterial and fungi under the natural environment. These polymers degrade in microbial environments either by anaerobic or aerobic biodegradation and sometimes, a combination of both. In these processes, the original polymer is transformed to carbon dioxide, methane or other hydrocarbons, biomass and non-toxic products (Jayasekara et al., 2005; Tiwari et al., 2018).

As discussed earlier, there are many possibilities of polymers' degradation induced by numerous environmental or external factors. Among these degradative processes, this chapter focuses on photo-oxidative degradation in order to evaluate alternatives to minimize the undesirable effects, not only on plastic packaging, but also on food products.

3. Stabilizing Additives for Polymers

Light exposure may cause significant photo-degradation of polymers, which results in breaking of polymer chains, production of free radicals and reduction of molecular weight (Yousif and Haddad, 2013). Consequently, the final properties, such as strength, malleability, appearance and color deteriorate, leading to material damage after an unpredictable time. These changes have a dramatic effect on polymer properties and materials' service life and can be prevented or slowed down by adding stabilizing agents to polymeric matrices. Stabilizers are chemical substances that are added to polymers in small amounts (0.05–2 wt %) and are capable to inhibit or retard their degradation (Zweifel, 1998), minimizing not only packaging material degradation but also food deterioration. The amount of light stabilizer required to provide an adequate and economical protection is related to many aspects, such as plastic thickness and additive-polymer compatibility. Among other factors, additive concentration is a function of the inherent polymer light stability, the specific application under consideration and the presence of other additives in formulations, e.g., antioxidants, pigments, and fillers (Singh et al., 2012). Polymer stabilizers might be incorporated directly or through the addition of a masterbatch during material processing. Besides good stabilizer activity, these additives must have a number of properties to be widely used in polymer applications. Thus, they should be highly effective, preferably non-toxic, stable and compatible with the polymer as well as cost-effective (Marzec, 2014).

Polymers' photo-stabilization may be achieved in many ways. Usually three alternatives can be mentioned to protect polymers against photo-degradation: (a) blocking or screening out the incident radiation, (b) absorbing damaging radiation and dissipating the energy in a harmless way and (c) deactivating species or intermediates in the polymer as it undergoes degradation. Different commercial stabilizers can be found, depending on their action mode. They can be grouped in primary or chain-breaking and secondary or preventive antioxidants (Mamta et al., 2014). The first group is capable of interrupting free radical processes donating labile hydrogen atoms that react with free radical species more quickly than polymers.

Thus they are effective in reducing or halting the chain length shortening during propagation reactions (Spalding and Chatterjee, 2018). On the other hand, preventive antioxidants destroy hydroperoxides in an ionic reaction; so radicals are not produced or they protect hydroperoxides from decomposition by absorbing UV light (Scott and Gilead, 1995). A more specific classification divides stabilizers in light screeners, UV absorbers, excited-state quenchers, peroxide decomposers and radical scavengers. Among this last classification, excited-state quenchers, peroxide decomposers and radical scavengers are the most effective systems (Yousif and Haddad, 2013). Regardless of the stabilizing agent, the presence of a photo-stabilizer significantly reduces polymer oxidation rate, extending material service life and giving to the packed product a good protection against UV and visible light.

According to Hawkins (1984), a true light screener is a surface-coating which forms an impenetrable shield between the polymer and the radiation source, involving only physical effects. Light screeners are substances that cover or reflect UV radiation so that light cannot get inside the polymer and thus plays an important role in protecting the materials. To avoid or reduce polymer photo-degradation, inorganic and organic UV absorbers have also been used (Forsthuber et al., 2013; Devi and Maji, 2012). The mechanism of UV absorbers against UV light varies from one material to another; inorganic UV absorbers often scatter light, while organic UV absorbers convert UV light to heat (Nikafshar et al., 2017). Both, inorganic and organic UV filters are chemical products differing in their molecular composition and in the UV protection mechanism (Morabito et al., 2011). Organic UV filters are called chemical filters as their action mode is related to chemical changes in their molecules that prevent radiation reaching the material. Meanwhile, inorganic UV filters are called physical filters because their protection mode against radiation is associated with physical phenomena, such as scattering and reflection UV radiation (Antoniou et al., 2008; Jin et al., 2008; Seite et al., 2000; van der Pols et al., 2006; Wang et al., 2010). Organic UV absorbers have an increased ability to absorb light in the UV range, transmitting thermal energy and reducing the amount of radiation that can be absorbed by chromophores present in the polymer (Rudko et al., 2015). It is important to remember that a UV absorber must be light stable because otherwise it would be destroyed in the stabilizing reactions. Compared with inorganic, organic UV absorbers have higher efficiency but less durability due to evaporation and migration to polymer surface (Bail et al., 2016). Inorganic stabilizers, such as zinc oxide and titanium dioxide, absorb in the visible region and in a UV wide range, while organic additives mainly absorb UV wavelengths but not the visible ones. For this reason, organic UV absorbers are highly demanded and are suitable for clear and transparent packaging (Parejo et al., 2006). Additionally, polymeric packaging, containing organic UV absorbers, compared with materials with inorganic additives, are more flexible and have a lower glass transition temperature (Nikafshar et al., 2017). Organic UV additives are usually aromatic compounds with a carbonyl group. According to Manaia et al. (2013), when these compounds receive the energy of UV photons, they can act in three ways: (a) undergo conformational molecular changes, (b) emit radiation at higher wavelength, or (c) release incident energy as heat (Antoniou et al., 2008; Kiss et al., 2008). The action mode of organic protector molecules is reversible, allowing molecules to function repeatedly.

The most effective UV filter is carbon black that absorbs all wavelength of UV light and converts it into heat, while physically blocking UV penetration (Tolinski,

2015). It consists of very fine particles fused together to form primary aggregates, as can be seen in Fig. 8 (Yousif and Haddad, 2013). The main carbon black properties are blackness and dispersability when mixed with inks, paints, or polymeric resins. Carbon black gives a blacker appearance to polymer products; this is the main disadvantage of this additive to be used in food packaging. It is important to highlight that nowadays consumers emphasize the product aesthetics, thus packaging color and transparency have relevance in the selection of stabilizers agents. It is believed that carbon black acts as a simple physical screen, a UV absorber, a radical trap and a terminator of the free radical chains through which the photo-oxidative reactions are propagated (Liu and Horrocks, 2002). Stabilizing capacity against UV light of carbon black is associated with particle type and size as well as to their concentration and dispersion within the polymer matrix (Deveci et al., 2018). Small particles are known to have the greatest UV stabilizing efficiency, but they tend to agglomerate into aggregates or clusters, which cannot be easily dispersed (Ghasemi-Kahrizsangi et al., 2015). An optimum concentration of carbon black is 3–5 wt %. At higher than this concentration, the polymer loses its tensile strength and other mechanical properties (Suits and Hsuan, 2003).

Among UV absorbers, it can be mentioned that substituted benzophenones or benzotriazoles act by dissipating absorbed UV radiation harmlessly as heat (Ammala et al., 2002). Chemical structures of different derivatives of these UV absorbers are presented in Fig. 9. Benzophenones are effective UV stabilizers in thin films, giving

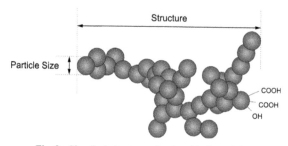

Fig. 8. Chemical structure of carbon black particles.

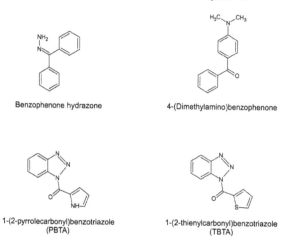

Benzophenone hydrazone

4-(Dimethylamino)benzophenone

1-(2-pyrrolecarbonyl)benzotriazole
(PBTA)

1-(2-thienylcarbonyl)benzotriazole
(TBTA)

Fig. 9. Chemical structure of different derivatives of substituted benzophenones or benzotriazoles.

slight color to the polymer and have low toxicity. These UV absorbers have relatively low price but discolor during processing and weathering. Regarding benzotriazoles, these compounds have higher UV absorption capacity than benzophenones, at equivalent concentrations. Besides, they impart less color, but are more expensive.

Nowadays, there has been extensive interest in ultrafine or nanoparticle fillers and pigments with regard to their properties as a UV 'blockers' in coating applications (Allen et al., 2002). According to Reiss et al. (2011), the incorporation of semiconductor nanoparticles into polymers leads to dramatic changes in the optical properties of the polymer. Nanoparticles usually have higher photostability than organic molecules and are, therefore, of great interest as additives to polymers. Moreover, it is possible to change the absorption edge of the composite via tuning the particle size. Among the inorganic UV absorbers, titanium dioxide (TiO_2) and zinc oxide (ZnO) have been extensively used as polymer stabilizers against photo-degradative reactions. Chemical structures of both pigments are shown in Fig. 10. TiO_2 occurs naturally in three crystalline structures: rutile, anatase and brokite (Allen et al., 2018). Rutile is the most common and stable form of this pigment. An important optical property of this birefringent crystal is its refractive index in the UV and visible wavelength range. The whiteness of TiO_2 pigments is partly due to these high refractive indices (Smijs and Pavel, 2011). ZnO exists in three main crystalline forms: hexagonal (wurtzite), cubic zinc blende and cubic rock salt (Bodke et al., 2018). The wurtzite structure is the most common and stable form. Compared with TiO_2, the whitening effect of ZnO is lower. The light attenuation results from reflection and scattering of UV and visible

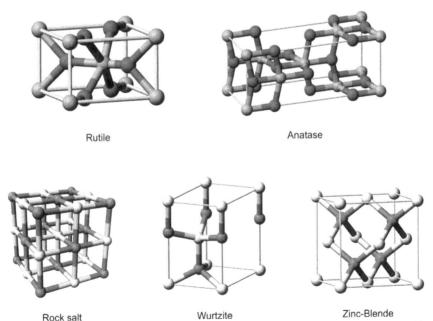

Rutile   Anatase

Rock salt   Wurtzite   Zinc-Blende

Fig. 10. Crystalline structures of titanium dioxide (rutile and anatase) and zinc oxide (wurtzite, cubic zinc blende and cubic rock salt).

radiation and from UV absorption. UV attenuation properties of these two particles are complementary, with TiO_2 being primarily a UVB-absorbing compound, while ZnO is more efficient in UVA absorption.

The first approach to prevent degradation is to attempt to filter off the harmful irradiation with suitable absorbers before it can form excited chromophores. However, this is not 100 percent efficient and some excited chromophores will nevertheless be formed. Energy acceptor molecules, usually called quenchers, can be used to deactivate these excited species (singlet and/or triplet) of polymers chromophore groups before bond scission (Leppard et al., 2002). Energy transfer can occur efficiently only if the quencher energy level is below that of the chromophore. Quenchers dissipate the excess energy harmlessly, such as heat, fluorescence, or phosphorescence—methods that do not result in polymer degradation (Drobny, 2014). Most widely used quenchers are organic nickel complexes, such as 2,2-thiobis (4-octylphenolto)-n-butylamine nickel, nickel salts of thiocarbamate and nickel complexes with alkylated phenol phosphates (Marzec, 2014; Yousif and Haddad, 2013). Figure 11 shows some of the most commonly employed quenchers for polymers. Quenchers are often used in combination with UV absorbers.

Free-radical scavengers are also useful additives to inhibit polymers' light degradation. They react with free radicals in polymer material, reducing them to stable unreactive products (Drobny, 2014). Among these kinds of additives, hindered amine light stabilizers (HALS or HAS) represent the latest development in this field (Singh et al., 2012). There is a wide variety of HALS with different molecular weight. In Fig. 12 are shown different chemical structures of commercially available HALS. It is a common practice to combine hindered HALS with UV absorbers for optimal protection of coatings and plastics. The stabilization mechanism of HALS is extensively studied and reported to be a cyclic chain-breaking antioxidant process, called Denisov cycle (Schaller et al., 2009).

Another type of polymer photostabilizers is known as hydroperoxide decomposers. These compounds can decompose hydroperoxide groups before they are photolysed by absorbed photons (Rabek, 1995). According to Yousif and Haddad (2013), many metal complexes of sulphur containing ligands, such as dialkylthiocarbonate and dialkylthiophosphate not only decompose hydroxide but are also effective in UV stabilization (as UV absorbers and excited state quencher). In Fig. 13 is shown an example of hydroperoxide decomposer.

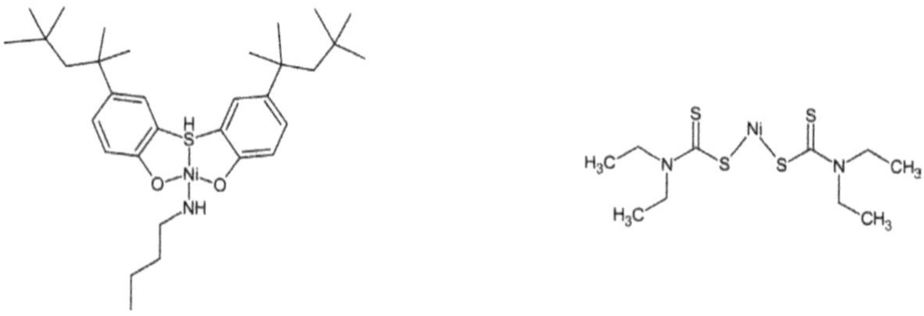

2,2'-Thiobis(4-tert-octylphenolato)-n-butylamine nickel(II)   Nickel diethyldithiocarbamate

Fig. 11. Chemical structure of most commonly employed quenchers for polymers.

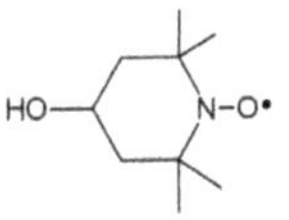

HALS-1

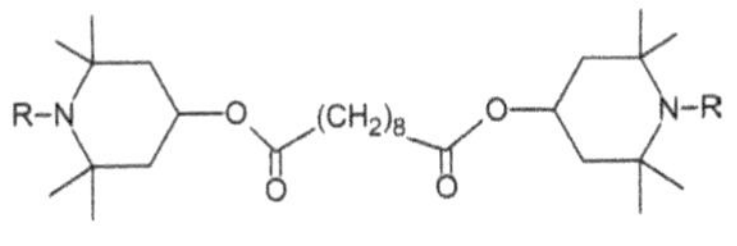

HALS-2	R = H
HALS-3	R = CH$_3$
HALS-4	R = OC$_8$H$_{17}$

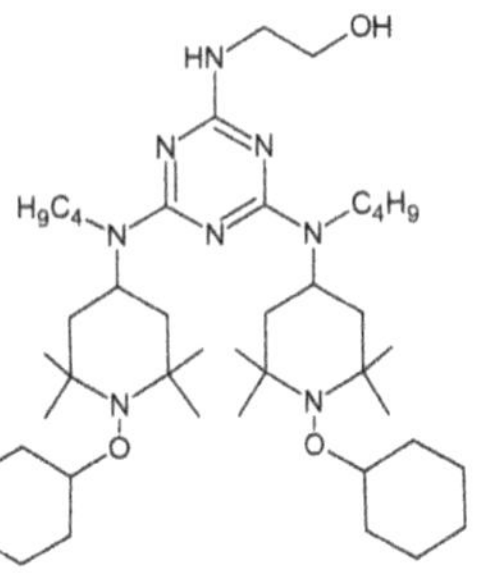

HALS-5

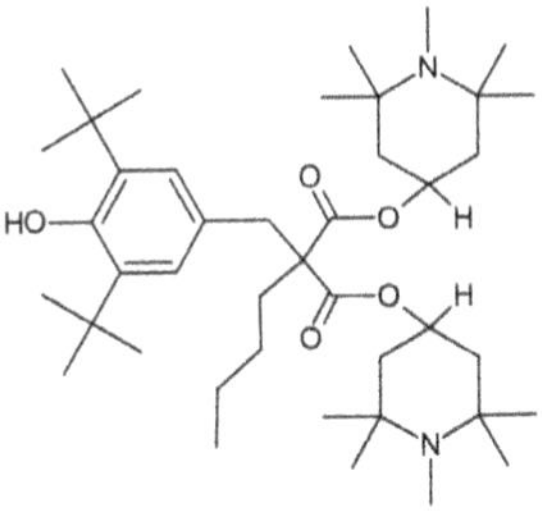

HALS-6

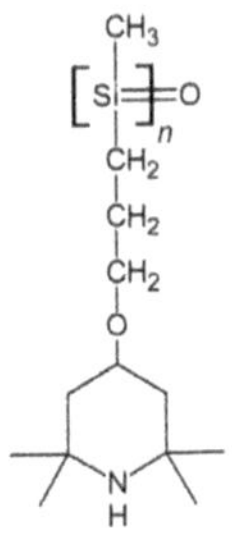

Uvasil992

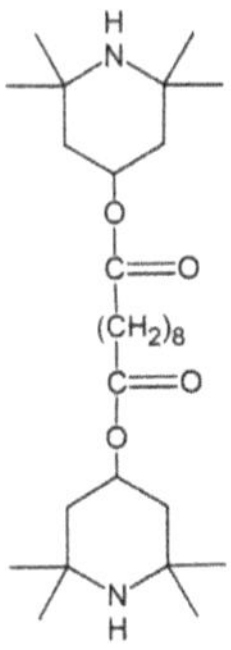

Tinouvin770

Fig. 12. Chemical structures of commercially available HALS.

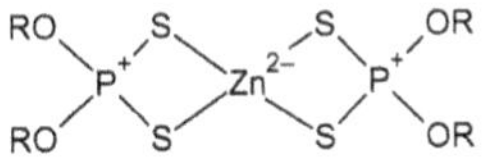

Zinc dithiophosphate

Fig. 13. Chemical structure of a hydroperoxide decomposer.

4. Plastics for Food Packaging

4.1 Types and Applications

Packaging is a relevant aspect of food technology and is involved with protection and preservation of all type of food products (Jabeen et al., 2015). Due to economical abundance, petrochemical plastics have been largely used as packaging material. They are made by condensation polymerization (polycondensation) or addition polymerization (polyaddition) of monomer units (Marsh and Bugusu, 2007). Differences in chemical characteristics of monomers in the structure of polymer chains, and in the interrelationship of chains, determine the different properties of various polymeric materials. Final properties are determined by the chemical and physical nature of polymers used in their manufacture; meanwhile, polymers' properties are determined by their molecular structure, molecular weight, degree of crystallinity and chemical composition (Robertson, 2006).

Plastics are used in food packaging because they offer a wide range of appearance and performance properties, which are derived from the inherent features of the individual material and how it is processed and used (Kirwan, 2009). The main advantages are related with their flowability, mouldability, inertness, cost effectiveness, lightweight, optical properties, heat sealability, mechanical resistance and gas barrier properties. Plastics are selected for specific applications considering their requirement in packaging process, distribution, storage and end use of the product. Other relevant aspects at the moment of choosing the more adequate plastic material are associated with marketing and environment awareness (Emblem, 2012).

Plastic versatility helps in obtaining diverse types of food packaging, which can be rigid containers, such as bottles, jars, tubs and trays. Besides, plastic flexible films are also employed to obtain packaging for food products like bags, sachets and pouches (Cooper, 2013; Rahman, 2007).

According to Emblem (2012), amongst synthetic polymers, low-density polyethylene (LDPE), high-density polyethylene (HDPE), polypropylene (PP), polyvinyl chlorine (PVC), polystyrene (PS) and polyethylene terephthalate (PET) dominate the packaging market. All these traditional plastics are derived from non-renewable petroleum resources and give non-biodegradable plastic materials after their useful life. Therefore, in order to recover the material and valorizing plastic wastes, a commonly used practice is to recycle them (Arrieta et al., 2017).

The ASTM International Resin Identification Coding System, often abbreviated as the RIC, is a set of symbols appearing on plastic products that identify the plastic resin out of which the product is made (ASTM D7611/D7611M-18 2018). It was developed in 1988, by the Society of the Plastics Industry (now the Plastics Industry Association) in the United States, but since 2008, it has been administered by ASTM, an international standards organization. In its original form, the symbols used as part of the RIC consisted of arrows that cycled clockwise to form a triangle that enclosed a number. The number broadly referred to the type of plastic used in the product:

- #1: polyethylene terephthalate (PET)
- #2: high-density polyethylene (HDPE)
- #3: polyvinyl chloride (PVC)
- #4: low-density polyethylene (LDPE)
- #5: polypropylene (PP)
- #6: polystyrene (PS)
- #7: other plastics, such as acrylic, nylon, polycarbonate and polylactic acid (PLA)

PET is semi-crystalline, thermoplastic polyester of characteristic high strength, transparency and safety (López-Fonseca et al., 2011). PET bottle was patented in 1973 by Nathaniel Wyeth and began to be used popularly to produce disposable soft drink bottles in the 1980s (Al-Sabagh et al., 2016). In the last few years, PET consumption has increased significantly due to the strong demand for food packaging and bottle markets for glass replacement (Rogers and Long, 2003). HDPE is valuable for its numerous indoor and outdoor applications due to its remarkable properties, such as high toughness and impact strength, low thermal conductivity, good barrier to humidity, resistance to abrasion and corrosion, inertness to most chemicals, while holding a reasonable pricing and low energy demands for processing (Grigoriadou et al., 2018). One of the most important applications of HDPE is in food packaging in the form of films. LDPE exhibits very poor printability, wetability and barrier properties and its industrial applications are mainly developed in the domains of food packaging (Peyroux et al., 2015). PVC based wrap films are widely used to package a wide range of foodstuffs owing to its flexibility, permeability to water vapor and oxygen transmission (Ohata, 2016). PP is one of the most widely used thermoplastics in the world due to its combination of easy processability, good balance of mechanical properties and low cost. However, PP has certain shortcomings that limit its use in some applications, such as its poor oxygen barrier that prevents the widespread use of this material in the packaging industry (Khalaj et al., 2016). PS is a versatile plastic used to make a wide variety of consumer products. As a hard, solid plastic, it is often used in products that require clarity, such as food packaging.

The Food and Drug Administration (FDA) regulates the safety of all plastics currently used in food packaging and food-contact materials. Food-safe plastic resins are those approved for contact with edibles by FDA and they are known as food contact substances (FCS). The FDA defines food contact substances as "any substance that is intended for use as a component of materials used in manufacturing, packing, packaging, transporting or holding food." In Table 1 are shown examples of plastic food packaging obtained from synthetic polymers.

4.2 Polymers Processing

4.2.1 Methods

Polymers are often divided according to their possibility of melting and reshaping through the application of heat and pressure. Polymer processing can be defined as the process whereby raw materials are converted into products of desired shape and properties. Thermoplastic polyolefins are generally supplied as pellets of varying sizes, and they may contain some or all of the desired additives. When heated above their glass transition and/or melting temperature, thermoplastic materials soften and flow as viscous liquids that can be shaped by using a variety of techniques (Fakirov, 2017). The most commonly employed thermoplastics are PE, PP, PS, PVC, PC, PMMA, PET, polyamide (PA) and nylon, among others (Vlachopoulos and Strutt, 2003).

In most extrusion-based processes, the machine consists essentially of a screw that fits closely into a cylindrical barrel, with just sufficient clearance to allow its rotation. Tasks of an extruder include compacting polymer pellets into a solid bed, melting the solid bed, pumping the melt and mixing the melt with any additives before extrusion. In addition, when the formulation includes volatile matter, it is necessary to have a

Table 1. Examples of Food Packaging Obtained from Synthetic Polymers and the Corresponding Resin Identification Code (RIC) and the Associated Symbol.

RIC	Polymer base	Symbol	Examples of packaging
#1	Polyethylene Terephthalate—PET		• Microwaveable containers • Take-out containers • Single-serving food trays • Bottles for: soft and sport drinks, single-serve water, ketchup, salad dressing, and vegetable oil
#2	High-Density Polyethylene—HDPE		• Juice and milk jugs • Squeeze butter and vinegar bottles • Chocolate syrup containers • Butter containers • Cereal box liner
#3	Polyvinyl Chloride—PVC		• Blister for breath mints or gum • Trays for sweets and fruits • Plastic packing (bubble foil) and foils to wrap foodstuff
#4	Low-Density Polyethylene—LDPE		• Coffee can lids • Bread bags • Pack soda can rings • Fruit and vegetable bags
#5	Polypropylene—PP		• Yogurt containers • Syrup containers • Cream cheese and sour cream containers
#6	Polystyrene—PS		• Cups • Deli and bakery trays • Fast food containers • Lids • Hot cups • Egg cartons
#7	Plastic resin other than the six types of resins listed above		Depends on the polymer

venting port on the barrel (Vogel, 2013). Then, polymeric materials are subsequently processed into finished and semi-finished parts by compression moulding, injection, blowing or thermoforming.

• *Compression moulding*

Films of polymers, that are thermally stable above their melting or softening temperature, can be obtained through a combination of heat and pressure by employing a hydraulic heat press machine. The melt-pressing technique is more often used at the laboratory scale because large films are difficult to prepare by this batch process (Fakirov, 2017).

- *Injection moulding*

  Injection moulding is a two-step cyclical process: first, the polymer is melted in the injection moulding machine and second, it is injected into the mould, where it cools and solidifies into the final part (Vlachopoulos and Strutt, 2003; Introduction to Plastics, 2018). Injection moulding is used to produce thin-walled plastic parts for a wide variety of applications: household appliances, consumer electronics, power tools, automotive dashboards, open containers, toothbrushes, small plastic toys and medical devices.

- *Blowing*

  In this process, polymer melt is extruded through an annular slit die, usually vertically, to form a thin-walled tube. Air is introduced via a hole in the center of the die to blow up the tube like a balloon. Mounted on top of the die, a high-speed air ring blows on to the hot film to cool it. The tube of film then continues upwards, continually cooling, until it passes through nip rolls where the tube is flattened to create what is known as a *lay-flat* tube of film (Köhl et al., 2012).

- *Thermoforming*

  This technique involves a process of transforming a thermoplastic sheet into a three-dimensional shape by using heat, vacuum and pressure. The sheet may be stretched over a male mould (positive forming) or into a female mould (negative forming). On contact with the mould, heat is lost and the material regains stiffness as it cools. Geometries of thermoformed products are usually simple (boxes, food trays, various containers, refrigerator liners, computer cases) (Vlachopoulos and Strutt, 2003).

4.2.2 Additives Incorporation

There has been a growing trend in recent years for polymer product manufacturers to use natural polymers and additive masterbatches. Additives can offer more than one effect on polymer compounds: plasticizing, light stabilizing, pigmentation and reinforcement among others (Murphy, 2001).

According to Prashantha et al. (2008), a key issue is related to the handling of additive incorporation in plastic manufacturing process. Using a commercial masterbatch in the production of polymer composites would be a better choice as this includes elimination of dispersion difficulties, formulation development and easy handling. Nevertheless, distribution of masterbatch and subsequent dispersion of additives in the polymer matrix after processing needs to be ascertained.

UV masterbatches protect polymers against photodegradation, avoid discoloration and improve polymer processability. Numerous UV masterbatches are commercially found and some of them are listed in Table 2.

Table 2. Commercial Masterbatches.

Master batch code/name	Activity/application	Company
1146.1	Antistatic masterbatch migrates to the polymer surface, allowing dissipate static generated during processing.	INCOPLAS SRL http://incoplas.com.ar/masterbatches/
1143.1	Sliding masterbatch is recommended to provide sliding properties to polyethylene films, by reduction of friction coefficient.	
1101.2	Masterbatch to retardation of combustion by flame.	
1698.1	Masterbatch to UV protection of polyolefins.	
POLYCOM	Improves the elongation strength in blown film bags.	PLASTINES S.A.S https://www.plastines.com/en_US/productos/masterbatch-aditivos/
ANTI-FOG	Anti-fog masterbatch is an additive, which prevents the film to forming the fog on the surface of plastic.	
ANTI-OXIDANT	Antioxidant masterbatch avoids discoloration, change in viscosity, surface crazing, loss of physical and mechanical and optical properties.	
ANTI-STATIC	Antistatic masterbatch incorporation reduces the accumulation of statics elements to polymer surface, enhancing the appearance of final product.	
WEATHERPROOFING (UV STABILIZER)	UV Stabilizer masterbatch provides UV radiation protection to polyolefin plastics, can be used in films and tapes and is approved for use in contact with foodstuffs.	
NANOMODIFIED	Nano Modified is used as reinforce agent applied to HDPE and LLDPE blown film, PP Injection molding, PP sheet materials, PP wire drawing and HDPE, among others.	
CROMOFIX 50070	UV absorber. General use in low-thickness articles (films, raffia, sheet).	IQAP MASTERBATCH http://www.iqapgroup.com/iqap/sites/default/files/descargas-masterbatch/iqap-group-masterbatch-additives-english.pdf
CROMOFIX 50024	Antioxidant. Injection-moulded articles.	
CROMOFIX 50B017	UV absorber/Antioxidant. Agricultural films in contact with chemical agents.	
CROMOFIX 50441	UV absorber. Polyethylene bottles/Protection of content.	
CROMOFIX 794360	UV absorber. Injection-moulded articles, polyolefin bottles / protection of content.	
CROMOFIX 50234	UV absorber. PET bottles, PET or PETG sheet/ protection of content.	

Product	Description	Reference
CROMOFIX 50B005	UV absorber. PET bottles, PET or PETG sheet/protection of content and non-yellowing.	
CROMOFIX 50100	Phenolic antioxidant. Film, bottles and injection-moulded articles/long-term stabilizer.	
CROMOFIX 50179	Phenolic antioxidant. Film, bottles and injection-moulded articles/long-term stabilizer.	
CROMOFIX 750C042	Phenolic antioxidant/Organic phosphite. Film, bottles and injection-moulded articles/Process and long-term stabilizer.	
CROMOFIX 50175	Phenolic antioxidant/Organic phosphite. Film, bottles and injection-moulded articles/Process and long-term stabilizer.	
CROMOFIX 500078	Natural silica. Polyethylene film.	
CROMOFIX 50B400	Natural silica. Polyethylene film.	
CROMOFIX 50455	Synthetic silica. Film.	
CROMOFIX 50772	Natural silica, purified erucamide. Film, polyolefin sheet.	
CROMOFIX 50840	Partial polyol ester and talc. PET sheet, maximum transparency.	
CROMOFIX 50845	Partial polyol ester. PET sheet, injection-moulded articles.	
CROMOFIX 50127	Inorganic opaquing compound. Sheet or multi-layer films.	
CROMOFIX 50B00	Acrylic compound. PET bottles.	
CROMOFIX 750C027	Increase transparency. Clarifying additive. Injection-moulded articles, extrusion blow moulding, thermoforming.	
CROMOFIX 750C082	Increase transparency. Nucleating additive. Injection-moulded articles.	
PELLETHANE® 80A	Maintaining of polymer shelf-life, minimizing unanticipated changes in chemistry, molecular weight and color over time and designed for medical applications. Simulated Heat Aging/	FOSTER http://www.fostercomp.com/products/specialty-additives
PELLETHANE® 55D	Light Aging /Yellowness after Light Aging.	
PELLETHANE® 72D		
MAXITHEN® HP79860UV	Light stabilizers masterbatch. Suitable for the UV stabilization of several polyolefin films. Can be used for food packaging articles.	GABRIEL-CHEMIE https://gabriel-chemie.com/sites/default/files/inline-files/GC-FlexiblePackaging-en.pdf

Table 2 cont. ...

...Table 2 contd.

Master batch code/name	Activity/application	Company
MAXITHEN® HP796230UV	Provides excellent stabilization for articles exposed to higher temperatures. Can be used for food packaging articles.	
MAXITHEN® HP7790UV	UV absorber for PE films, in order to protect the packaged goods against UV radiation. Gives UV-blocking effects at UV wavelengths of between 250 and 340 nm.	
MAXITHEN® HP793700C12UV	UV absorber for PE films, in order to protect the packaged goods against UV radiation, with long-lasting UV absorption properties. Provide UV-blocking effects at UV wavelengths of between 250 and 380 nm.	
92011	UV-stabilizer. Reduces exposure to UV radiation and increases the duration of the use of the products without losing chemical and mechanical properties.	POLISTOM http://polistom.com/en/products/ masterbatches/additive-masterbatches. html
96011	Antistatic. Prevents the formation of static electricity on the surface of the product.	
00050	Chalk additive. LDPE base. Used in the production of films.	
95031	Reduces the coefficient of friction of the surface.	
EP UV 500 CL	Performance HALS-based UV stabilizers for all polyolefins.	ENERPLASTICS http://www.enerplastics.com/products/ uv-masterbatches/
EP UV 588 CP	PP tape extrusion & non-woven application.	
EP UV BARRIER PE 15	UV absorber for PE applications especially designed for food packaging to prolong shelf-life.	
EP UV 915 GH	HALS based UV Masterbatch for greenhouse films with green tint where medium pesticide-resistant is required.	
ZI019	Strong UV stabilizing properties. No discoloration of final product. recommended for transparent layers	SAMTECH http://www.samtech.kr/
UPL901		
UPL902	Strong UV protection properties. Color protection on final product. Recommended for colored layers.	
UPL903		

Product	Description	Source
COLOR MATRIX™ ULTIMATE 390	Allows less than 10% UV light transmission up to 390 nm accuracy. Can be combined with liquid colorants to achieve protection up to 575 nm	POLYONE http://www.polyone.com/products/polymer-additives/uv-and-light-blocking-additives/colormatrix-ultimate-uv-light-barrier-pet
COLORMATRIX™ ULTIMATE 370	Allows less than 10% UV light transmission up to 370 nm.	
KRITILEN UV 23	Special UV-absorber.	PLASTIKA KRITIS SA https://www.plastikakritis.com/assets/uploads/files/Additives_2013_March_2014.pdf
KRITILEN IR 550	Inorganic Infra-red absorber.	
KRITILEN HT 555	Special Infra-red absorber.	
KRITILEN DIFFUSER 557	Special inorganic diffuser.	
KRITILEN BROWN 70964	Pigments and Infra-red absorber.	
UV MASTERBATCH 1588	Composed of hindered amine light stabilizer that is dispersed in a poly-olefin carrier. Agriculture film. Chair moulding. Outdoor application.	UNIVERSAL MASTERBATCH LLP http://www.universalmasterbatch.com/additive-masterbatches.html
SOLASORB™ UV200F	Provides a UV absorption effect over a wide range with minimal loss of clarity. Recommended for food contact applications.	CRODA https://www.crodapolymeradditives.com/en-gb/products-and-applications/product-finder/product/129/Solasorb_1_UV200F

Acknowledgements

We express our gratitude to the Agencia Nacional de Promoción Científica y Técnica (ANPCyT, Argentina), the Consejo Nacional de Investigaciones Científicas y Técnicas (CONICET, Argentina), the Universidad Nacional de Cuyo (UNCu, Argentina) and the Universidad Nacional del Sur (UNS, Argentina) for their financial support.

References

Agboola, O., R. Sadiku, T. Mokrani, I. Amer and O. Imoru (2017). Polyolefins and the environment, pp. 89–133. *In*: S.C.O Ugbolue (Eds.). *Polyolefin Fibres: Structure, Properties and Industrial Applications*, Elsevier, Cambridge, USA.

Allen, N.S., M. Edge, A. Ortega, C.M. Liauw, J. Stratton and R.B. McIntyre (2002). Behaviour of nanoparticle (ultrafine) titanium dioxide pigments and stabilisers on the photooxidative stability of water-based acrylic and isocyanate-based acrylic coatings, *Polym. Degrad. Stab.*, 78: 467–478.

Allen, N.S., N. Mahdjoub, V. Vishnyakov, P.J. Kelly and R.J. Kriek (2018). The effect of crystalline phase (anatase, brookite and rutile) and size on the photocatalytic activity of calcined polymorphic titanium dioxide (TiO_2), *Polym. Degrad. Stab.*, 150: 31–36.

Allwood, M.C. and H.J. Martin (2000). The photodegradation of vitamins A and E in parenteral nutrition mixtures during infusion, *Clin. Nutr.*, 19: 339–342.

Al-Sabagh, A.M., F.Z. Yehia, G. Eshaq, A.M. Rabie and A.E. ElMetwally (2016). Greener routes for recycling of polyethylene terephthalate, *Egyp. J. Petrol.*, 25: 53–64.

Ammala, A., A.J. Hill, P. Meakin, S.J. Pas and T.W. Turney (2002). Degradation studies of polyolefins incorporating transparent nanoparticulate zinc oxide UV stabilizers, *J. Nanopart. Res.*, 4: 167–174.

Ammala, A., S. Bateman, K. Dean, E. Petinakis, P. Sangwan, S. Wong, Q. Yuan, L. Yu, C. Patrick and K.H. Leong (2011). An overview of degradable and biodegradable polyolefins, *Progr. Polym. Sci.*, 36: 1015–1049.

Andersen, M.L. and L.H. Skibsted (2010). Light-induced quality changes in food and beverages, pp. 113–139. *In*: L. Skibsted, J. Risbo and M. Andersen (Eds.). *Chemical Deterioration and Physical Instability of Food and Beverages*, Elsevier, New York, USA.

Antoniou, C., M. Kosmadaki, A. Stratigos and A. Katsambas (2008). Sunscreens—What's important to know, *J. Eur. Acad. Dermatol.*, 22: 1110–1119.

Arrieta, M.P., L. Peponi, D. López, J. López and J.M. Kenny (2017). An overview of nanoparticles role in the improvement of barrier properties of bioplastics for food packaging applications, pp. 391–424. *In:* A.M. Grumezescu (Ed.). *Food Packaging*, Elsevier, New York, USA.

Arslan, D. (2015). Effects of degradation preventive agents on storage stability of anthocyanins in sour cherry concentrate, *Agron. Res.*, 13: 892–899.

ASTM D7611/D7611M-18 (2018). Practice for coding plastic manufactured articles for resin identification, *ASTM International*, West Conshohocken, PA.

Ball, G.F.M. (2006). *Vitamins in Foods: Analysis, Bioavailability and Stability*, Taylor & Francis, Boca Raton, FL. USA.

Barba, F.J., L.R.B. Mariutti, N. Bragagnolo, A.Z. Mercadante, G.V. Barbosa-Cánovas and V. Orlien (2017). Bioaccessibility of bioactive compounds from fruits and vegetables after thermal and non-thermal processing, *Trends Food Sci. Tech.*, 67: 195–206.

Barrett, D.M., J.C. Beaulieu and R. Shewfelt (2010). Color, flavor, texture and nutritional quality of fresh-cut fruits and vegetables: Desirable levels, instrumental and sensory measurement and the effects of processing, *Crit. Rev. Food Sci.*, 50: 369–389.

Bhuvaneswari, H.G. (2018). Degradability of polymers, pp. 29–44. *In*: S. Thomas, A. Vasudeo Rane, K. Kanny, A. VK and M.G. Thomas (Eds.). *Recycling of Polyurethane Foams*, Elsevier, New York, USA.

Bodke, M.R., Y. Purushotham and B.N. Dole (2018). Comparative study on zinc oxide nanocrystals synthesized by two precipitation methods, *Cer.*, 64: 91–96.

Boon, C.S., D.J. McClements, J. Weiss and E.A. Decker (2010). Factors influencing the chemical stability of carotenoids in foods, *Crit. Rev. Food Sci.*, 50: 515–532.

Brothersen, C., D.J. McMahon, J. Legako and S. Martini (2016). Comparison of milk oxidation by exposure to LED and fluorescent light, *J. Dairy Sci.*, 99: 2537–2544.

Caponio, F., M.T. Bilancia, A. Pasqualone, E. Sikorska and T. Gomes (2005). Influence of the exposure to light on extra virgin olive oil qualityduring storage, *Eur. Food Res. Technol.*, 221: 92–98.

Cataldo, F. (2001). On the ozone protection of polymers having non-conjugated unsaturation, *Polym. Degrad. Stabil.*, 72: 287–296.

Chiste, R.C., A.S. Lopes and J.G. Lenio (2010). Thermal and light degradation kinetics of anthocyanin extracts from mangosteen peel (*Garcinia mangostana* L.), *Int. J. Food Sci. Tech.*, 45: 1902–1908.

Cooper, T.A. (2013). Developments in plastic materials and recycling systems for packaging food, beverages and other fast-moving consumer goods, pp. 58–107. *In:* N. Farmer (Eds.). *Trends in Packaging of Food, Beverages and other Fast-Moving Consumer Goods (FMCG)*, Woodhead Publishing Series in Food Science, Technology and Nutrition.

Cortez, R., D.A. Luna-Vital, D. Margulis and E. Gonzalez de Mejia (2017). Natural pigments: Stabilization methods of anthocyanins for food applications, *Compr. Rev. Food Sci. F.* 16: 180–198.

Dalsgaard, T.K., J. Sørensen, M. Bakman, L. Vognsen, C. Nebel, R. Albrechtsen and J.H. Nielsen (2010). Light-induced protein and lipid oxidation in cheese: Dependence on fat content and packaging conditions, *Dairy Sci. Technol.*, 90: 565–577.

Deveci, S., N. Antony and B. Eryigit (2018). Effect of carbon black distribution on the properties of polyethylene pipes - Part 1: Degradation of post-yield mechanical properties and fracture surface analyses, *Polym. Degrad. Stabil.*, 148: 75–85.

Devi, R.R. and T.K. Maji (2012). Effect of nano-ZnO on thermal, mechanical, UV stability, and other physical properties of wood polymer composites, *Ind. Eng. Chem. Res.* 51: 3870–3880.

Drobny, J.G. (2014). *Handbook of Thermoplastic Elastomers*, William Andrew, Amsterdam.

Duncan, S.E. and H.H. Chang (2012). Implications of light energy on food quality and packaging selection, pp. 25–73. *In:* S. Taylor (Eds.). *Advances in Food and Nutrition Research*, Elsevier.

Duncan, S.E. and S. Hannah (2012). Light-protective packaging materials for foods and beverages, pp. 303–322. *In:* K.L. Yam and D.S. Lee (Eds.). *Emerging Food Packaging Technologies*, Woodhead Publishing Limited

Eldesouky, A., A.F. Pulido and F.J. Mesias (2015). The role of packaging and presentation format in consumers preferences for food: An application of projective techniques, *J Sens. Stud.*, 30: 360–369.

Emblem, A. (2012). Plastics properties for packaging materials, pp. 287–309. *In:* A. Emblem and H. Emblem (Eds.). 2012. *Packaging Technology: Fundamentals, Materials and Processes*, Elsevier.

Fakirov, S. (2017). *Fundamentals of Polymer Science for Engineers*, Wiley-VCH Verlag GmbH & Co. KGaA, Weinheim, Germany.

Forsthuber, B., C. Schaller and G. Grüll (2013). Evaluation of the photo stabilising efficiency of clear coatings comprising organic UV absorbers and mineral UV screeners on wood surfaces, *Wood Sci. Technol.*, 47: 281–297.

Francis, J. and D. Dickton (2015). Effects of light on riboflavin and ascorbic acid in freshly expressed human milk, *J. Nutr. Health Food Eng.*, 2: 00083.

Gadioli, I.L., L. de L. de O. Pineli, J.D.S.Q. Rodrigues, A.B. Campos, I.Q. Gerolim and M.D. Chiarello (2013). Evaluation of packing attributes of orange juice on consumers intention to purchase by conjoint analysis and consumer attitudes expectation, *J Sens. Stud.*, 28: 57–65.

Ghasemi-Kahrizsangi, A., J. Neshati, H. Shariatpanahi and E. Akbarinezhad (2015). Improving the UV degradation resistance of epoxy coatings using modified carbon black nanoparticles, *Prog. Org. Coat.*, 85: 199–207.

Gomes, L.B., J.M. Klein, R.N. Brandalise, M. Zeni, B.C. Zoppas and A.M.C. Grisa (2014). Study of oxo-biodegradable polyethylene degradation in simulated soil, *Mater. Res.*, 17: 121–126.

Grand, A.F. and C.A. Wilkie (2000). *Fire Retardancy of Polymeric Materials*, CRC Press, New York. USA.

Grant-Preece, P., C. Barril, L.M. Schmidtke, G.R. Scollary and A.C. Clark (2015). Light-induced changes in bottled white wine and underlying photochemical mechanisms, *Crit. Ver. Food Sci.*, 57: 743–754.

Grigoriadou, I., E. Pavlidou, K. M. Paraskevopoulos, Z. Terzopoulou and D.N. Bikiaris (2018). Comparative study of the photochemical stability of HDPE/Ag composites, *Polym. Degrad. Stabil.*, 153: 23–36.

Hansen, E. and L.H. Skibsted (2000). Light-induced Oxidative Changes in a Model Dairy Spread. Wavelength Dependence of Quantum Yields, *J. Agric. Food Chem.*, 48: 3090–3094.

Hawkins, W.L. (1984). *Polymer Degradation and Stabilization*, Springer Berlin Heidelberg, Berlin, Heidelberg.

Hussmann Corporation (2017). *Retail Lighting Effects on Fresh Product Stability.*

Huvaere, K., M.L. Andersen, L.H. Skibsted, A. Heyerick and D. De Keukeleire (2005). Photooxidative degradation of beer bittering principles: A key step on the route to lightstruck flavor formation in beer. *J. Agr. Food. Chem.*, 53: 1489–1494.

Introduction to plastics (2018). *Practical Testing and Evaluation of Plastics*, pp. 1–44, Wiley-VCH Verlag GmbH & Co. KGaA, Weinheim, Germany.

Jabeen, N., I. Majid and G.A. Nayik (2015). Bioplastics and food packaging: A review, *Cogent Food Agric.*, 1: 1–6.

Jayasekara, R., I. Harding, I. Bowater and G. Lonergan (2005). Biodegradability of a selected range of polymers and polymer blends and standard methods for assessment of biodegradation, *J. Polym. Environ.*, 13: 231–251.

Jin, C.Y., B.S. Zhu, X.F. Wang and Q.H. Lu (2008). Cytotoxicity of titanium dioxide nanoparticles in mouse fibroblast cells, *Chem. Res. Toxicol.*, 21: 1871–1877.

Kaimainen, M. (2014). Stability of natural colorants of plant origin, doctoral thesis in Food Sciences, University of Turku, Turku, Finland.

Khalaj, M.J., H. Ahmadi, R. Lesankhosh and G. Khalaj (2016). Study of physical and mechanical properties of polypropylene nanocomposites for food packaging application: Nano-clay modified with iron nanoparticles, *Trends in Food Sci. Tech.*, 51: 41–48.

Kirwan, M.J. (2009). Plastics in food packaging, *Food Packaging Technology*, Wiley, Williston.

Kiss, T., T. Shimojima, K. Ishizaka, A. Chainani, T. Togashi, T. Kanai, X.Y. Wang, C.T. Chen, S. Watanabe and S. Shin (2008). A versatile system for ultrahigh resolution, low temperature, and polarization dependent laser-angle-resolved photoemission spectroscopy, *Rev. Sci. Instrum.*, 79: 023106.

La Mantia, F.P., M. Morreale, L. Botta, M.C. Mistretta, M. Ceraulo and R. Scaffaro (2017). Degradation of polymer blends: A brief review, *Polym. Degrad. Stabil.*, 145: 79–92.

Leppard, D., P. Hayoz, T. Schäfer, T. Vogel and F. Wendeborn (2002). Light stabilisers, *Int. J. Chem.*, 56: 216–224.

Limbo, S., L. Torri and L. Piergiovanni (2007). Light-induced changes in an aqueous β-carotene system stored under halogen and fluorescent lamps, affected by two oxygen partial pressures, *J. Agr. Food Chem.*, 55: 5238–5245.

Liu, M. and A. Horrocks (2002). Effect of carbon black on UV stability of LLDPE films under artificial weathering conditions, *Polym. Degrad. Stabil.*, 75: 485–499.

López-Fonseca, R., I. Duque-Ingunza, B. de Rivas, L. Flores-Giraldo and J.I. Gutiérrez-Ortiz (2011). Kinetics of catalytic glycolysis of PET wastes with sodium carbonate, *Chem. Eng. J.*, 168: 312–320.

López-Rubio, A., E. Almenar, P. Hernandez-Muñoz, J.M. Lagarón, R. Catalá and R. Gavara (2004). Overview of active polymer-based packaging technologies for food applications, *Food Rev. Int.*, 20: 357–387.

Lu, B. and Y. Zhao (2017). Photooxidation of phytochemicals in food and control: A review, photooxidation of phytochemicals and control, *Ann N.Y. Acad. Sci.*, 1398: 72–82.

Mamta, K. Misra, G.S. Dhillon, S.K. Brar and M. Verma. 2014. Antioxidants, pp. 117–138. *In:* S.K. Brar, G.S. Dhillon and C.R. Soccol (Eds.). *Biotransformation of Waste Biomass into High Value Biochemicals*, Springer New York, New York, USA.

Manaia, E.B., R.C.K. Kaminski, M.A. Corrêa and L.A. Chiavacci (2013). Inorganic UV filters, *Braz. J. Pharm. Sci.*, 49: 201–209.

Marsh, K. and B. Bugusu (2007). Food packaging - Roles, materials and environmental issues, *J. Food Sci.*, 72: R39–R55.

Marzec, A. (2014). The effect of dyes, pigments and ionic liquids on the properties of elastomer composites, *Doctor of Philosophy in Polymers and Composite Materials*, Technical University of Łódź and University Claude Bernard, Łódź, Poland.

Massey, S., A. Adnot, A. Rjeb and D. Roy (2007). Action of water in the degradation of low-density polyethylene studied by X-ray photoelectron spectroscopy, *Express Polym. Lett.*, 1: 506–511.

Mestdagh, F., B. De Meulenaer, J. De Clippeleer, F. Devlieghere and A. Huyghebaert (2005). Protective influence of several packaging materials on light oxidation of milk, *J. Dairy Sci.*, 88: 499–510.

Min, D.B. and J.M. Boff (2002). Chemistry and reaction of singlet oxygen in foods, *Compre. Rev. Food Sci. F.*, 1: 58–72.

Mitchell, B.S. (2003). *An Introduction to Materials Engineering and Science*, John Wiley & Sons, Inc., Hoboken, NJ, USA.

Moller, J.K.S. and L.H. Skibsted (2006). Myoglobins – The link between discoloration and lipid oxidation in muscle and meat, *Quim. Nova*, 29: 1270–1278.

Morabito, K., N.C. Shapley, K.G. Steeley and A. Tripathi (2011). Review of sunscreen and the emergence of non-conventional absorbers and their applications in ultraviolet protection: Emergence of non-conventional absorbers, *Int. J. Cosmetic Sci.*, 33: 385–390.

Murphy, J. (2001). *Additives for Plastics Handbook*, Elsevier, Oxford. UK.

Nair, N.R., V.C. Sekhar, K.M. Nampoothiri and A. Pandey. 2017. Biodegradation of biopolymers, pp. 739–755. *In:* A. Pandey, S. Negi, and C.R. Soccol (Eds.). *Current Developments in Biotechnology and Bioengineering*, Elsevier.

Nerin, C., L. Tovar, D. Djenane, J. Camo, J.S. Salafranca, J.A. Beltran and P. Roncales (2006). Stabilization of beef meat by a new active packaging containing natural antioxidants, *J. Agric. Food Chem.*, 54: 7840–7846.

Niaounakis, M. (2017). Degradation of plastics in the marine environment, pp. 127–142, *Management of Marine Plastic Debris*, Elsevier.

Nikafshar, S., O. Zabihi, M. Ahmadi, A. Mirmohseni, M. Taseidifar and M. Naebe (2017). The effects of UV light on the chemical and mechanical properties of a transparent epoxy-diamine system in the presence of an organic UV absorber, *Mat.*, 10: 180.

Ohata, M. (2016). Effect of long-time heating for polyvinyl chloride and polypropylene resin pellet certified reference materials for heavy metal analysis, *Anal. Sci.*, 32: 1003–1005.

Ojeda, T., A. Freitas, K. Birck, E. Dalmolin, R. Jacques, F. Bento and F. Camargo (2011). Degradability of linear polyolefins under natural weathering, *Polym. Degrad. Stabil.*, 96: 703–707.

Parejo, P.G., M. Zayat and D. Levy (2006). Highly efficient UV-absorbing thin-film coatings for protection of organic materials against photodegradation, *J. Mater. Chem.*, 16: 2165.

Petersen, K., P.V. Nielsen, G. Bertelsen, M. Lawther, M.B. Olsen, N.H. Nilsson and G. Mortensen (1999). Potential of biobased materials for food packaging, *Trends in Food Sci. Technol.*, 10: 52–68.

Peyroux, J., M. Dubois, E. Tomasella, L. Frézet, A.P. Kharitonov and D. Flahaut (2015). Enhancement of surface properties on low density polyethylene packaging films using various fluorination routes, *Eur. Polym. J.*, 66: 18–32.

Prashantha, K., J. Soulestin, M.F. Lacrampe, M. Claes, G. Dupin and P. Krawczak (2008). Multi-walled carbon nanotube filled polypropylene nanocomposites based on masterbatch route: Improvement of dispersion and mechanical properties through PP-g-MA addition, *Express Polym. Lett.*, 2: 735–745.

Rabek, J.F. (1995). *Polymer Photodegradation: Mechanisms and Experimental Methods*, Springer, Stockholm, Sweden.

Rahman, S. (2007). *Handbook of Food Preservation*, CRC Press, Boca Raton, FL, USA.

Rather, I.A., W.Y. Koh, W. Paek and J. Lim (2017). The sources of chemical contaminants in food and their health implications, *Front. Pharmacol.*, 8: 1–8.

Ray, S. and R.P. Cooney (2018). Thermal degradation of polymer and polymer composites, pp. 185–206. *In:* Myer Kutz (Ed.). *Handbook of Environmental Degradation of Materials*, Elsevier.

Reiss, P., E. Couderc, J. De Girolamo and A. Pron (2011). Conjugated polymers/semiconductor nanocrystals hybrid materials—Preparation, electrical transport properties and applications, *Nanoscale*, 3: 446–489.

Robertson, G.L. (2006). *Food Packaging: Principles and Practice*, Taylor & Francis/CRC Press, Boca Raton, FL, USA.

Rogers, M.E. and T.E. Long (2003). *Synthetic Methods in Step-Growth Polymers*, Wiley-Interscience, Hoboken, N.J.

Rudko, G., A. Kovalchuk, V. Fediv, W.M. Chen and I.A. Buyanova (2015). Enhancement of polymer endurance to UV light by incorporation of semiconductor nanoparticles, *Nanoscale Res. Lett.*, 10: 1–6.

Schaller, C., D. Rogez and A. Braig (2009). Hindered amine light stabilizers in pigmented coatings, *J. Coat. Technol. Res.*, 6: 81–88.

Schwartz, S.J., J.I. Cooperstone, M.J. Cichon, J.H. von Elbe and M.M. Giusti (2017). Colorants, pp. 1107. *Fennema's Food Chemistry*, Fifth ed., CRC Press, Boca Raton, FL.

Scott, G. (1999). Antioxidant control of polymer biodegradation, *Macromol. Sy.*, 144: 113–125.

Scott, G. and D. Gilead (1995). *Degradable Polymers: Principles and Applications*, Chapman & Hall, London.

Seite, S., A. Colige, P. Piquemal-Vivenot, C. Montastier, A. Fourtanier, C. Lapiere and B. Nusgens (2000). A full-UV spectrum absorbing daily use cream protects human skin against biological changes occurring in photoaging, *Photodermatol. Photo.*, 16: 147–155.

Sevim, K. and J. Pan (2018). A model for hydrolytic degradation and erosion of biodegradable polymers, *Acta Biomater*, 66: 192–199.

Singh, B. and N. Sharma (2008). Mechanistic implications of plastic degradation, *Polym. Degrad. Stabil.*, 93: 561–584.

Singh, P., S. Saengerlaub, A.A. Wani and H.C. Langowski (2012). Role of plastics additives for food packaging, *Pigm. Resin Technol.*, 41: 368–379.

Smijs, T.G. and S. Pavel (2011). Titanium dioxide and zinc oxide nanoparticles in sunscreens: focus on their safety and effectiveness, *Nanotechnol. Sci. Appl.*, 4: 95–112.

Spalding, M.A. and A.M. Chatterjee (2018). Handbook of Industrial Polyethylene and Technology: Definitive Guide to Manufacturing, Properties, Processing, Applications and Markets, Scrivener Publishing, Beverly, MA.

Sudhakar, M., A. Trishul, M. Doble, K. Suresh Kumar, S. Syed Jahan, D. Inbakandan, R.R. Viduthalai, V.R. Umadevi, P. Sriyutha Murthy and R. Venkatesan (2007). Biofouling and biodegradation of polyolefins in ocean waters, *Polym. Degrad. Stabil.*, 92: 1743–1752.

Suits, L.D. and Y.G. Hsuan (2003). Assessing the photo-degradation of geosynthetics by outdoor exposure and laboratory weatherometer, *Geotext. Geomembranes*, 21: 111–122.

Suman, S.P. and P. Joseph (2013). Myoglobin chemistry and meat color, *Annual Rev. Food Sci. Technol.*, 4: 79–99.

Tiwari, A.K., M. Gautam and H.K. Maurya (2018). Recent development of biodegradation techniques of polymer, *Int. J. Res.*, Granthaalayah 6: 414–452.

Tolinski, M. (2015). *Additives for Polyolefins: Getting the Most out of Polypropylene, Polyethylene and TPO*, Elsevier, Kidlington, Oxford, UK.

Tyler, D.R. (2004). Mechanistic aspects of the effects of stress on the rates of photochemical degradation reactions in polymers, *J. Macromol. Sci. C Polym. Rev.*, 44: 351–388.

Utracki, L.A., P. Mukhopadhyay and R.K. Gupta (2014). Polymer blends: Introduction, pp. 3–170. *In:* L.A. Utracki and C.A. Wilkie (Eds.). 2014. *Polymer Blends Handbook*, Springer, Netherlands, Dordrecht.

Van der Pols, J.C., C. Xu, G.M. Boyle, P.G. Parsons, D.C. Whiteman and A.C. Green (2006). Expression of p53 tumor suppressor protein in sun-exposed skin and associations with sunscreen use and time spent outdoors: A community-based study, *Am. J. Epidemiol.*, 163: 982–988.

Vlachopoulos, J. and D. Strutt (2003). Polymer processing, *Mat. Sci. Technol.*, 19: 1161–1169.

Vogel, H. (2013). Processing: Structural polymeric materials, pp. 151–166. *In:* M. Köhl, M.G. Meir, P. Papillon, G.M. Wallner and S. Saile (Eds.). *Polymeric Materials for Solar-Thermal Applications*, Wiley-VCH Verlag GmbH & Co. KGaA, Weinheim, Germany.

Wang, S.Q., Y. Balagula and U. Osterwalder (2010). Photoprotection: A review of the current and future technologies, *Dermatol. Ther.*, 23: 31–47.

Wheeler, J.S.R., S.W. Reynolds, S. Lancaster, V.S. Romanguera and S.G. Yeates (2014). Polymer degradation during continuous ink-jet printing, *Polym. Degrad. Stabil.*, 105: 116–121.

Wold, J.P., A. Veberg Dahl, F. Lundby, A.N. Nilsen, A. Juzeniene and J. Moan (2009). Effect of oxygen concentration on photo-oxidation and photosensitizer bleaching in butter, *Photochem. Photobiol.*, 85: 669–676.

Yousif, E. and R. Haddad (2013). Photodegradation and photostabilization of polymers, especially polystyrene: Review, Springer Plus, 2: 398.

Zhang, C., N. Waksmanski, V.M. Wheeler, E. Pan and R.E. Larsen (2015). The effect of photodegradation on effective properties of polymeric thin films: A micromechanical homogenization approach, *Int. J. Eng. Sci.*, 94: 1–22.

Zweifel, H. (1998). *Stabilization of Polymeric Materials*, Springer, Berlin, Heidelberg.

CHAPTER 2

Packaging Biodegradability
Trends in Food Industry Applications

Olga B. Alvarez-Pérez,[1] *Cristian Torres-León,*[1]
Rene Diaz-Herrera,[1] *Rafael Gomes-Araujo,*[1]
Jorge A. Aguirre-Joya,[2] *Janeth M. Ventura-Sobrevilla,*[2]
Miguel A. Aguilar-González[3] and *Cristóbal N. Aguilar*[1,*]

1. Introduction

With the increase in population during the last decade, an increasing problem of contamination due to the indiscriminate use of plastics and their disposal has been observed around the world, causing serious problems to the environment. For this reason, a large number of investigations focus on the study of compounds and materials of natural origin for the generation of plastics with short degradation time or for their incorporation into existing ones and modify the degradation times. Within the total production of plastics in the world amounting to 140 million tons, those that are used in the food industry make up two-thirds of the total volume and the difficulty in achieving their recycling and/or avoiding their use is still a challenge today. Biodegradable packaging forms a viable alternative for this sector of the industry, due to the multiple varieties of materials used in its manufacture. Among the most used are some polymers and polysaccharides, such as alginate, starch, gums, galactomannans, extracts of various plants rich in bioactive compounds, some of synthetic origin, such as polycaprolactone and polyhydroxybutyrate, which give packaging properties similar to plastic synthetics, potentially expanding its use in this industry. We can

[1] Food Research Department, School of Chemistry, Autonomous University of Coahuila, Coahuila, México.
[2] School of Health Sciences, Autonomous University of Coahuila, Coahuila, México.
[3] Center for Research and Advanced Studies of the National Polytechnic Institute. Parque Industrial Saltillo-Ramos Arizpe, Coahuila, México.
* Corresponding author: cristobal.aguilar@uadec.edu.mx

define biodegradable materials as those that are degraded by enzymatic action of bacteria, fungi and/or yeasts, which can be compostable and have a great potential to replace plastics based on petrochemical derivatives. These provide a barrier between the external environment and the food products, guaranteeing their innocuousness and extending their shelf-life, especially for those in which deterioration depends on oxidation and microbiological quality (Suderman et al., 2018). The materials used in this type of packaging play a very important role in determining the quality of these bioplastics and can be used alone or in combination with other materials to improve their properties (Fig. 1).

Packaging materials are classified into different categories by various criteria, either by the majority component in them, which can be biodegradable polymers and biopolymers, or by the origin and synthesis of the materials, which can be natural, synthetic and microbial polymers.

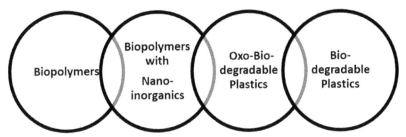

Fig. 1. Representative interrelation of biodegradable materials.

2. Natural Biopolymers

2.1 Proteins

Proteins are one the most abundant natural raw materials that can be used for creating novel food structures as films for food packaging, which is one of the larger consumers of industrial plastics (Wijaya et al., 2017). Proteins are known specifically for their surface activity, which allows them to play a major role in the formation and stabilization of emulsions (Rodriguez et al., 2011). Proteins are biomolecules formed by carbon, nitrogen, hydrogen and oxygen and which can also contain sulfur, phosphorus, iron, magnesium and copper among other elements. They consist of 20 to 500 amino acids, which in addition to the alpha amino and alpha carboxyl groups are involved in the peptide bonds. Differences in protein function result from differences in composition and amino acid sequence. The structure of proteins can be described at various levels of complexity. Four levels of protein structure are commonly defined (Nelson and Cox, 2013). The primary structure consists of a peptide-bonded amino acid sequence and includes any disulfide bonds. The secondary structure refers particularly to stable arrays of amino acid residues, giving rise to recurrent structural patterns (Nelson and Cox, 2013). The tertiary structure reflects the three-dimensional organization of the polypeptide chain, based on the hydrogen bond junctions, Van der Waals forces, electrostatic and hydrophobic interactions and disulfide bridges, to form globular, fibrous or random protein structures. Finally, the quaternary structure occurs as a consequence of the association of different polypeptide chains, equal or not to each other, that interact through non-covalent unions, originating unique molecules (Thompson et al., 2008).

Proteins provide the opportunity to use raw materials to produce biodegradable packaging. These biomaterials show promising properties because of their capacity to form three-dimensional networks stabilized and strengthened by hydrogen bonds, hydrophobic interactions and disulfide bonds (Romani et al., 2019). The films are formed by different types of amino acids that allow the development of intermolecular interactions (as ionic interactions), that combined with other compounds and carried out with different temperatures, allow obtaining of coatings with different chemical and physical properties. Optimizing interactions between amino acids, the formation of polymers with improved stability, barrier, mechanical and solubility properties are favored (Cinelli et al., 2014). Currently, in the food industry proteins are used from two sources: derived from animal sources (such as whey protein, casein, among others) and derived from plants (such as zein, soy, gluten, among others).

Proteins, such as wheat gluten, corn zein, soy protein, myofibrillar proteins and whey proteins have been successfully formed into films, using thermoplastic processes, such as compression molding and extrusion (Hernandez-Izquierdo and Krochta, 2008). Furthermore, plasticizers (such as glycerol, sucrose, and sorbitol) are generally added to the protein matrix to improve its processing and facilitate deformation and processing without thermal degradation. Plasticizers extend, soften the film structure and reduce cohesion within the film network by entering between polymeric molecular chains (Sharma and Singh, 2016). The resulting packages express excellent mechanical and barrier properties to different gases, such as water vapor, CO_2, O_2 and N_2 (Cinelli et al., 2014).

In recent years, most research works from the edible components field have focused on composite or multi-component films. Manufacturing composite (heterogeneous) films and coatings improves the barrier and mechanical properties of the formulations (Hassan et al., 2018). Biodegradable packaging has been developed from protein-polysaccharide interactions by cross-linking or other treatments that promote the attractive interactions between proteins and polysaccharides (Osés et al., 2009). These two classes of biopolymers can be used to create different types of colloidal systems with unique properties, such as nano or microparticles, hydrogels, films, emulsions, oil-filled hydrogels, foams and oleogels, to name a few (Wijaya et al., 2017). Generally, interactions between proteins and polysaccharides may occur non-covalently (attraction and repulsion between unlike biopolymers) and covalently (establishing permanent binding).

Protein-polysaccharide complexes also have been used in the stabilization of hydrophilic-hydrophobic interfaces. Eghbal et al. (2016) produced biopolymer-based films via complex coacervation of low methoxyl pectin (LMP) and sodium caseinate (SC) at pH 3.0 where the two macromolecules carry net opposite charges. The authors discovered that the complexation of polymers with opposite charge caused a decrease in water absorption and an increase in mechanical strength. The highest values of Young's modulus (182.97 ± 6.48 MPa) and tensile strength (15.64 ± 1.74 MPa) with a slight increase of elongation at break (9.35 ± 0.10 per cent) were obtained for films prepared at a CAS/LMP ratio equal to 0.05. This phenomenon may be due to the electrostatic interactions between the biopolymers of the opposite charge. The authors concluded that coacervation can be used to improve the properties of the films. This field of study is very promising for the development of active packaging.

Tsai and Weng (2019) manufactured biodegradable packaging based on whey and zein protein, finding that the packages have different properties, such as transparency, solubility among others when they are made individually and when the material is mixed, then the properties improve. Galus (2018) studied the properties of soy-based packaging and how they are affected with different concentrations of canola oil. The properties of a barrier to water vapor and sorption improved by increasing its functionality at low concentrations, showing that these packages can be used in food systems and can be dissolved during cooking. Piccirilli et al. (2019) used concentrated whey protein to evaluate the effect of protein and liquid smoke variations. They found that the packages show different alterations when subjected to different storage conditions. The temperature of 8°C showed the best results. Table 1 shows the barrier, mechanical, and molecular properties of protein-based films, according to the most recent studies in their manufacture with different components.

Depending on the desired application, it is required that the films have specific functionalities, for example, permeability and mechanical characteristics among others, and with this objective it is possible to modify their functionality at various stages of their manufacturing process, such as chemical or physical modification of the initial structure, use of additives, modification of chemical processes, among others (Deeth and Bansal, 2019).

Despite the results which demonstrate the advantages of biodegradability and good gas barrier and mechanical properties, the fact that proteins were not totally hydrophobic and contain predominantly hydrophilic amino acid residues limits their moisture-barrier properties (Shit and Shah, 2014). New alternatives to improve these properties are being investigated. Romani et al. (2019) investigated the effect of cold plasma as a surface modification strategy for the treatment of films prepared from fish myofibrillar proteins. The authors found that cold plasma glow affected the physicochemical properties, microstructure and thermal stability of fish protein films. This treatment is a promising alternative to improve the properties of biodegradable films of proteins.

Encapsulation of bioactive with protein-polysaccharide electrostatic complexes can also ensure the controlled-release in a biological system (Wijaya et al., 2017). One of the latest trends in food science and technology is the production of edible films enriched with various bioactive ingredients (antioxidants and antimicrobials) to improve their functional properties. These active molecules can be extracted from medicinal plants or from agro-industrial waste and byproducts (Adilah et al., 2018; Torres-León et al., 2017). Bioactive extracts represent an interesting ingredient for biodegradable food packaging material, mainly due to their functional properties which help in obtaining active materials aimed at extending the shelf-life of the food (Mir et al., 2018). Additionally, the interaction of extracts influences the techno-functional properties of the biopolymer, such as thickness, color, water vapor permeability, tensile strength and antioxidant properties (Torres-León et al., 2018). Wang et al. (2012) modified the properties of edible soy protein isolate films with anthocyanin-rich red raspberry (*Rubus strigosus*) extract. The bioactive extract significantly improved the tensile strength and percentage of elongation at break, while increasing the water-

Table 1. Different Properties of Protein Packaging Materials.

Protein-based film	Moisture content (%)	Solubility (%)	Thickness (mm)	TS (MPa)	EB (%)	WVP	Color (L*)	References
Whey protein and liquid smoke	15.4	33.5	0.133	1.8	2.6			(Piccirilli et al., 2019)
Soy protein			0.255–0.361	0.91–1.93	3.95–4.19			(Galus, 2018)
Whey protein and Zein	17.86–20.25		0.079–0.082	0.966–3.387	18.27–95.30			(Tsai and Weng, 2019)
Whey protein and almond		41.4–46.9	0.7	5.4–11.8	14.5–53.7			(Galus and Kadzińska, 2016)
Sesame protein		51.06–87	0.12	2.73	16.12		44.33	(Sharma and Singh, 2016)
Pectin and sodium caseinate			0.039	15.64	9.35	0.166a	42.34	(Eghbal et al., 2016)
Protein hydrolysate-chitosan			0.07	23.46	40		88–89	(Zhang et al., 2019)
Protein and mango seed extract		33	0.066	2.08–7.74		7.93b		(Adilah et al., 2018)
Fish gelatin and mango peels extract		20–23	0.039–0.042	8.22–15.78	42.09–50.68	1.98–2.31 c	85–87	(Adilah et al., 2018)

WVP: Water vapor permeability (a: x10⁻¹¹g.m/Pa.s.m²; b: x10⁻⁸g/m.s.Pa; c: x10⁻⁸g.mm/kPa.s.m²)

TS: Tensile strength

EB: Elongation at break

swelling ratio. The films showed significant decrease in water solubility and water vapor permeability. Adilah et al. (2018) discovered that incorporation of mango kernel extracts in soy protein isolate (SPI) and fish gelatin (FG) films increased the thickness, tensile strength and antioxidant activity significantly.

Protein materials have a great possibility of application in the food industry, although their introduction will be gradual due to pressure from consumers and legislation regarding the greater use of biodegradable materials. These can be used in future if necessary to continue working on the techniques and procurement processes, as well as in finding appropriate applications.

2.2 Polysaccharides

Polysaccharides are, in most cases, the generating matrix of biodegradable packaging, due to the multiple advantages they have due to barrier properties (Elham et al., 2016). In general, they produce packages with good mechanical properties and are efficient barriers against low polarity compounds. However, their hydrophilic nature means that they have a low resistance to water loss (Parra Huertas, 2009). Its selectivity, in terms of oxygen permeability and carbon dioxide, conditions the creation of modified atmosphere inside the food. This translates into an increase in the useful life of the product and also has selectivity between oxygen and carbon dioxide. Among the polysaccharides commonly used are derivatives of cellulose, starches, chitosan, alginates, carrageenans and pectins. Many of these materials are usually easily accessible, economical, abundant and completely biodegradable, making them ideal starting materials to develop fully degradable products that meet the needs of different markets (Bin et al., 2004). Starch is characterized as the first polysaccharide used in industrial applications. However, its hydrophilic nature has led to the investigation of other materials with similar properties and to the improvement thereof with modification and/or addition of compounds.

Starch is an important polysaccharide of energy reserves in plants and the second most abundant biopolymer on earth. It is natural, abundant, available, renewable, versatile and biodegradable polymer, composed mainly by amylose and amylopectin and some minor components as protein, lipids and minerals. Amylose is a linear polymer of α-1,4-linked glucans, and amylopectin is a larger molecule with highly α-1,6 branched chains (Chel-guerrero et al., 2016; Liu et al., 2017; Araújo et al., 2018; Khlestkin et al., 2018). The ratio between amylose and amylopectin, structure, morphology and size of starch granules varies according to the botanic source, stage of development of the plant and environmental conditions. The normal starches contain 75–80 per cent of amylopectin and 20–25 per cent of amylose, with some exceptions as the modified starch has only amylopectin. The starch is an important sustainable resource for various industries and can be obtained easily from fruits, seeds and tubers. The most common sources of commercial starch are cereals, like corn and wheat and the roots or tuberous, like cassava and potato. The tuberous plants have about 16–24 per cent of starch in weight and the rest is water with traces of lipids and proteins. The content of starch in cereals can be more than 60 per cent and 10–20 per cent of fibers, proteins and lipids. This is an important factor to keep in consideration in processes of manufacturing and further transformations (Zhu, 2017; Khlestkin et al., 2018). The food industry uses about 60 per cent of all starch extracts in products like sugar syrups, ice cream, snack foods, meat products, baby foods, fat replacers, coffee whitener, soups, sauces, beer and bakery products. The pharmaceutical industry uses the rest to produce packing material, seed coatings, paper, cardboard, adhesives, textiles, diapers, fertilizers, bioplastics, building materials, oil drilling and cement. Starch is an ingredient extensively used in many industries, but native starch cannot always withstand the extreme processing conditions, like freeze thaw cycles, high shear rates, high temperatures, strong acid and alkali treatments. Thus its use is limited (Hermansson and Svegmark, 1996; Liu et al., 2017; Hsieh et al., 2019). Some modifications can be induced in starch by physical (e.g., osmotic pressure treatment, extrusion, hydrothermal processes), chemical (e.g., oxidation, hydrozypropylation, etherification) and biological modifications (e.g., enzymatic modifications, isoamylase and pullulanase) to obtain different structures and consequently, different or new proprieties in gelatinization, pasting and retrogradation as per the requirements of specific applications (Henríquez et al., 2008; Zavareze et al., 2011; Schirmer et al., 2013; Matignon and Tecante, 2017). Currently, a large part of the packaging materials is used as one-time packaging material produced from petrochemical polymers, which are not friendly to environment due to improper disposal, non-degradability and incompatibility with living organisms (Thakur et al., 2016; Podshivalov et al., 2017). The increasing consumers' awareness concerning their health, environment, food nutritional value and food safety has generated the necessity to find an alternative to overcome the consequences of the non-renewable packaging. The edible films and coatings of starch from renewable sources have been considered as an alternative of packaging due to their being in abundance, biodegradable, environmental-friendly and inexpensive (Luchese et al., 2018). The major applications of biodegradable films are in the food industry for their prolongation of shelf-life of food and fruit, based on creating a protective membrane, performing the correct gas transfer between the environment and in medical applications to product delivery and preservation of drugs, in technology of feeding livestock in agriculture as well as in creating new food products (Falguera et al., 2011; Dhall, 2013; Galus and Kadzinska, 2016; Guo et al., 2017). Starch is a complex biopolymer used as a material of choice for the development of biodegradable films. The low cost and good film-forming proprieties of starch provides a good blending combination with other biopolymers, like carrageenan, chitosan, gelatin, xanthan gum and carnauba wax (Arismendi et al., 2013; Podshivalov et al., 2017; Talón et al., 2017). Combinations of different polysaccharides provide excellent miscibility and interactions and have been reported to overcome their individual limitations and produce desirable functional properties by combining the advantages of each component. Glycerol is the most common plasticizer used in starch films to obtain flexibility, due its low molecular weight which modifies the interactions between the macromolecules, increases the mobility of the polymer chains and reduces the glass transition temperature of the film. However, glycerol is very hydrophilic and hygroscopic and affects the water vapor and gas permeability of the films. Nevertheless, these proprieties can be improved by adding essential oils, stearic acid or waxes. These natural ingredients are hydrophobic components that improve the mechanical and barrier properties, and sometimes produce beneficial proprieties in the development of films, such as antioxidant and antimicrobial activity as a natural alternative to food preservation (Chiumarelli and Hubinger, 2014; Thakur et al., 2016; López-Córdoba et al., 2017; Caetano et al., 2018).

Recently, there has been development in the investigation of starch films and has sought to use some non-conventional sources of starch and bioactive ingredients from agro-industrial residues. The bioactive ingredients most used are essential oils or extracts, for example thyme extract, yerba mate extract, tea polyphenols, rosemary extract, basil extracts and many more bioactive compounds that can be added directly in the formulation or in controlled liberation in nanoparticles. Table 2 shows the recent formulations of starch films (Galus and Kadzinska, 2016; Jaramillo et al., 2017; Caetano et al., 2018). The formulation of films, coatings or bioplastics with natural compounds and with added bioactive ingredients generates novel and biodegradable packaging. These packagings are considered biodegradable materials because they can be degraded by enzymatic action of living organisms, such as fungi, bacteria and yeasts to generate end-products of the bioprocess of degradation, such as biomass, H_2O and CO_2 under aerobic conditions or methane, hydrocarbons and biomass under anaerobic conditions (Jamarillo et al., 2016; Carissimi et al., 2018; Khalid et al., 2018).

2.3 Lipids

2.3.1 Definition

The word 'lipid' derived from the Greek *lipos* means fats. Lipids are the fourth major constituent to be present in all the cells. There are several definitions for the word 'lipid' and one of them defines lipids as small hydrophobic or amphipathic molecules, which are insoluble in water, but soluble in non-polar organic solvents. Lipids are classified as important biomolecules since they fulfill both structural and physiological functions within cellular processes, such as energy storage, signal transduction and cell membrane component (Voet et al., 2007; Li-Beisson et al., 2016). Lipids have a great structural variety, which is why their properties depend on several factors (degree of saturation, length of fatty acid chain and lipid composition) (Kadhum and Shamma, 2017). The United Nations Food and Agriculture Organization (FAO) and Ibero-American Nutrition Foundation (FINUT) classify lipids into eight different categories, which include fatty acids, glycerolipids, glycerophospholipids, sphingolipids, sterols, isoprenoids, glycolipids and polyketides (FAO and FINUT, 2012).

2.3.2 Industrial Level

Due to their wide structural variety and properties, these compounds have many applications in different industrial sectors, such as food, nutraceuticals, pharmaceutical, cosmetic, chemical industry; in the area of detergents and in recent years in the field of nanomedicine, bioenergy and biofuel with the production of biodiesel (Zhu and Jackson, 2015; Li-Beisson et al., 2016; Castañeda et al., 2017; Guerrand, 2017). Oils are the most used lipids in the industrial sector, mainly due to two reasons: the first is because oils are one of the lipids with high concentrations within the cell and second, due to the processes of oil extraction, they are affordable and easy to reproduce on an industrial scale (Li-Beisson et al., 2016).

2.3.3 Food Industry

Lipid compounds are widely used and produced in different sectors of the food industry mainly as edible oils and animal and vegetable fats (FAO and FINUT, 2012;

Table 2. Properties of Starch Biodegradable Packaging.

Components	Thickness (mm)	Moisture content (%)	Solubility (%)	Tensile strength (MPa)	Elongation (%)	References
Starch film/tea polyphenol	0.1	12.6–13.7	-	16.3–20	64.3–74	(Feng et al., 2018)
Cassava starch/oregano essential oil and pumpkin extract	0.122–0.193	12.67–29.33	18.28–21.75	0.32–1.74	114–247	(Caetano et al., 2018)
Cassava starch/blueberry pomace	0.0913–0.1243	-	-	-	-	(Luchese et al., 2018)
Cassava starch/green tea and basil extracts	0.25	25.3–28.6	28–30	-	-	(Jaramillo et al., 2017)
Pea starch/chitosan	0.050–0.087	12.4–21.6	19.3–23.7	-	-	(Talón et al., 2017)
Gelatin/potato starch	0.10–0.14	15–18	-	-	70–145	(Podshivalov et al., 2017)
Rice starch-l-carrageenan	0.084–0.114	9.51–17.06	43.35–63.22	0.116	15.7–49.6	(Thakur et al., 2016)
Corn starch/buttermilk	0.045–0.057	-	-	0.66–31	0.44–16	(Moreno et al., 2014)
Cassava starch/carnauba wax and stearic acid	0.13	-	27.5–43.14	0.211–1.067	17.67–31.07	(Chiumarelli and Hubinger, 2014)
Tapioca starch/xanthan gum	-	0.75–0.97	21.9–42	-	71.6–280.6	(Arismendi et al., 2013)

Guerrand, 2017; Kadhum and Shamma, 2017). The use of these lipids in food coating is a very common application in the industry. Some of the applications are in the use of wax as a coating on fruits and vegetables to control food drying, in the wine industry, as an alternative to replace the plastic plugs, to cover corks with a layer of wax to prevent moistening, to avoid the release of flavors from the cork into the wine. Butter or cocoa-based coatings are widely used in the confectionery and biscuit industry and food-grade lacquers and varnishes to improve the appearance of food. Films are also used in dough-based products, such as pizza and cereal-based products to avoid migration of moisture and loss of flavor (Debeaufort and Voilley, 2009). Some of the lipids used for coating or as a barrier in foods are of animal and vegetable origin, fractionated and reconstituted oils and fats, natural and non-natural waxes, natural origin resins, essential oils and surfactants (Debeaufort and Voilley, 2009; Henriques et al., 2016).

2.3.4 Customer Demand

The use of biopolymers as a raw material in the preparation of packaging is an attractive alternative to reduce the use of plastic packaging. Since these compounds are biodegradable, unlike their counterparts of synthetic origin, they are of particular importance in the industrial sector (Peña-Rodriguez et al., 2014; Breda et al., 2017; Escamilla-García et al., 2017; Wang et al., 2019). Pollution caused due to the accumulation of plastic-based materials is expected to decrease in the near future and that is why researchers are focussing on the development of new packaging that is biodegradable and not toxic to the environment (Chevalier et al., 2018; Nouraddini et al., 2018). Therefore, it seems that the studies on edible films or coatings are the answer to the search for packaging materials of natural origin that are friendly to the environment and can provide more than food protection. This is something that results in addition to its importance for the consumer (Kaewprachu and Rawdkuen, 2016; Chevalier et al., 2018).

2.3.5 Importance of Lipid Application in Edible Films and Coatings

A large range of lipid compounds (fats and oils) can be used in coating food. The decision depends on the use it would have (Debeaufort and Voilley, 2009; Galus and Kadzińska, 2016). Films and edible coatings based on lipids can be applied alone as a layer on the food. However, the lipid films do not show good mechanical properties by themselves. An alternative is the addition of lipids to coatings made from polysaccharides and proteins as matrix, thereby forming a composite film or coating. This is done with the intention of providing hydrophobicity because the coatings with lipids show better properties of water vapor barrier, decrease the migration of moisture into the food and have a greater moisture resistance than coatings or films made with hydrocolloids. This is an important property in fresh foods, such as fruits and vegetables (Debeaufort and Voilley, 2009; Zaritzky, 2010; Sharma et al., 2016; Umaraw and Verma, 2017). In the last few years, numerous works on composite edible coatings have been carried out, since the interaction of the ingredients is a very important topic in research (Hassan et al., 2018). The films and edible coatings can be made according two methodologies—the first is to apply it in the form of a film over the polymeric coating to obtain a bilayer; second is to mix the lipid with the film-forming solution as an emulsion (Debeaufort and Voilley, 2009; Zaritzky, 2010).

The emulsion technique compared to bilayer presents some advantages since it takes less time in the process. The preparation of the emulsion only requires one drying process, which is less complex and the coating combines the properties of the components in it (Galus and Kadzińska, 2016). Therefore, the properties of the film or coating depend on the functionality of the components. For lipids, their physicochemical, functional, organoleptic and mechanical properties can be influenced by different factors (composition, melting and solidification temperatures, crystalline structure, interactions with water, oxygen, ingredients and food) (Debeaufort and Voilley, 2009; Alvarez-Pérez, 2013; Téllez-Rangel et al., 2018).

2.3.6 Edible Films and Coatings

In the food-coating area, many lipid compounds have been used, such as animal and vegetable waxes and vegetable oils. Table 3 shows a list of reported works of edible films and coatings with lipids (Moore and Akoh, 2017; Zhang et al., 2018).

Oils

The incorporation of oils in edible films and coatings turns out to be an interesting option for the food industry, since in addition to contributing hydrophobicity to the coating, the oils have a great nutritional impact on the human health because they help in prevention of numerous diseases through their antioxidant activity. Several sources of oils are reported to be used for food coating, like edible oils (soybean, corn, sunflower, olive, and palm oils) and essential oils (thyme oil, oregano oil and lemongrass oil) (Galus and Kadzińska, 2016; Bustos et al., 2016; Shahidi and Camargo, 2016; Zhang et al., 2018). Essential oils are secondary metabolites recognized generally as safe (GRAS). These oils have been used for years in edible films, essentially because their compounds improve the physical barrier and sensory properties of the coating; besides, the addition of essential oils improves the quality and shelf-life of food by decreasing the risk of pathogen growth, thanks to their antimicrobial and antioxidant properties (Atarés and Chiralt, 2016; Avila-Sosa et al., 2016; Bustos et al., 2016; Aminzare et al., 2017; Escamilla-García et al., 2017). Essential oils are ingredients of great interest in the area of biodegradable packaging (Atarés and Chiralt, 2016).

Waxes

Edible waxes have a greater resistance to water transport than most lipid-based coatings and are highly effective in blocking the migration of moisture. Waxes in nature can act as a coating for the protection of the fruit's skin (Alvarez-Pérez, 2013; Zhang et al., 2018). Some of the waxes, that are applied as coatings to reduce transpiration are beeswax, that is produced by the honeybees, the wax of carnauba, that is collected from the leaves of palms, sugarcane wax and candelilla wax, with candelilla wax being most effective (Olivas and Barbosa-Cánovas, 2009; Zaritzky, 2010; Barbosa et al., 2013; Umaraw and Verma, 2017). Candelilla wax is of vegetable origin, insoluble in wáter, but soluble in organic solvents. It is an ester with long-chain fatty acids derived from the leaves of a shrub which grows in northern Mexico and southern United States. This serum material has several applications, particularly in

Table 3. Reported Works of Edible Films and Coatings with Lipids.

Lipidic compound	Film matrix	Effect	References
Beeswax and carnauba wax	Gelatin	Enhances thermal stability and UV/visible light and water vapor barriers properties	(Zhang et al., 2018)
Candelilla wax	Candelilla wax	Improves shelf-life of Papaya (*Carica papaya* L.)	(Télles-Pichardo et al., 2013)
Candelilla wax and jojoba oil	Mixture of biopolymers (pectin, arabic and xanthan gums)	Extends shelf-life of green bell pepper	(Ochoa-Reyes et al., 2013)
Beeswax, candelilla wax and carnauba wax	Rennet casein native poder	Decreases WVP values	(Chevalier et al., 2018)
Anise essential oil	Chitosan	Improves moisture, solubility, water vapor permeability, tensile strength and elasticity properties of film, delays lipid oxidation and microbial spoilage and improves the chemical properties in chicken burger (4°C)	(Mahdavi et al., 2018)
Anise, orange, and cinnamon essential oils	Chitosan-Zein mixture	Improves physical properties, decrease WVP values and inhibit the growth of *Penicillium* sp. and *Rhizopus* sp.	(Escamilla-Garcia et al., 2017)
Microencapsulated lemongrass oil	Alginate	Decreases the growth of *Listeria monocytogenes*	(Bustos et al., 2016)
Combined essential oils of cinnamon and ginger	Chitosan	Retards total microbial growth in pork packaging	(Wang et al., 2016)
***Bunium persicum* and *Zataria multiflora* essential oils**	Corn Starch	Antioxidant properties	(Aminzare et al., 2017)
Sunflower oil	Quinoa protein/chitosan	Improves WVP properties	(Abugoch et al., 2016)
Emulsified oil	Collagen and egg white albumin	Decreases WVP values	(Téllez-Rangel et al., 2018)
Almond/walnut oils	Whey protein	Enhances light and water vapor barrier properties	(Galus and Kadzinska, 2016)
Butyric acid, lauric acid, palmitic acid, oleic acid, stearic acid and sucrose fatty acid ester	Rice starch and carrageenan	Improves physical and water barrier properties	(Thakur et al., 2017)

*WVP: Water vapour permeability

hardening other waxes, in manufacturing polishes as well as brighteners, in storage and transport of products, as well as in various industrial sectors, such as food, cosmetics, electrical and mechanical. It is a substance recognized as GRAS by the Food and Drug Administration (FDA) for safe application in the food sector (Saucedo-Pompa et al., 2009; Barbosa et al., 2013; Alvarez-Pérez et al., 2015).

3. Encapsulation of Compounds Incorporated in Biodegradable Packaging

Food packaging systems (FPM) have an important role in the food supply chain, such as in storage, handling, transport and protection of food from external contamination and in product preservation. Thus it is necessary to attend to their demands by developing novel, cost-effective, eco-friendly and versatile materials and processes for protecting and monitoring the quality of food products, to guarantee food safety and to improve the traceability of products (Han et al., 2018). Actually, most of the materials used in food packaging are non-biodegradable petroleum-based plastic materials, whose environmental impact leads to generation of 8 per cent of global gas and fossil feedstock. So bio-based (biodegradable and biocompatible) active and smart packaging using nanotechnology and encapsulation is an effective alternative (Kuswandi, 2017). One strategy is to modify these materials by adding features (antimicrobial, antioxidant, marking properties, etc.), in the matrix of the packaging. Encapsulation occurs when the active material is surrounded within protective wall materials used to protect or to delivery active material (McClements and Li, 2010). There are several encapsulation systems, such as liposomes, microemulsions, emulsion droplets, multilayer emulsions, multiple emulsions, solid lipid particles, molecular inclusion complexes and biopolymers nanoparticles (Steiner et al., 2018). Nevertheless, the three principal encapsulation applications used in green food packaging are nanocomposites, microgels and edible films.

The mixture of two or more components of different nature and significantly different properties in the composition of the packaging material generates a composite. A composite that integrates nanoparticles (NP) inside its matrix, embedded into its layers, or at least one of its components is in the nanometer range, is known as a nanocomposite (NC) (Dlamini et al., 2019). So nanocomposites have two phases, namely matrix (composite) and reinforcement (nanoparticles). The initial objective of a mixed-matrix nanocomposite was to resolve a recurrent issue in the gas separation process in the decade of 1990s (Chung et al., 2007). The different phases can be combined by several techniques (Xu et al., 2018) to obtain materials with properties that make them useful in different areas, like biomedical fields, water treatment, mixes separation, rechargeable batteries construction, sensor applications and fuel cells among others (Esfahani et al., 2019).

Bionanocomposites (BNC) are those which comprise a biopolymer of natural origin in combination with an inorganic molecule to enhance the biocompatibility and biodegradability of the obtained material (Shchipunov, 2012; Ahmad et al., 2017) and require specific methods for their preparation. They are also classified according to their structure into particulate, elongated and layered particle-reinforcement BNC (Zafar et al., 2016). BNC and NP are used: (a) to improve the mechanical and barrier properties, such as elasticity and gas barrier against oxygen, carbon dioxides and flavor compounds' diffusion and stability under temperature and moisture conditions (Youssef and El-Sayed, 2018); (b) to give information on real-time of packaged food product qualities in smart packaging (Suh et al., 2016; Wang et al., 2017); and (c) to protect the food products from internal and external factors, extending their shelf-life in active packaging (Wyrwa and Barska, 2017). The principal polymer-based nanomaterials used are polylactic acid, polyhydroxyalkanoates, starch, cellulose, chitosan, zein, whey protein; metal and non-metal oxides including MgO, Fe_3O_4, SiO_2, TiO_2; clay and silicates, such as chlorite, montmorillonite, illite and kaolinite (Huang et al., 2018).

Microgels are a special type of colloids that are constituted by a polymeric network with advantageous properties, like high porosity, adjustable dimensions and architecture and have important applications in industry (including food industry), basic research (as model system for condensed matter physics) and biomedicine (Agrawal and Agrawal, 2018; Rovigatti et al., 2019). These properties give microgels a unique interfacial behavior when interacting with other substances or environmental conditions (like temperature, pH, among others). The preparation of a functional microgel needs a special synthetic strategy, with most popular methodologies used in microgel preparation being microfluidic method, miniemulsion polymerization and precipitation polymerization. Each type of technique offers specific advantages in preparing microgels with specific properties and structure (Agrawal and Agrawal, 2018). Some of the most remarkable applications of microgels are in the designing of smart drugs and bioactive delivery systems and for immobilization and addition of functional ingredients into food matrixes and lately, in designing of smart food packaging (Ali and Ahmed, 2018; Batista et al., 2019).

The large amounts of food waste produced in the world is a priority problem as it impacts the individual and his social health. The most waste products are those that are regularly consumed fresh, like fruits, vegetables, meat, or fish (Martín-Sánchez et al., 2014). The use of edible covers or coatings has been a useful tool in improving protection of food products in recent years. Beside the benefits in the shelf-life of foodstuff, these materials have environmental, economic and biological advantages. In the area of food preservation, edible coatings are used as a replacement or as a complement of the natural layers of the product (like the husk in fruits) to avoid, prevent, or regulate the exchange of gases and other matter involved in the food respiration and aging processes in the environment (Tsironi and Taoukis, 2018). These materials are commonly made of polysaccharides/hydrocolloids, polypeptides, lipids and synthetic or mixed matrixes (Shit and Shah, 2014; Ali and Ahmed, 2018; Dhumal and Sarkar, 2018). The design and production of innovative edible covers have an intrinsic relationship with the nanobio-technological techniques mentioned above and can be prepared as two types: coatings (a layer formed by a solution where the food is dipped) and films (a layer that wraps the food) (Dhumal and Sarkar, 2018). Many animal and plant materials present good properties for use as edible films; an example is the poly(lactic acid), a thermoplastic material built with monomers derived from a wide variety of plants (Tawakkal et al., 2018). Other materials suitable for production of edible coating are cellulose and derivatives, like hydroxypropyl cellulose, hydroxypropyl methylcellulose, etc., but their poor water barrier needs to be improved by addition of hydrophobic substances (Pathare et al., 2013). Zein is a corn-derived, hydrophobic, prolamin protein with convenient thermoplastic properties and has been used in the reduction of moisture loss and color change in fresh fruits (Shukla and Cheryan, 2001; Bai et al., 2003; Shit and Shah, 2014). Candelilla wax is obtained from the Candelilla plant which is a complex substance manly of lipids that have been used to make edible coatings to improve shelf-life, gas and moisture exchange in avocado, strawberry and apple (Aguirre-Joya et al., 2017; Oregel-Zamudio et al., 2017; De León-Zapata et al., 2018).

By itself, the use of biomaterials in preparation of biodegradable food packaging presents benefits of diverse nature, like environmental, economic and commercial uses, but these features can be further improved by adding bioactive compounds. Nowadays, several approaches of this strategy have been developed to fuse bioactive compounds in the matrix of edible coatings mainly to preserve food quality for a longer time. The techniques to produce encapsulated food components are vesicles, emulsion preparations, liposomes, hydrogels, and solid nanoparticles mainly (Yin and Tsai, 2015). An application of these techniques entails addition of antioxidants, antimicrobial or biocontrol agents in the film-forming dispersion that will be used directly to cover foods. These additives can also be encapsulated to protect food from degrading by prolonging its availability or controlling its release. An example of this is the preparation of a nanogel of poly(N-isopropylacrylamide) copolymerized with acrylic acid to encapsulate pimaricin—a fungicide-type preservative of common use in food industry but it presents low solubility and poor stability at low pH and light exposure, to be applied in the surface of Arzúa-Ulloa DOP cheese for preventing spoilage. This technique did not affect the ripening parameters or the organoleptic properties of cheese during the maturation process or the shelf-life period, but prevented the appearance of spoilage microorganisms and reduced the direct release of pimaricin into the cheese, compared with the traditional application of the preservative as an aqueous solution directly on the product surface (Fuciños et al., 2017). Another approach is the nanoencapsulation of plants' essential oils with different objectives; for example, the essential oil of *Zataria multiflora* Boiss (Shirazi thyme) was nanoemulsified with sunflower oil, water and Tween 80 and applied to fresh tout in the cold storage to extend its shelf-life and preserve its organoleptic properties (Shadman et al., 2017); on the other hand, chitosan nanoparticles when loaded with essential oil of summer savory (*Satureja hortensis*) presented high antioxidant activity and antimicrobial function against *Escherichia coli* O157:H7, *Listeria monocytogenes* and *Staphylococcus aureus* (Zanetti et al., 2015).

Issues like the environmental impact of non-degradable food packaging, food spoilage and other concerns related to storage of food products may be, at least, partially addressed by the use of innovative technologies of bio-based polymers and the addition of bioactive compounds like essential oils or polyphenols. The inclusion of edible materials prepared from GRAS components (even non-food components) offers a wide range of possibilities for food protection, traceability, smart labeling, etc.

4. Characterization Techniques of Biodegradable Packaging

Characterization by means of spectrometric and microscopic techniques provides a status to the food-packaging manufacturing process. Simultaneously with these analysis techniques, it is reviewed that these materials are not the means of contamination by chemical compounds used in the elaboration or that there is any contamination of any type. For example, in the production of polylactic acid, zeolites (aluminosilicates based on Al, Si, Na and O mainly) are used. This material acts as a catalyst in the chemical process of conversion of lactic acid into lactide. Polylactic acid, PLA, is one of the main materials with which biodegradable edible packaging is made.

This section presents an overview of the updates in the microscopic techniques and some alternative techniques inherent in microscopy and which have been carried out recently for a complete characterization of biodegradable food packaging and some properties related to it have been reported.

There are different fabrication routes for biodegradable food packaging—chemical method (Qu et al., 2010), melt blending (Li et al., 2018), extrusion (Aranda-Garcia et al., 2015), blending film (Wang et al., 2015), solution casting (Farrag et al., 2018a), conventional solvent-casting technique and dried and multi-methods (Tunç and Duman, 2014).

4.1 Characterization of Food Packaging Morphology

Structural characterization is a very important part of the analysis because it allows the morphology-size-structure-swelling properties and solubility of the materials obtained with the new properties acquire a new composition (or in a pure manner) (Farrag et al., 2018b; Tunç and Duman, 2018).

The defects that can affect the barrier properties in films of samples of edible biodegradable packaging can be, among others, micropores and micro-cracks, open or closed porosity, irregular holes, irregular dispersion (in the case of biphasic or multiphase systems), extrusion marks, segregations, roughness, etc. Also factors inherent in its manufacturing process, such as concentration, chain length, morphology and crystalline structure.

The main objectives of food packaging are preservation of the level of quality similar to when they were produced/packaged and increase in the shelf-life while maintaining the beneficial effects. The characteristics of the packing must be resistant to chemical or biological alterations in addition to physical changes that may occur. It is very important to evaluate the safety conditions of packaging materials used for food products. Environmental factors can also alter the optimal condition of the packaging temperature, humidity or effect of oxygen or any other gas.

According to many researchers, packages based on plastic materials are more permeable than glass or metal. At present, there are different review works which report a successful application of nano-composites, where the advantages, disadvantages, expectations and precautions of nanomaterials are analyzed (Quirino et al., 2014; Yu et al., 2014; Tunc and Duman, 2014; Qu et al., 2010; McLauchlin and Thomas, 2012; Arora and Padua, 2010).

4.2 X-Ray Diffraction Analysis

X-ray diffraction (XRD) is the indicated technique to verify crystalline structure totally or partially for each component, specially in polymer–polymer, polymer-ceramic, polymer-organic, polymer-inorganic and biodegradable–biodegradable systems. Each component has the possibility of being analyzed, both in proportion as well as in its identification. In addition, in refinement analysis of diffractograms

(Rietveld Method), crystallographic information can be obtained. Currently there is a modality in the diffraction analysis of R-X, specific to analysis of thin films. The proportion of the industry of biodegradable products is categorized as metallic (11 per cent), vitreous (13 per cent), flexible (53 per cent), rigid (15 per cent) and others, including biodegradable (7 per cent).

4.3 High Resolution Optical Microscope with Image Analyzer

4.3.1 Phase Identification by Optical Microscopy

The definition of 'image analysis' refers to the physical quantification of some microstructural parameter contained in a sample/image (total or proportional), in a digital photomicrograph or image. This means obtaining a real, non-approximate value of some typical characteristic of some material, with the objective of relating the measurements to the elaboration process or an important event of some material. It is a graphic representation with a finite numerical value of the interior or details that constitute a regular form of some morphology.

It is important to note that according to estimates the industry of biodegradable packaging materials will increase in the year 2022 to 22 trillion dollars globally. The definition of biodegradable is attributed to those materials that present natural chemical processes and can also be decomposed into natural chemical elements by contact with natural agents, such as water, air, sun or the composition of soils. Examples of these materials are bacteria, plants or animals. Plastic materials with a synthetic polymer base and their derivatives are not biodegradable. This property is inherently linked to its crystalline structure.

There are different research works in which different methodologies are used for the characterization of both the properties of materials and their procedures. Some of them related to biodegradable food packaging are antimicrobial/antioxidant (Atares and Chiralt, 2016; Escamilla-García et al., 2017; Dutta et al., 2009) and biodegradable (Herrera et al., 2016; Wilfred et al., 2018, Li et al., 2018), color and barrier properties (Aguirre-Loredo et al., 2014; Atares and Chiralt, 2016; Mohan et al., 2017; Regalado-Gonzalez et al., 2013), mechanical properties and water absorption (Qu et al., 2010; Silverajah et al., 2012; Aranda-García et al., 2015; Aguirre-Loredo et al., 2014) and thermal properties (Aranda-García et al., 2015; Li et al., 2018).

4.3.2 Scanning Electron Microscopy of Ultra-high Resolution

To carry out efficient surface exploration, it is necessary to resort to the combination of *electromagnetic lenses with electromechanical lenses (in a mix)*. This ensures that the electron-sample interaction does not generate thermal damage and also retains the volume without any alteration. In addition, with the help of a new generation detector, called Gentle-beam, and the arrangement of detectors above and below of the stage—UED (upper electron detector) and LED (Lower electron detector) respectively—it leads to deceleration of the electrons to take place in the form of an electrical circuit or as a battery (1 keV with 1 mm of working distance). A study based on starch-polyphenols will be the reason for our upcoming publication, in which these techniques are used.

Recently, a series of scanning electron microscopes for field emission were introduced with which it is possible to verify biodegradable packaging since the 'degradation' of materials can be observed. Wilfred et al. (2018) conducted a very thorough study of the degradation of the materials and related it to their properties and carried out a very interesting characterization by FEG-SEM Field Emission Gun-Scanning Electron Microscopy.

According to published literature, the use of higher energy voltages causes the information acquired to be from regions farther away than the surface. Conventionally, for metal-based samples, it is recommended to use 20 kV at a working distance (WD) of 10 mm. In the case of samples of biodegradable food packaging, as they have low electrical conductivity, damages and short circuits occur. Uncertainty persists about the existence or not of the effect of the vacuum on the samples when they use conductive type coatings to create vacuum (High Vacuum =, Ion Sputering =). With the new mode of analyzing non-conductive samples at low energy voltages and working distances as short as 1 or 2 millimeters, the aforementioned gets decreased.

In general, the refined crystal structures contribute to an improvement in the mechanical properties of the nanocomposites (Yu et al., 2014).

4.4 Analysis of the Thickness in Edible Biodegradable Thin Films

The study of the thickness by the microscope contributes very important information from two fundamental points of view: firstly, observing the fracture corresponding to the packaging material, it can be related to the acquired mechanical properties (fracture mechanism) and secondly, measurements in the thickness can be related to values of the flexural strength. Cryogenic fracture implies that the sample of the film is introduced into liquid nitrogen and fractured later. After this, the sample is dried (in air at room temperature) and then coated in a sputtering ion device with a coating based on Gold-pd (Fig. 2).

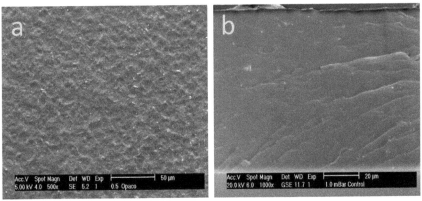

Fig. 2. SEM photomicrographs (a) of a section in an apple starch film and (b) surface section of the same film. Electron mode (SE) secondary electrons and gaseous secondary electrons (GSE), work at a distance of 11.7–5.2 and 5 and 20 keV.

4.5 Optical Microscopy of Depth Composition and 3D Visualization

The new microscopic techniques allow the observation of dispersion of particles both as reinforcers and as matrices. Also, it is possible to analyze, micro-cracks, porosity, cleanliness, inclusions, roughness, defects, height-depth profiles, fractures, thicknesses, etc. The best and most extraordinary technique is the manipulation of files obtained from the images and their processing in 3D. In this work, images of starch films with poly-phenols observed at 2500 and 5000x high resolution optical microscopy are reported for the first time (Fig. 3).

The new analysis techniques for biodegradable materials include different modes of use, such as stereographic of conventional reflected light and analysis with transmitted light. The range for observation has been increased to 5,000x in any of the three modes (with different lenses but in the same microscope). Bi-phasic or multi-phasic composites can be analyzed both superficially as well as in their depth of field from different angles (including 3D).

The development of reinforcing materials and modulators of flexibility as well as nano-particulates in biodegradable food packaging has been reported.

Then microscopic tests help to observe the total-migration behavior before and after making edible packages. Additionally, migration has also been regarded as a health-risk factor and/or a negative influence since many substances in contact with the content could alter their composition (Fig. 4).

Some researchers have paid special attention to the section on characterization of the sealing materials and their influence as well as on the interfacial compatibility of the elements of the packaging (Yu et al., 2014; Wang et al., 2015).

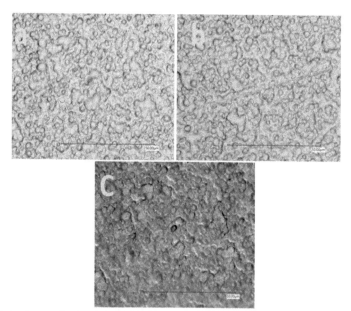

Fig. 3. Photomicrographs of digital optical microscopy of a thin film based on polyphenols and starch (a) with reflected light, (b) with transmitted light, and (c) with electronic filters.

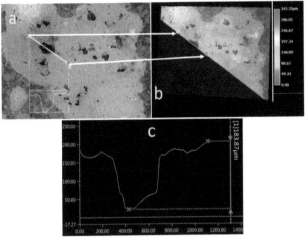

Fig. 4. Photomicrographs of a digital optical microscope. Samples were analyzed to 1000x in software of image analysis. Analysis of the depth pore composition is done in a composite material: (a) surface stroke, (b) depth analysis in microns, and (c) representative graph.

In most of the works reviewed here, conventional analysis techniques were used, such as thermogravimetric analysis (DSC-TGA), conventional and high-resolution microscopic studies, thermomechanical, biodegradable, antimicrobial, atomic force microscopy, visual inspection, titrations, chromatography of gases, DRX, FTIR, dynamic-mechanical, uv-vis and physico-chemical.

5. Mechanism of Biodegradation

Biodegradation refers to the disintegration, degradation and/or loss of attributes of packaging materials using microorganisms and hydrolysis followed by oxidation. All polymers are degraded in one way or another; therefore, all polymers should be considered degradable. Biodegradation rates are based mainly on temperature, type of microorganisms and humidity. Accelerated degradation is possible in the composting process where biopolymers are degraded to water, CO_2 and biomass for an average of six to 12 weeks. The effects of degradation can be manifested by different changes in the structure, in the surface of the materials, loss of mechanical properties, reduction of molecular weight and toxicity among others. On the other hand, degradation can be anaerobic and aerobic in nature, resulting in the formation of methane and hydrogen (biogas). The process of oxobiodegradation is carried out in natural and synthetic polymers. However, biodegradation needs an instantaneous mineralization, where the oxobiodegradation of carboxylic groups results in alcohol, aldehydes and ketone molecules, which are degradable and generate little waste during the peroxidation that begins by the action of light or heat. Majid et al. (2018) mentioned that this is the main reason why hydrocarbon polymers lose their mechanical properties, and after that, bioassimilation begins with enzymes, bacteria or fungi, resulting in the generation of

CO_2 and biomass. The presence of compounds, such as antioxidants or stabilizers, can interfere in the biodegradation process due to the inhibition of the oxidation process (Leela et al., 2018). But, in spite of this, it is necessary to incorporate them in the packaging to increase the shelf-life of the materials (Yu et al., 2014).

The biodegradability tests can be classified into two groups—breathing or using active sludges. These are based on measuring the oxygen consumption by microorganisms or the speed of assimilation while referring to the gaseous products of metabolism under different conditions, or they are put in contact with microorganisms in high concentrations to observe the degradation time.

The challenge for the successful use of biodegradable polymer products is achieving a controlled lifetime. Products must remain stable and function properly during storage and intended use, but biodegrade efficiently later. Edible films based on natural biopolymers are assumed to be biodegradable. However, the rate of biodegradation may decrease in the biopolymer films formed through covalent bonding. Intermolecular covalent cross-linking, i.e., the formation of new chemical bonds between individual macromolecules maybe considered the opposite of degradation, as it leads to an increase in molecular size and, at higher conversions, to certain kinds of superstructures (spatial networks) with characteristic physical properties.

6. Application of Biodegradable Packaging in Food Industry

It has been an increased demand to substitute traditional plastic packaging due to the negative impact on the entire ecosystem, such as plastic conglomeration, releasing toxic gases during incineration and water contamination with oil, whereas the main difference between conventional polymers and polymers from renewable sources is biodegradability. Conventional plastics are not biodegradable due to their long chain molecules that are complex and big enough to inhibit microbial degradation. However, they are not that way in the case for renewable polymers or biodegradable ones (Ivanković et al., 2017).

According to the *Biodegradable Polymers Market Report*, published in United Kingdom (Esch et al., 2010), biodegradable polymers are useful in a variety of applications, such as agriculture, bags and sacks and food industry. This last one has the largest market demand for bio-plastics or biodegradable plastics. This demand corresponds to a variety of factors, such as consumers' exigency and the regulation of governances, particularly in developed countries, to reduce the plastic waste and the negative ecological impact. For example, in actuality there are at least 14 countries with prohibition to use of plastic bags in supermarkets, including Kenya, United States, Spain, Mexico, Argentina, Chile, China and Bangladesh.

The classification of food packaging industry is done as flexible, rigid, glass, metal and others. In this last classification, biodegradable plastics are included and represent 7 per cent of the total food-packaging market (Government of Gujarat, 2017).

On the other hand, agricultural products are the main source for natural and renewable biopolymers for biodegradable plastics with food applications, such as starch, protein, cellulose and plant oils. Biomass from agricultural products represents the raw material to fabricate biodegradable packaging in the food industry and it can undergo any kind of biological physical or chemical treatment to create biodegradable packaging (FAO, 2016).

Biomass represents an attractive industrial option as raw material for biodegradable polymers, since it has the beneficila impact on the cost of biodegradable plastic manufacture due to its low cost. It is for this reason that the development of biodegradable packaging from renewable sources has acquired relevance in recent years (Mostafa and Tayeb, 2018). From agricultural biomass, by direct extraction, starch- and cellulose-based materials can be obtained by microbial fermentation of polyhidroxyalkanotes (PHA). Table 4 summarizes some of the main natural biodegradable polymers, the raw materials of origin; application in food industry and the traditional non-biodegradable polymer that is intended to be substituted.

According to Galgano et al. (2015) the main biopolymers are classified into three groups divided by their source; the groups are (1) polymers from renewable monomers synthetized by chemical reactions (e.g., polylactic acid), (2) polymers synthetized by microorganisms (e.g., polyhydroxyalkanoates) and (3) polymers obtained by extraction from animal or vegetal sources (e.g., lipids, proteins, etc.).

The main applications for biodegradable polymers in the food industry are as bio-packaging in films, stand-alone films and for modified atmospheric packages (MAP) where most of the times the films are edible. The packaging materials made from natural biopolymers represent a key innovation to reduce the negative environmental impact of plastic production and its waste management (Maharana et al., 2009; Ajay et al., 2018).

Edible films are coatings formed directly on the surface of the food. They are an integrated part of the food and are consumed along with the coated product, while the stand-alone packaging represents a food wrap that is not consumed with the food (Khalil et al., 2018). On the other hand, MAP are a type of food packaging for intentionally modifying the inner atmospheric conditions in the food packaged. Most of the times, fruits and vegetables are minimally processed or ready to eat and modification is to decrease O_2 and/or increase CO_2 to prolong the shelf-life or storability by metabolism retardation. Further, MAP can improve moisture retention (Ajay et al., 2018; Peelman et al., 2014).

Other forms of biodegradable food packaging include gels, that are used most of the time to control or prevent microbial contamination. As an example, there are the hydrogels that are a polymeric material conformed by a three-dimensional network with the capacity to absorb water or any water-soluble fluid, where the porpoise of the material in food packaging industry is humidity control inside the packaging (Batista et al., 2019).

Industrial application of biodegradable hydrogels includes removal of excess of moisture generated by fruit transpiration, water loss by physicochemical changes or by permeation of water vapor. It is also applicable to ready-to-eat meals and hygroscopic and crispy products (Ahmed, 2015; Azeredo et al., 2013).

Biodegradable bags represent a package technology based on the use of biomaterials, particularly the use of dextrose from corn. Biodegradable bags are resistant to breakage, are strong, moisture resistant and also temperature resistant and this makes them useful for food storage and packaging (Nampoothiri et al., 2010). On the other hand, biodegradable boxes with lids are made from corn, that are biodegradable in 47 days but do not release harmful substances into the environment.

Table 4. Some of the Main Raw Sources of Natural Biopolymers and Their Industrial Application.

Raw material	Natural polymer	Application in food industry	Polymer to substitute	References
Potatoes, corn, tapioca am other starch-based crops (residues can be used)	Starch based polymers	Bottles, coffee machine capsules, disposable cutlery and tableware	Polystyrene (PS)	Wagner (2014)
Wood and herbaceous crops, fiber crops	Cellulose based polymers	Dried products, meats, bread and fruits	Polystyrene (PS), Low density polyethylene (LDPE), High density polyethylene (HDPE), and Polypropylene (PP)	Wagner (2014)
Sugar cane and oil crops like olive, sunflower and soy	Polyhydroxyalkanoate (PHA)	Food containers	Polypropylene (PP) and polyethylene (PE)	Bilbow (N.D.)
Fish muscle	Protein	Coatings, films, food contact material	polyethylene (PE)	(Romani et al., 2019)

Its application in food industry is to maintain the fruit quality, as demonstrated by Almenar et al. (2008).

Another type of biodegradable food packaging is silver biodegradable packaging. Incorporation of silver in small concentrations in the biodegradable package maintains the quality of food and improves its storage capacity. Silver has the capacity to damage cell wall, cell membrane and cytoplasm of bacteria, and also affect the replication of DNA, rendering it antibacterial (Otoni et al., 2016).

6.1 Commercially Available Biodegradable Packaging

There are several industries that offer biodegradable packaging with the use of various and diverse polymers with diverse applications in food industry. The intention of producing biodegradable polymers on an industrial scale is to produce environmentally-responsible packaging for the food industry. Around the world exist diverse examples of companies dedicated to production on a commercial scale of biodegradable food packaging that are listed in Table 5.

One of the most used polymers in addition to starch is polylactic acid (PLA), a thermo-biopolymer derived from sugarcane and corn starch, which are renewable products for biodegradable film manufacturing. They can be produced in two ways: (1) polymerization of ring-opening of lactic with metal catalyst, and (2) condensation of lactic acid monomers at less than 200°C. However, PLA presents two main disadvantages in biodegradable packaging: firstly, its low glass transition temperature makes PLA-based plastics inefficient for hot liquids and also due to its inherent brittleness. An important consideration in use if PLA-based biodegradable food packaging is that PLA is classified as Generally Recognized As Safe (GRAS) by the American Food and Drug Administration (FDA) and thus can be used with no more restrictions than the Good Manufacture Practices (GMP) (Chiesa, 2017). It is authorized by the European Commission (Commission Regulation No. 10/2011) and for these characteristics, PLA is a good option for food-contact material (Bishai et al., 2014).

Table 5. Examples of Some Companies Around the World with Commercially Available Biodegradable Packaging for foods.

Company	Product	Application	Characteristics	References
Novamont (Novara, Italy)	Mater-Bi®	Films for vegetables or bags	Made from starch, biodegradable and compostable	(Nobile et al., 2008)
Wilkinson Industries Inc. (Fort Calhoun, New England)	VersaPack®	Containers for minimally processed or ready to eat food	Prolong shelf-life of blue berries. Recyclable PET	(Almenar et al., 2008)
United Biopolymers (Germany)	BIOPAR®	Vegetable, shopping and garbage bags including hot liquids (coffee, tea)	Potato starch based, completely biodegradable	(Khalil et al., 2018)
Plantic Technologies, (United States of America)	Plantic™	Diverse foods including pasta, chocolate, sea food, meat and bakery, etc.	Starch based packaging	(Ajay et al., 2018)

Nowadays there exists a diversity of options in the market of biodegradable food packaging, made with the intention to substitute traditional non-biodegradable plastics from fossil sources because of their negative effects on the environment. Biodegradable materials for food packaging are made from renewable sources, such as plants, microorganisms and agroindustry wastes. The intention to substitute non-biodegradable plastics with biodegradable ones is also to decrease pollution problems around the world by reducing the environmental impact and dependency on oil-based products (Muller et al., 2017). The food industry exhibits its compromise by offering eco-friendly options for food packaging, like shelf-life prolongation, moisture retention, organoleptic characteristics improvement, food safety and sustainability. Biodegradable food packaging represents a key technology for the responsible industrial development around the world.

7. Regulations Established for the Use of Biodegradable Packaging

All the materials that are used in the formulation of films and edible coatings must be regulated and cover all the specifications established for food products. Therefore, they must be used within the quantities or units specified in the guidelines for each country. All those who use them commercially must include all the ingredients of the formulations in the final label of the product. For its part, the manufacturers of films and coatings should be responsible in seeking the approval of all the concerned food regulatory agencies to ensure the health of consumers.

7.1 The European Food Safety Authority (EFSA)

EFSA offers independent scientific advice on risks related to food. Its advice is applied in European legislation and policies. In this way, it helps to protect the consumers from facing risks in the food chain. Their competence includes the feed safety, nutrition, animal health and welfare, plant protection and plant health.

Among all its tasks, the most important are to collect scientific data and knowledge, offer independent and updated scientific advice on food safety issues, disseminate its scientific work and cooperate with EU countries, international organizations and other interested parties. Trust lies in the EU's food safety system by offering reliable advice.

This legislation benefits the European consumers, who are among the best protected and informed in the world on the risks of the food chain. European and national institutions are responsible for the management of public health and authorize the consumption of food and feed.

Films and coatings are classified as food products, ingredients or additives. According to the legislation, an additive can be defined as a substance used in food for various reasons, such as sweetening, coloring or prolonging the shelf-life. An additive must present benefits for the consumers by preserving the nutritional quality of the food, helping in its manufacture, transformation, preparation, treatment, packaging, transport or storage and meeting special dietary needs.

7.2 The FDA Food Safety Modernization Act (FSMA)

The purpose of this Act is to improve and strengthen the public health protection schemes, ensuring a safe and reliable supply according to the sanitary requirements for food in the United States. Under this legislation, all components that make up an edible coating or film must have the GRAS (Generally Recognized As Safe) designation by its acronym in English and be on the list of food additives by the FDA (Food and Drugs Administration).

Acknowledgements

Authors thank National Council of Science and Technology (CONACYT, Mexico) for finnancial support. This document is part of the research program in Food Packaging and Edible Films and Coatings of the Candelilla Research Group and the postgraduate program in Food Science and Technology of the Universidad Autónoma de Coahuila, México.

References

Abugoch, L., C. Tapia, D. Plasencia, A. Pastor, O. Castro-Mandujano, L. Lopez and V.H. Escalona (2016). Shelf-life of fresh blueberries coated with quinoa protein/chitosan/sunflower oil edible film, *J. Sci. Food Agr.*, 96: 619–626.

Adilah, A.N., B. Jamilah, M.A. Noranizan and Z.A.N. Hanani (2018). Utilization of mango peel extracts on the biodegradable films for active packaging, *J. Food Pack. Shelf Life*, 16: 1–7.

Adilah, Z.A.M., B. Jamilah and Z.A.N. Hanani (2018). Functional and antioxidant properties of protein-based films incorporated with mango kernel extract for active packaging, *Food Hydrocolloids*, 74: 207–218.

Agrawal, G. and R. Agrawal (2018). Stimuli-responsive microgels and microgel-based systems: Advances in the exploitation of microgel colloidal properties and their interfacial activity, *Polymers*, 10: 418.

Aguirre-Joya, J.A. (2017). Effects of a natural bioactive coating on the quality and shelf-life prolongation at different storage conditions of avocado *Persea americana* Mill. cv. Hass, *J. Food Pack. Shelf-Life*, 14: 102–107.

Ahmad, M. (2017). Chitosan centered bionanocomposites for medical specialty and curative applications: A review, *Int. J. Pharm.*, 529: 200–217.

Ahmed, E.M. (2015). Hydrogel: Preparation, characterization, and applications: A review, *Journal of Advanced Research*, 6: 105–121.

Ali, A. and Ahmed, S. (2018). Recent advances in edible polymer-based hydrogels as a sustainable alternative to conventional polymers, *J. Agric. Food Chem.*, 66: 6940–6967.

Almenar, E., H. Samsudin, R. Auras, B. Harte and M. Rubino (2008). Postharvest shelf-life extension of blueberries using a biodegradable package, *Food Chem.*, 110: 120–127.

Alvarez-Pérez, O.B. (2013). *Desarrollo de una película comestible bioactiva a base de pectina y cera de candelilla*, thesis, *Universidad Autonoma de Coahuila*, Saltillo, Coahuila.

Alvarez-Pérez, O.B., J.C. Montanez, C.N. Aguilar and R. Rojas (2015). Pectin-candelilla wax: An alternative mixture for edible films, *J. Microbiol. Biotech. Food Sci.*, 5: 167–171.

Aminzare, M., E. Amiri, Z. Abbasi, H. Hassanzadazar and M. Hashemi (2017). Evaluation of *in vitro* antioxidant characteristics of corn starch bioactive films impregnated with *Bunium persicum* and *Zataria multiflora* essential oils, *Annual Research & Review in Biology*, 15: 1–9.

Aranda-García, F.J., R. González-Núñez, C.F. Jasso-Gastinel and E. Mendizábal (2015). Water absorption and thermomechanical characterization of extruded starch/polylactic acid/agave bagasse fiber bioplastic composites, *Int. J. Polym. Sci.*, 1–7.

Araújo, R.G. et al. (2018). Avocado by-products: Nutritional and functional properties, *Trends Food Sci. Tech.*, 80: 51–60.

Arismendi, C. et al. (2013). Optimization of physical properties of xanthan gum/tapioca starch edible matrices containing potassium sorbate and evaluation of its antimicrobial effectiveness, *LWT Food Sci. Tech.*, 53: 290–296.

Arora, A. and G. Padua (2010). Review: Nanocomposites in food packaging, *J. Food Sci.*, 75: 43–49.

Atarés, L. and A. Chiralt (2016). Essential oils as additives in biodegradable films and coatings for active food packaging, *Trends Food Sci. Tech*, 48: 51–62.

Avila-Sosa, R., E. Palou and A. López-Malo (2016). Essential oils added to edible films. *In: Essential Oils in Food Preservation, Flavor and Safety*, pp. 149–154.

Azeredo, H.M.C. (2013). Antimicrobial nanostructures in food packaging, *Trends Food Sci. Tech.*, 30: 56–69.

Bai, J. et al. (2003). Formulation of zein coatings for apples *Malus domestica* Borkh, *Postharvest Biol. Tec.*, 28: 259–268.

Barbosa, J., J. Dutra, M. Nucci, D. Barrera, L. Ricardo and R. Lopes (2013). Thermal and rheological properties of organogels formed by sugarcane or candelilla wax in soybean oil, *Food Res. Int.*, 50: 318–323.

Batista, R.J., P.J.P. Espitia, J.S.S. Quintans, M.M. Freitas, M.Â. Cerqueira, J.A. Teixeira and J.C. Cardoso (2019). Hydrogel as an alternative structure for food packaging systems, *Carbohyd. Polym.*, 205: 106–116.

Bilbow, D.N.D. (2019). Biopolymer Company MHG becomes First Company to Receive Vincotte certification for Marine Biodegradability, http://www.mhgbio.com/biopolymer-company-mhgbecomes-first-company-to-receive-vincotte-certification-for-marine-biodegradability/ Accessed Jan 28 2019.

Bin, Li and Bijun Xie (2004). Synthesis and characterization of konjac glucomannan/polyvinyl alcohol interpenetrating polymer networks, *J. Appl. Polym. Sci.*, 93: 2775–2780.

Bishai, M., S. De, B. Adhikari and R. Banerjee (2014). A comprehensive study on enhanced characteristics of modified polylactic acid based versatile biopolymer, *Eur. Polym. J.*, 54: 52–61.

Breda, C.A., D.L. Morgado, O.B.G. Assis and M.C.T. Duarte (2017). Processing and characterization of chitosan films with incorporation of ethanolic extract from 'pequi' peels, *Macromol. Res.*, 25: 1049–1056.

Bustos, R.O., F.V. Alberti and S.B. Matiacevich (2016). Edible antimicrobial films based on microencapsulated lemongrass oil, *J. Food Sci. Tech.*, 53: 832–839.

Caetano, S. et al. (2018). Characterization of active biodegradable films based on cassava starch and natural compounds, *J. Food Pack. Shelf-Life*, 16: 138–147.

Carissimi, M., S.H. Flôres and R. Rech (2018). Effect of microalgae addition on active biodegradable starch film, *Algal Research*, 32: 201–209.

Castañeda, M.T., S. Nuñez, C. Voget, F. Garelli and H. De Battista (2017). *Análisis De Balance De Flujos Metabólicos Aplicado a La Producciòn de Lìpidos Microbianos. In: IV Jornadas de Investigación, Transferencia y Extensión de la Facultad de Ingeniería, La Plata*, pp. 308–313.

Chel-guerrero, L. et al. (2016). Some physicochemical and rheological properties of starch isolated from avocado seeds, *Int. J. Biol. Macromol.*, 86: 302–308.

Chevalier, E., A. Chaabani, G. Assezat, F. Prochazka and N. Oulahal (2018). Casein/wax blend extrusion for production of edible films as carriers of potassium sorbate—A comparative study of waxes and potassium sorbate effect, *J. Food Pack. Shelf- Life*, 16: 41–50.

Chiesa, L.M. (2017). Department of environmental science and policy, *Università degli Studi di Milano*, Milan.

Chiumarelli, M. and M.D. Hubinger (2014). Evaluation of edible films and coatings formulated with cassava starch, glycerol, carnauba wax and stearic acid, *Food Hydrocolloids*, 38: 20–27.

Chung, T.S. et al. (2007). Mixed matrix membranes MMMs comprising organic polymers with dispersed inorganic fillers for gas separation, *Progress in Polymer Science*, Oxford, pp. 483–507.

Cinelli, P., M. Schmid, E. Bugnicourt, J. Wildner, A. Bazzichi, I. Anguillesi and A. Lazzeri (2014). Whey protein layer applied on biodegradable packaging film to improve barrier properties while maintaining biodegradability, *Polym. Degrad. Stabil.*, 108: 151–157.

De León-Zapata, M.A. et al. (2018). Changes in the shelf-life of candelilla wax/tarbush bioactive-based nanocoated apples at industrial level conditions, *Sci. Hortic.*, 231: 43–48.

Debeaufort, F. and A. Voilley (2009). Lipid-based Edible Films and Coatings. *In: Edible Films and Coatings for Food Applications*, pp. 135–168, Springer, New York, NY.

Deeth, H. and N. Bansal (2019). Whey Proteins: An Overview, *Whey Proteins*, 1–50.

Dhall, R.K. (2013). Advances in edible coatings for fresh fruits and vegetables: A review, *Crit. Rev. Food Sci.*, 53: 435–450.

Dhumal, C.V. and P. Sarkar (2018). Composite edible films and coatings from food-grade biopolymers, *J. Food Sci. and Technology*, 55: 4369–4383.

Dlamini, D.S., B.B. Mamba and J. Li (2019). The role of nanoparticles in the performance of nano-enabled composite membranes—A critical scientific perspective, *Sci. Total Environ.*, 656: 723–731.

Dutta, P.K., Shipra, Tripathi, G.K. Mehrotra and Joydeep Dutta (2009). Perspectives for chitosan based antimicrobial films in food applications, *Food Chem.*, 114: 1173–1182.

Eghbal, N., M.S. Yarmand, M. Mousavi, P. Degraeve, N. Oulahal and A. Gharsallaoui (2016). Complex coacervation for the development of composite edible films based on LM pectin and sodium caseinate, *Carbohyd. Polym.*, 151: 947–956.

Escamilla-García, M., G. Calderón-Domínguez, J.J. Chanona-Pérez, A.G. Mendoza-Madrigal, P. Di Pierro, B.E. García-Almendárez and C. Regalado-González (2017). Physical, structural, barrier and antifungal characterization of chitosan-zein edible films with added essential oils, *Int. J. Mol. Sci.*, 18: 2370.

Esch, P.M., C. Moor, B. Schmid, S. Albertini, S. Hassler, G. Donzé and H.P. Saxer (2010). *Biodegradable Polymers Market Report*, Qual. Assur. J.13.

Esfahani, M.R. et al. (2019). Nanocomposite membranes for water separation and purification: Fabrication, modification and applications, *Separation and Purification Technology*, 213: 465–499.

Falguera, V. et al. (2011). Edible films and coatings: Structures, active functions and trends in their use, *Trends Food Sci. Tech.*, 22: 292–303.

FAO & FINUT (2012). *Grasas y ácidos grasos en nutrición humana: Consulta de expertos, Estudio FAO alimentación y nutrición*.

Farrag, Y., S. Malmir, B. Montero, M. Rico, S. Rodríguez-Llamazares, L. Barral and R. Bouza. (2018a). Starch edible films loaded with donut-shaped starch microparticles, *LWT Food Sci. Tech.*, 98: 62–68.

Farrag, Y., W. Ide, B. Montero, M. Rico, S. Rodríguez-Llamazares, L. Barral and R. Bouza (2018b). Preparation of starch nanoparticles loaded with quercetin using nanoprecipitation technique, *Int. J. Biol. Macromol.*, 114: 426–433.

Feng, M. et al. (2018). Development and preparation of active starch films carrying tea polyphenol, *Carbohyd. Polym.*, 196: 162–167.

Fuciños, C. et al. (2017). Evaluation of antimicrobial effectiveness of pimaricin-loaded thermosensitive nanohydrogel coating on Arzúa-Ulloa DOP cheeses, *Food Control*, 73: 1095–1104.

Galus, S. and J. Kadzińska (2016). Whey protein edible films modified with almond and walnut oils, *Food Hydrocolloids*, 52: 78–86.

Galus, S. (2018). Functional properties of soy protein isolate edible films as affected by rapeseed oil concentration, *Food Hydrocolloids*, 85: 233–241.

Government of Gujarat (2017). *Establishment of Biodegradable Packaging for Food Products Manufacturing Unit Agro and Food Processing*.

Guerrand, D. (2017). Lipases industrial applications: Focus on food and agroindustries, *Oilseeds and Fats Crops and Lipids*, 24: 1–7.

Guo, M., M.P. Yadav and T.Z. Jin (2017). Antimicrobial edible coatings and films from micro-emulsions and their food applications, *Int. J. Food Microbiol.*, 263: 9–16.

Han, J.-W. et al. (2018). Food Packaging: A Comprehensive Review and Future Trends, *Comprehensive Reviews in Food Science and Food Safety*, 17: 860–877.

Hassan, B., S.A.S. Chatha, A.I. Hussain, K.M. Zia and N. Akhtar (2018). Recent advances on polysaccharides, lipids and protein based edible films and coatings: A review, *Int. J. Biol. Macromol.*, 109: 1095–1107.

Henriques, M., D. Gomes and C. Pereira (2016). Whey protein edible coatings : recent developments and applications, pp. 177–196. *In: Emerging and Traditional Technologies for Safe, Healthy and Quality Food*, Food Eng. Series.

Henríquez, C. (2008). Characterization of piñon seed (*Araucaria araucana* (Mol) K. Koch) and the isolated starch from the seed, *Food Chem.*, 107: 592–601.

Hermansson, A. and K. Svegmark (1996). Developments in the understanding of starch functionality, *Trends Food Sci. Tech.*, 7: 345–353.

Hernandez-Izquierdo, V.M. and J.M. Krochta (2008). Thermoplastic processing of proteins for film formation—A review, *J. Food Sci.*, 73: 30–39.

Hsieh, C. et al. (2019). Structure, properties, and potential applications of waxy tapioca starches— A review, *Trends Food Sci. Tech.*, 83: 225–234.

Huang, Y. et al. (2018). Recent developments in food packaging based on nanomaterials, *Nanomaterials*, 8: 830.

Ivanković, A., K. Zeljko, S. Talić, Martinović Bevanda, Anita and Lasić, M. (2017). Biodegradable packaging in the food industry, *Archiv Für Lebensmittelhygiene*, 23–52.

Jamarillo, C. et al. (2016). Biodegradability and plasticizing effect of yerba mate extract on cassava starch edible films, *Carbohyd. Polym.*, 151: 150–159.

Jaramillo, C. et al. (2017). Active and smart biodegradable packaging based on starch and natural extracts, *Carbohyd. Polym.*, 176: 187–194.

Kadhum, A.A.H. and M.N. Shamma (2001). Edible lipids modification processes: A review, *Crit. Rev. Food Sci.*, 57: 48–58.

Kaewprachu, P. and S. Rawdkuen (2016). Application of active edible film as food packaging for food preservation and extending shelf-life, *Micro. Food Hea.*, 185–205.

Khalid, S. et al. (2018). Development and characterization of biodegradable antimicrobial packaging films based on polycaprolactone, starch and pomegranate rind hybrids, *J. Food Pack. Shelf-Life*, 18: 71–79.

Khalil, H.P.S.A., A. Banerjee, C.K. Saurabh, Y.Y. Tye, A.B. Suriani and A. Mohamed (2018). Biodegradable films for fruits and vegetables packaging application: Preparation and properties, *Food Eng. Rev.*, 10: 139–153.

Khlestkin, V.K., S.E. Peltek and N.A. Kolchanov (2018). Review of direct chemical and biochemical transformations of starch, *Carbohyd. Polym.*, 181: 460–476.

Kuswandi, B. (2017). Environmental-friendly food nano-packaging, *Environmental Chemistry Letters*, 15: 205–221.

Leela, S.D., J.L. Lane, T. Grant, S. Pratt, P.A. Lant and B. Laycock (2018). Environmental impact of biodegradable food packaging when considering food waste, *J. Clean Prod.*, 180: 325–334.

Li-Beisson, Y., Y. Nakamura and J. Harwood (2016). Lipids: From chemical structures, biosynthesis and analyses to industrial applications, pp. 1–18. *In: Lipids in Plant and Algae Development.*

Liu, G. et al. (2017). Structure, functionality and applications of debranched starch: A review, *Trends Food Sci. Tech.*, 63: 70–79.

López-Córdoba, A. et al. (2017). Cassava starch films containing rosemary nanoparticles produced by solvent displacement method, *Food Hydrocolloids*, 71: 26–34.

Luchese, C. et al. (2018). Development and characterization of cassava starch films incorporated with blueberry pomace, *Int. J. Biol. Macromol.*, 106: 834–839.

Maharana, T., B. Mohanty and Y.S. Negi (2009). Melt–solid polycondensation of lactic acid and its biodegradability, *Progress in Polymer Science*, 34: 99–124.

Mahdavi, V., S.E. Hosseini and A. Sharifan (2018). Effect of edible chitosan film enriched with anise *Pimpinella anisum* L. essential oil on shelf-life and quality of the chicken burger, *Food Sci. Nutr.*, 62: 269–279.

Majid, I., M. Thakur and V. Nanda (2018). Biodegradable Packaging Materials. *In: Reference Module in Materials Sci. Mat. Eng.*, pp.1–11.

Martín-Sánchez, A.M. et al. (2014). Phytochemicals in date co-products and their antioxidant activity, *Food Chem.*, 158: 513–520.

Matignon, A. and A. Tecante (2017). Starch retrogradation: From starch components to cereal products, *Food Hydrocolloids*, 68: 43–52.

McClements, D.J. and Y. Li (2010). Structured emulsion-based delivery systems: Controlling the digestion and release of lipophilic food components, *Advances in Colloid and Interface Science*, 159: 213–228.

McLauchlin, A.R. (2012). *Biodegradable Polymer Nanocomposites*, University of Exeter, UK and N. Thomas, Loughborough University, UK, Woodhead Publishing Limited.

Mir, S.A., B.N. Dar, A.A. Wani and M.A. Shah (2018). Effect of plant extracts on the techno-functional properties of biodegradable packaging films, *Trends Food Sci. Tech.*, 80: 141–154.

Mohan, T.P., Kay Devchand and K. Kanny (2017). Barrier and biodegradable properties of corn starch-derived biopolymer film filled with nanoclay fillers, *Journal of Plastic Film & Sheeting*, 33: 309–336.

Moore, M.A. and C.C. Akoh (2017). Enzymatic interesterification of coconut and high oleic sunflower oils for edible film application, *J. Am. Oil Chem. Soc.*, 94: 567–576.

Moreno, O. et al. (2014). Physical and bioactive properties of corn starch - Buttermilk edible films, *J. Food Eng.*, 141: 27–36.

Mostafa, N.A. and A.M. Tayeb (2018). Production of biodegradable plastic from agricultural wastes, *Arabian J. Chemistry*, 11: 546–553.

Muller, J., C. González-Martínez and A. Chiralt (2017). Combination of poly(lactic) acid and starch for biodegradable food packaging. *Materials*, 10: 952.

Nampoothiri, K.M., N.R. Nair and R.P. John (2010). Bioresource Technology: An overview of the recent developments in polylactide, *PLA Research*, 101: 8493–8501.

Nelson, D. and M. Cox (2013). Lehninger Principles of Biochemistry, fifth ed., *J. Chem. Inf. Model.*

Nobile, M.A. Del, A. Conte, M. Cannarsi and M. Sinigaglia (2008). Use of biodegradable films for prolonging the shelf-llife of minimally processed lettuce, *J. Food Eng.*, 85: 317–325.

Nouraddini, M., M. Esmaiili and F. Mohtarami (2018). Development and characterization of edible films based on eggplant flour and corn starch, *Int. J. Biol. Macromol.*, 120: 1639–1645.

Ochoa-Reyes, E., G. Martínez-Vazquez, S. Saucedo-Pompa, J. Montañez, R.A. Rojas-Molina, M. de Leon-Zapata, R. Rodriguez-Herrera and C. Aguilar (2013). Improvements of shelf-life quality of green bell peppers using edible coating coatings formulations, *J. Microbiol. Biotechn. Food Sci.*, 2: 2448–2451.

Olivas, G.I. and G. Barbosa-Cánovas (2009). Edible films and coatings for fruits and vegetables, pp. 211–244. *In: Edible Films and Coatings for Food Applications.*

Oregel-Zamudio, E. et al. (2017). Effect of candelilla wax edible coatings combined with biocontrol bacteria on strawberry quality during the shelf-life, *Sci. Hortic.*, 214: 273–279.

Osés, J., M. Fabregat-Vázquez, R. Pedroza-Islas, S.A. Tomás, A. Cruz-Orea and J.I. Maté (2009). Development and characterization of composite edible films based on whey protein isolate and mesquite gum, *J. Food Eng.*, 92: 56–62.

Otoni, C.G., P.J.P. Espitia, R.J. Avena-bustillos and T.H. Mchugh (2016). Trends in antimicrobial food packaging systems: Emitting sachets and absorbent pads, *Food Research International*, 83: 60–73.

Pankaj, S.K., C. Bueno-Ferrer, N.N. Misra, L. O'Neill, A. Jiménez, P. Bourke and P.J. Cullen (2014). Characterization of polylactic acid fi lms for food packaging as affected by dielectric barrier discharge atmospheric plasma, *Innov. Food Sci. Emerg.*, 21: 107–113.

Pathare, Y.S., V.S. Hastak and A.N. Bajaj (2013). Polymers used for fast disintegrating oral films: A review, *Int. J. Pharm. Sci. Rev. Res.*, 21: 169–178.

Peelman, N., P. Ragaert, A. Vandemoortele, E. Verguldt, B. Meulenaer, De and F. Devlieghere (2014). Use of bio-based materials for modified atmosphere packaging of short and medium shelf-life food products, *Innov. Food Sci. Emerg.*, 26: 319–329.

Peña-Rodriguez, C., J. Martucci, L. Neira, A. Arbelaiz, A. Eceiza and R. Ruseckaite (2014). Functional properties and *in vitro* antioxidant and antibacterial effectiveness of pigskin gelatin films incorporated with hydrolysable chestnut tannin, *Food Sci. Technol. Int.*, 21: 221–231.

Piccirilli, G.N., M. Soazo, L.M. Pérez, N.J. Delorenzi and R.A. Verdini (2019). Effect of storage conditions on the physicochemical characteristics of edible films based on whey protein concentrate and liquid smoke, *Food Hydrocolloids*, 8: 221–228.

Podshivalov, A. et al. (2017). Gelatin/potato starch edible biocomposite films: Correlation between morphology and physical properties, *Carbohyd. Polym.*, 157: 1162–1172.

Qu, P., Y. Gao, G. Wu and Z. Zhang (2010). Nanocomposites of poly lactic acid reinforced with cellulose nanofibrils, *Bio-Resources*, 5: 1811–1823.

Quirino, R.L., T.F. Garrison and M.R. Kessler (2014). Matrices from vegetable oils, cashew nut shell liquid, and other relevant systems for biocomposite applications, *Green Chem.*, 16: 1700–1715.

Regalado, C., C. Pérez-Pérez, E. Lara-Cortés and B. García-Almendarez (2006). Whey protein-based packaging films and coatings, *Research Signpost*, 237–261.

Rocío Yaneli Aguirre-Loredo, Adriana Inés Rodríguez-Hernández and Norberto Chavarría-Hernández (2014). Physical properties of emulsified films based on chitosan and oleic acid, *CyTA—J. Food*, 12: 305–312.

Rodriguez Patino, J.M. and A.M.R. Pilosof (2011). Protein-polysaccharide interactions at fluid interfaces, *Food Hydrocolloids*, 25: 1925–1937.

Romani, V.P., B. Olsen, M.P. Collares, J.R.M. Oliveira, C. Prentice-Hernández and V.G. Martins (2019). Improvement of fish protein films properties for food packaging through glow discharge plasma application, *Food Hydrocolloids*, 87: 970–976.

Rovigatti, L. et al. (2019). Numerical modelling of non-ionic microgels: An overview, *Soft Matter, Royal Soc.*, Ch, 15, pp. 1108–1119.

Saucedo-Pompa, S., R. Rojas-Molina, A.F. Aguilera-Carbó, A. Saenz-Galindo, H. de la Garza, D. Jasso-Cantú and C.N. Aguilar (2009). Edible film based on candelilla wax to improve the shelf-life and quality of avocado, *Food Res. Int.*, 42: 511–515.

Schirmer, M. et al. (2013). Physicochemical and morphological characterization of different starches with variable amylose/amylopectin ratio, *Food Hydrocolloids*, 32: 52–63.

Shadman, S. (2017). Evaluation of the effect of a sunflower oil-based nanoemulsion with *Zataria multiflora* Boiss. Essential oil on the physicochemical properties of rainbow trout *Oncorhynchus mykiss* fillets during cold storage, *Food Sci. Techn.*, 79: 511–517.

Shahidi, F. and A.C. de Camargo (2016). Tocopherols and tocotrienols in common and emerging dietary sources: occurrence, applications and health benefits, *Int. J. Mol. Sci.*, 17: 1745.

Sharma, D., P.K. Sharma, D. Singh and P.K. Sharma (2016). Edible membranes containing antimicrobial compounds: current approach and future prospects, pp. 207–223. *In*: Garg, N., Abdel-Aziz, S. and Aeron, A. (Eds.). *Microbes in Food and Health. Springer, Cham.*

Sharma, L. and C. Singh (2016). Sesame protein-based edible films: Development and characterization, *Food Hydrocolloids*, 61: 139–147.

Shchipunov, Y. (2012). Bionanocomposites: Green sustainable materials for the near future, *Pure and Applied Chemistry*, 84: 2579–2607.

Shit, S.C. and P.M. Shah (2014). Edible polymers: Challenges and opportunities, *Journal of Polymers*, 2014: 1–13.

Shukla, R. and M. Cheryan (2001). Zein: The industrial protein from corn, *Ind. Crop Prod.*, 13: 171–192.

Silverajah, V.S., N.A. Ibrahim, N. Zainuddin, W.M. Yunus and H.A. Hassan (2012). Mechanical, thermal and morphological properties of poly(lactic acid)/epoxidized palm olein blend, *Molecules*, 17: 11729–11747.

Steiner, B.M., D.J. McClements and G. Davidov-Pardo (2018). Encapsulation systems for lutein: A review, *Trends Food Sci. Tech.*, 82: 71–81.

Suh, S., X. Meng and S. Ko (2016). Proof of concept study for different-sized chitosan nanoparticles as carbon dioxide CO_2 indicators in food quality monitoring, *Talanta*, 161: 265–270.

Talón, E. et al. (2017). Release of polyphenols from starch-chitosan based films containing thyme extract, *Carbohyd. Polym.*, 175: 122–130.

Tawakkal, I.S.M.A., M.J. Cran and S.W. Bigger (2018). The influence of chemically treated natural fibers in polylactic acid composites containing thymol, *Polymer Composites*, 39: 1261–1272.

Télles-Pichardo, R., K. Cruz-Aldaco, E. Ochoa-Reyes, C.N. Aguilar and R. Rojas (2013). Cubiertas comestibles de cera y polifenoles de candelilla: una alternativa de conservación de papaya *Carica papaya* L., *Acta Química Mexicana*, 5: 1–7.

Téllez-Rangel, E.C., E. Rodríguez-Huezo and A. Totosaus (2018). Effect of gellan, xanthan or locust bean gum and/or emulsified maize oil on proteins edible films properties, *Emir. J. Food Agr.*, 30: 404–412.

Thakur, R. (2016). Characterization of rice starch-l-carrageenan biodegradable edible film: Effect of stearic acid on the film properties, *Int. J. Biol. Macromol.*, 93: 952–960.

Thakur, R., P. Pristijono, J.B. Golding, C.E. Stathopoulos, C.J. Scarlett, M. Bowyer and Q.V. Vuong (2017). Amylose-lipid complex as a measure of variations in physical, mechanical and barrier attributes of rice starch-ι-carrageenan biodegradable edible film, *J. Food Pack. Shelf-Life*, 14: 108–115.

Thompson, A., M. Boland and H. Singh (2008). Milk Proteins: From expression to food. Academic Press.

Torres-León, C., R. Rojas, L. Serna-Cock, R. Belmares-Cerda and C.N. Aguilar (2017). Extraction of antioxidants from mango seed kernel: Optimization assisted by microwave, *Food Bioprod. Process*, 105: 188–196.

Torres-León, C., A.A. Vicente, M.L. Flores-López, R. Rojas, L. Serna-Cock, O.B. Alvarez-Pérez and C.N. Aguilar (2018). Edible films and coatings based on mango var. Ataulfo by-products to improve gas transfer rate of peach, *LWT Food Sci. Tech.*, 97: 624–631.

Tsai, M. and Y. Weng (2019). Novel edible composite films fabricated with whey protein isolate and zein: Preparation and physicochemical property evaluation, *LWT Food Sci. Tech.*, 101: 567–574.

Tsironi, T.N. and P.S. Taoukis (2018). Current practice and innovations in fish packaging, *J. Aquat. Food Prod. T.*, 27: 1024–1047.

Tunç S. and Duman (2010). Preparation and characterization of biodegradable methyl cellulose/montmorillonite nanocomposite films, *Applied Clay Science*, 48: 414–424.

Umaraw, P. and A.K. Verma (2017). Comprehensive review on application of edible film on meat and meat products: An eco-friendly approach, *Crit. Rev. Food Sci.*, 57: 1270–1279.

Voet, D., J. Voet and C. Pratt (2007). Fundamentos de bioquímica: La vida a nivel molecular 2da edicion, *Medica Panamericana*, 233–283.

Wagner, C. (2014). Bioplastics: Types, applications, toxicity and regulation of bioplastics in food contact materials. [online] Available at http://www.foodpackagingforum.org/food-packaging-health/bioplastics [accessed Jan. 28, 2019].

Wang, L., J. Xue and Y. Zhang (2019). Preparation and characterization of curcumin loaded caseinate/zein nanocomposite film using pH-driven method, *Ind. Crop Prod.*, 130: 71–80.

Wang, S., M. Marcone, S. Barbut and L.T. Lim (2012). The impact of anthocyanin-rich red raspberry extract ARRE on the properties of edible soy protein isolate spi films, *J. Food Sci.*, 77: 497–505.

Wang, Y., Y. Xia, P. Zhang, L. Ye, L. Wu and S. He (2016). Physical characterization and pork packaging application of chitosan films incorporated with combined essential oils of cinnamon and ginger, *Food Bioprocess. Tech.*, 10: 303–511.

Wang, Y.-C., L. Lu and S. Gunasekaran (2017). Biopolymer/gold nanoparticles composite plasmonic thermal history indicator to monitor quality and safety of perishable bioproducts, *Biosensors and Bioelectronics*, 92: 109–116.

Wijaya, W., A.R. Patel, A.D. Setiowati and P. Van der Meeren (2017). Functional colloids from proteins and polysaccharides for food applications, *Trends Food Sci. Tech.*, 68: 56–69.

Wilfred, O., H. Tai, R. Marriott, Q. Liu, V. Tverezovskiy, S. Curling, H. Tai, Z. Fan and W. Wang (2018). Biodegradation of polylactic acid and starch composites in compost and soil, *Int. J. Nano. Rech.*, 1: 43–52.

Wyrwa, J. and A. Barska (2017). Innovations in the food packaging market: Active packaging, *European Food Research and Technology*, 243: 1681–1692.

Xu, Z. et al. (2018). Preparation and biomedical applications of silk fibroin-nanoparticles composites with enhanced properties—A review, *Mat. Sci. Eng.*, 95: 302–311.

Yadav, A., S. Mangaraj, R.M.N.K. Singh and S. Arora (2018). Biopolymers as packaging material in food and allied industry, *Inte. Jouur. Chem. Stu.*, 6: 2411–2418.

Yin, H.-Y. and W.-C. Tsai (2015). Advances of Nanomaterials for Food Processing, pp. 1137–1159. *In:* P.C.K. Cheung and B.M. Mehta (Eds.). *Handbook of Food Chem.*, Berlin, Heidelberg: Springer, Berlin, Heidelberg.

Youssef, A.M. and S.M. El-Sayed (2018). Bionanocomposites materials for food packaging applications: Concepts and future outlook, *Carbohyd. Polym.*, 193: 19–27.

Yu, H., C. Yan and J. Yao (2014). Fully biodegradable food packaging materials based on functionalized cellulose nanocrystals/poly3-hydroxybutyrate-co-3-hydroxyvalerate nanocomposites, *RSC Adv.*, 4: 59792–59802.

Yu, Q., J. Wang, Y. Jiang, R.J. McCabe and C.N. Tomé (2014). Co-zone {1⁻012} twin interaction in magnesium single crystal, *Materials Research Letters*, 2: 82–88.

Zafar, R. et al. (2016). Polysaccharidebased bionanocomposites, properties and applications: A review, *Int. J. Biol. Macromol.*, 92: 1012–1024.

Zanetti, M. et al. (2015). Microbiological characterization of pure geraniol and comparison with bactericidal activity of the cinnamic acid in gram-positive and gram-negative bacteria, *J. Microb. Bio. Tech.*, 7: 186–193.

Zaritzky, N. (2010). Edible coatings to improve food quality and safety, pp. 631–659. *In: Food Engineering Interfaces.*

Zavareze, R., A. Renato and G. Dias (2011). Impact of heat-moisture treatment and annealing in starches: A review, *Carbohyd. Polym.*, 83: 317–328.

Zhang, C., Z. Wang, Y. Li, Y. Yang, X. Ju and R. He (2019). The preparation and physiochemical characterization of rapeseed protein hydrolysate-chitosan composite films, *Food Chem.*, 27: 694–701.

Zhang, Y., B.K. Simpson and M.J. Dumont (2018). Effect of beeswax and carnauba wax addition on properties of gelatin films: A comparative study, *Food Bioscience*, 26: 88–95.

Zhu, F. (2017). Structures, properties, modifications and uses of oat starch, *Food Chem.*, 229: 329–340.

Zhu, Q. and E.N. Jackson (2015). Metabolic engineering of Yarrowia lipolytica for industrial applications, *Curr. Opin. Biotech.*, 36: 65–72.

CHAPTER 3

New Trends in Smart and Intelligent Food Packaging

*Semih Otles** and *Buket Yalcin Sahyar*

1. Introduction

The current demand for 'ready to eat', 'ready to cook' and 'ready to use' food has strikingly increased with the increased necessity of the manufacturers to produce minimally processed food. One of the main issues in food processing is protection against food-borne diseases which still constitute a global problem of public health (Morris, 2011; Carbone et al., 2016). New technologies on food packaging are improving as a result of changing consumer demands or industrial production trends. As a result, mildly preserved, fresh, tasty and convenient food products with prolonged shelf-life and controlled quality can be defined as main consumer demands or industrial production trends. Additionally, changes in consumer lifestyles and retailing applications pose the main challenge to the food-packaging industry. Thus, these changes act as the driving force for the development of new packaging concepts which extend shelf-life while maintaining and monitoring food safety and quality (Dainelli et al., 2008). Innovations in packaging were up to now limited mainly to small changes in packaging materials which could be defined as passive packaging systems to meet the marketing demands in short term. New concepts of smart and intelligent packaging are due to play an increasingly important role by offering innovative solutions to extending the shelf-life or maintain, improve or monitor food quality and safety (Gontard, 2000). New concepts like ethanol emitters (such as for bakery products), ethylene absorbers (such as for climacteric fruits), carbon dioxide emitters/absorbers, time/temperature and oxygen indicators, chemical sensors and biosensors (freshness, pathogen, leakage, carbon dioxide, oxygen, pH, time and temperature), etc., have been developed to maintain and monitor food quality and safety of food from producers to consumers (Rooney, 2005; Kuswandi et al.,

Ege University, Engineering Faculty, Food Engineering Department, Izmir, Turkey.
* Corresponding author: semih.otles@ege.edu.tr

2011; Dobrucka and Cierpiszewski, 2014). Thus, application and usage of smart and intelligent packaging can extend the shelf-life of food or improve its organoleptic properties while preventing food losses (Dobrucka and Cierpiszewski, 2014).

Additionally, food waste is a major and global problem which depletes the environment of limited natural resources (i.e., water, energy, chemical substances and materials) (Alexandratos and Bruinsma, 2012). Although all actors in the food-chain have a role to play in reducing and preventing food waste—from processors and producers of food to consumers and retailers—the immediate concern regarding household waste demands that advances in packaging are given top priority. Packaging has become an important technology to provide safety in the food-chain, avoid undesired reactions, satisfy consumer expectations and increase food shelf-life. Among the new trends, smart and intelligent packaging systems represent a great potential to reduce food wastes. It was denoted in a report (Freedonia Group, 2015) that demand for intelligent packaging will grow increasingly in 2019 and is expected to reach $1.5 billion as products like time-temperature indicators and smart labels and tags become more common. Furthermore, it was denoted that research will be supported in smart intelligent/communicative packaging solutions to reduce cold/frozen chain use for fresh foods, which would consequently reduce energy consumption, carbon emissions, preservative intake and food waste (Poyatos-Racionero et al., 2018). According to the global food waste problem, food packaging materials have also been taken into consideration. Therefore, new food packaging materials are being under investigation, particularly for development of more eco-friendly packaging materials. Thus, food waste can be minimized while food packaging materials can consist of lower synthetic additive and preservative contents. Thus a low environmental footprint can also be achieved. As sustainability has acquired an increasingly critical relevance in food packaging, bio-based and biodegradable materials stand out as suitable alternatives to the synthetic ones (Espitia et al., 2019). There have been numerous samples for biodegradable materials, which have been applied or can be applied to the food-packaging industry. One of the examples is biodegradable films that contain antimicrobial agents that inhibit the growth of pathogen microorganisms (Nasab and Tabari, 2018). Another example is hydrogels that have a great potential to be used in food packaging systems or as carriers of bioactive components. Hydrogels can be defined as three-dimensional, hydrophilic networks, comprising polymeric chains linked through physical or chemical bonds. Hydrogels can be used as part of a packaging system to control the humidity generated by food products with high water content. In addition to this, hydrogels, that include nanoparticles, may be shown to have antimicrobial activity (Batista et al., 2019).

In this chapter, definitions of traditional food packaging and new food-packaging trends (intelligent and smart food packaging) are presented. Structure and working mechanisms of new food-packaging materials will also be defined and examples of industrial applications on food-packaging materials will be presented.

2. Food-packaging Trends

2.1 Traditional Food Packaging

Traditional food packaging can be defined as one of the four basic functions (protection, communication, convenience and containment) (Fig. 1). Firstly, protection can be

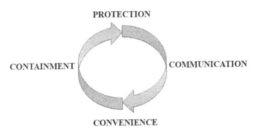

PROTECTION

CONTAINMENT

COMMUNICATION

CONVENIENCE

Fig. 1. Four basic functions of traditional food packaging.

easily defined as keeping food products in a limited volume and prevent its leak or break-up and protect it against possible contamination and changes. Communication can be defined as food-packaging communication with important information about the contained food product and its nutritional content, together with guidelines about preparation. Convenience can also be briefly defined as food packaging allowing consumers to enjoy food the way they want, at their convenience. Food packages might be designed towards individual lifestyles through, for example, portability and multiple single portions. Containment can be defined as the most basic function of package and its importance in easy transportation or handling (Vanderroost et al., 2014). Innovations in food packaging can be defined as improving, combining, or extending these four basic functions of traditional food packaging (Yam et al., 2005). Additionally, innovation in food-packaging technology is driven by not only the demand for safe and high-quality foods, but also changes in consumer preferences. Smart and intelligent food-packaging technologies are good examples of this development (Puligundla et al., 2012).

2.2 New Trends in Food Packaging

Intelligent packaging (Fig. 2) can be defined basically as packaging that contains an internal or external indicator to provide information about the history of the package and/or the quality of the food (Dobrucka and Cierpiszewski, 2014). According to another definition, intelligent packaging can also be defined as smart packaging, which can sense some properties of the food it encloses or the environment in which it is kept and which informs the manufacturer, retailer and consumer of the state of these properties (Hutton, 2003). Intelligent packaging devices are capable of sensing and providing information about the properties and function of packaged food and can provide assurances of the product safety and quality, pack integrity and are being utilized in applications, such as product authenticity and product traceability (Summers, 1992; Day, 2001). Intelligent packaging devices include sensors, time-temperature indicators, microbial growth indicators and gas-sensing dyes. Intelligent packaging also communicates and monitors information about food quality. Intelligent food-packaging technology also helps to trace the history of the product through critical points in the food supply chain (Puligundla et al., 2012).

Intelligent packaging should not be confused with active packaging. Active packaging is enhancing of the protection function of traditional food packaging and is so designed so that it includes a component that permits the absorption or release of substances into/from the packaged food or the environment surrounding the food

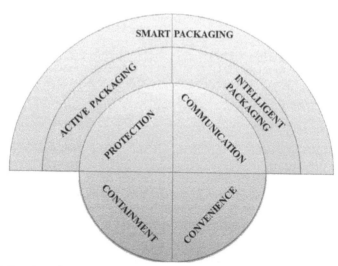

Fig. 2. Functions of smart and intelligent packaging (adapted from Maksimović et al., 2015).

(EC, 2009). Thus, active packaging can be defined as a system in which the product, the package and the environment interact to extend the shelf-life and improve the condition of packaged food so as to achieve some characteristics that cannot be obtained otherwise (Miltz et al., 1995). Intelligent packaging and active packaging systems can work synergistically to improve and perform smart packaging. Smart packaging enables a total packaging solution which, on the one hand monitors changes in the environment (intelligent) and/or in the product and on the other, acts upon these changes (active). Even though the concepts of smart and intelligent packaging are frequently used interchangeably in literature, they are not the same. Until today, three main technologies exist for applying intelligent packaging—sensors, indicators and radio frequency identification (RFID) systems (Kerry et al., 2006). These technologies differ from each other, not only in physical composition, but also in the amount and type of data that can be carried and how the data are captured and distributed (Heising et al., 2014; Vanderroost et al., 2014).

Smart packaging (Fig. 2) is an emerging area, which includes a number of functionalities, depending on the product being packaged, including food, beverage, pharmaceutical and various types of health and household products. There are some current and possible examples of smart packaging functions (Yezza, 2009; Kuswandi et al., 2011):

- Shelf-life to retain integrity and actively prevent food spoilage
- Enhance product attributes (such as color, taste, flavor, aroma, viscosity and texture)
- Respond actively to changes in the package and/or product environment
- Communicate product information, product history and/or condition to consumers
- Assist with opening and indicate seal integrity
- Confirm product authenticity

However, smart packaging practically focuses on the senses and informs the status of a product in term of its safety (food is safe or unsafe) and quality (representing the freshness, ripeness or firmness). In this direction, smart packaging can be defined as packaging that has the ability to track the product, sense the environment inside or outside the package and inform the manufacturer, retailer and consumer regarding the condition of the product (Yezza, 2009; Kuswandi et al., 2011).

The supply chain of the food industry can take advantage of the sensor technology in new trends in food technology. RFID programmes are directly applied to the sector by retailers in general merchandise, grocery, apparel and other categories. The benefit of RFID programme usage can be detailed by its improving sales with greater stock availability, cost savings and increased responsiveness, especially in receiving and inventory control operations (Delen et al., 2007). RFID tags are currently based on chips containing a memory that can be wirelessly read out to uniquely identify the products. Time-temperature, pressure, tilt monitoring and chemical sensing are emerging applications on RFID tags, which have proved their value with additional functionality (Yam et al., 2005). Moreover, monitoring of foods is especially relevant in the transport of perishable foods, such as fruit, vegetables, meat and fish. These products need to be kept under very stable conditions (such as specific temperature and humidity) to be transported as fresh as possible to their destination. On the contrary, food products could become unsellable products and a major health risk for the consumer. Meat and fish are very perishable food products due to the rapid growth of bacteria. Bacterial growth on meat and fish generally causes excretion of nitrogen and a sulfur-containing compound, which is accompanied by a strong and foul smell. Therefore trimethylamine (TMA) can be a good indicator to determine fish spoilage by using sensor technology (Olafsdóttir et al., 1997; Pacquit et al., 2007; Smits et al., 2012). Sensors integrated into food packages could also benefit the consumers by providing freshness and quality while allowing the retail industry to more efficiently manage food stocks and product authenticity. In a study, smart radio-frequency labels with sensors were presented, which were able to measure temperature, humidity and the presence of volatile amine compounds. These smart labels were used to quantify the freshness of fish (Smits et al., 2012).

On the other hand, food preservation, quality maintenance and safety are the main growing concerns of the food industry. It is obvious that the consumers demand safe and natural food products with stringent regulations to prevent food-borne infectious diseases. Antimicrobial packaging, which can be defined as a subset of new packaging trends and controlled release packaging, is one such promising technology which effectively impregnates the antimicrobes into the food packaging film material and subsequently delivers it over the anticipated period of time to kill the pathogenic microorganisms affecting food products. Thus the shelf-life of food products gets prolonged (Malhotra et al., 2015). One sample for antimicrobial packaging can be given from a study, which is directly related to essential oils (EOs). Eos, extracted from plants, have been the focus of numerous studies due to their potential in the food industry. Volatile compounds exist in EOs. These compounds are responsible for antimicrobial and antioxidant action. Several studies have emerged regarding applications of EOs and their incorporation into food packaging for their antimicrobial and antioxidant capacity. To the results of studies, these packages with EOs have shown efficiency against microorganisms and oxidants *in vitro* in tests with food and/ or food simulant evaluation tests. It can be concluded that EOs extend food stability during storage, inhibit the growth of pathogenic or spoilage microorganisms and protect against oxidation (Ribeiro-Santos et al., 2017).

Additionally, nanotechnology is needed to be emphasized to understand new trends in food packaging applications. New materials and devices can be easily set up by using nanotechnology (Pelaz et al., 2012). These nanomaterials and nanodevices can easily be found applicable in health and energy-related fields. Thus, these new technologies can also be implemented in the food packaging system and impact food packaging and quality control systems (Valdés et al., 2009; Kuswandi et al., 2011; Farahi et al., 2012). Before nanomaterials and devices were manufactured, many products in the supermarket were packed in standard plastic foils to directly prevent contact of the food with the environment and shield against the particular bacteria. It was denoted that use of nanomaterials in food packaging can provide feedback about the potential undesired contamination of food and the product still retains its optimal conditions or it should not be consumed any more. In this manner, sensors are the solution to monitor temperature, humidity, reduce and oxidize gases, and volatile organic compounds. Also sensors have been successfully integrated on plastic foil (Briand et al., 2011). There have been numerous reports directly related to the integration of sensing nanostructures on flexible substrates (McAlpine et al., 2007; Cao et al., 2008; Takei et al., 2010; Angione et al., 2011). These sensing nanostructures might have a groundbreaking impact on monitoring systems, for instance, in clothing and implants and clearly in the food industry.

Additionally, nano-based sensors can be very useful in rapid *in situ* analysis of food at critical locations (such as customs, storage facilities, *in situ* control of food's transportation). Also, nanomaterials might not be only useful in detecting spoilage but also in prevention. The prevention sample of nanomaterials usage can have antimicrobial coatings, which could be used as films derivatized Ag nanoparticles (NPs) (Fortunati et al., 2012). Broadly, biocompatibility, long-term stability and toxicity remain the main issues, which will have to be further characterized and optimized, particularly in the case of direct contact between nanomaterials and food (Pelaz et al., 2013; Jiang et al., 2015).

2.3 Structure and Working Mechanism of New Food Packaging Materials

Many examples have as been given about new food packaging studies. According to the food packaging material type that has been focused by many researchers, these examples have been given under main groups [biopolymers, valorization of bio-waste materials, phase change material (PCM), pH-sensitive dyes, organic acids, nanoparticles, nanomaterials, nanoemulsions, gas indicators, time-temperature indicators and films and coatings]. The definition of structure and working mechanism of new food packaging materials has also been identified.

2.3.1 Biopolymers

One of the food packaging materials is biopolymers. Till today, application and implementation of non-biodegradable materials or plastics into food packaging materials have significantly increased. These materials are mainly derived from petroleum products and cause the problem of waste disposal (Avella et al., 2005). New investigations mainly focus on the development of food packaging materials that could completely mineralize and rapidly degrade in the environment to meet the increasing demand for both environmental safety and sustainability (Jayaramudu et al., 2013; Majeed et al., 2013). Biopolymers have been one of the favorable samples to be developed and exploited into eco-friendly food packaging materials, thanks to its biodegradability (Tang et al., 2012). But, the use of biopolymers as food packaging materials has drawbacks, like poorer barrier, thermal and mechanical properties as compared to conventional non-biodegradable materials. In addition to this, nanocomposites present a promising route to enhance barrier and mechanical properties of biopolymers. Bio-nanocomposite is a multiphase material including two or more components which are a continuous matrix or phase, particularly biopolymer and nanofillers or discontinuous nano-dimensional phase (< 100 nm). The nanofillers represent a structural role in which they act as a reinforcement to improve the barrier and mechanical properties of the matrix. The matrix tension is transferred to the nanofillers through the boundary between them (Arafat et al., 2014; Azeredo et al., 2011; Trovatti et al., 2012). The integration of nanofillers as titanium dioxide (TiO_2), silicate and clay to biopolymers might improve not only the barrier and mechanical properties of biopolymers but also present applications and functions in food packaging as antimicrobial agent, biosensor, and oxygen scavenger (Azeredo, 2009; Azeredo et al., 2011; Rhim et al., 2013). On the other hand, the bio-nanocomposite could be used as a smart food packaging material whereby it can sense characteristics of the packaged food, such as microbial contamination, expiry date and use some mechanism to register and transfer information about the safety and quality of the food (Azeredo et al., 2011). The improvement of bio-nanocomposite materials for food packaging is important to counter the environmental problem and also improve the functions of food packaging materials (Othman, 2014).

2.3.2 Valorization of Bio-waste Material

In a study, aerogels from cellulose nanocrystals were produced by using rice and oat husks for food packaging applications. The analysis was done by comparing with commercial cellulose. Nanocrystals from cellulose were acquired by enzymatic hydrolysis and mechanical treatment at high pressure. Structures and crystallinity of cellulose showed different properties depending on the source of the cellulose and 16.0–28.8 nm were the average diameters for nanocrystals. A porous and uniform structure of the aerogels prepared with cellulose nanocrystals was obtained. The results denoted that aerogel of oat cellulose nanocrystals represented a larger pore size than eucalyptus cellulose nanocrystals. This may have influenced the lowest water absorption capacity of the aerogels of eucalyptus cellulose nanocrystals. It was concluded from these results that agro-industrial residues have promising applications in various industrial fields and could be used as aerogel absorbers of water in food packaging (Oliveira et al., 2019a).

In a study, bioactive aerogels were developed by using recycled *Gelidium sesquipedale* seaweed to apply in food packaging. Firstly, the raw seaweed was used to extract cellulose, highly crystalline and high-aspect ratio nanocellulose and an agar-based extract rich in polyphenols and with antioxidant capacity. Afterwards, pure polyvinyl alcohol (PVA) and hybrid aerogels, including cellulose and nanocellulose, were fabricated by a physical cross-linking method. It was noted that the presence of hydroxyl groups ensured by the high aspect ratio of nanocellulose promoted the interactions with water and facilitated the accessibility of moisture towards the interior of the aerogels, thereby generating high water-sorption capacity materials. It was also declared that the agar-based extract was then incorporated into selected formulations and the release in hydrophobic and hydrophilic food simulant media was examined. Bioactive was immediately released by dissolving pure PVA aerogels in aqueous media. Both the cellulose and nanocellulose hybrid aerogels preserved their integrity and ensured a more gradual release. Although the hybrid aerogels presented similar release profiles during the first 48 hours, the presence of nanocellulose led to greater release values after a more prolonged time. Thus it showed promising properties of hybrid PVA/cellulose/nanocellulose aerogels as matrices for the controlled release of bioactive compounds in food systems. This could be of interest in the development of bioactive packaging structures (Oliveira et al., 2019b).

In a study, bacterial cellulose (BC), which is gaining considerable attention due to its unique physico-chemical and mechanical properties, was produced by *Gluconacetobacter xylinus* PTCC 1734 in sugar-beet molasses, cheese whey and standard Hestrin–Schramm (HS) media. The BC was hydrolyzed by sulfuric acid to prepare bacterial cellulose nanocrystals (BCNC). The results showed that treated sugar-beet molasses led to the highest BC concentration and productivity, followed by treated cheese whey. It was concluded that food industrial byproducts can be used as cost-effective culture media to produce BC for large-scale industrial production. Isolated cellulose nanocrystals are useful in the fabrication of bio-nanocomposite films for food packaging applications (Salari et al., 2019).

Banana epidermis, which is a waste material and starchier, can be used to constitute the biodegradable composite film. Thus a biodegradable composite film had been developed by using modified banana epidermis starch. The modified banana epidermis starch (MBES) particle had been analyzed to identify the composite film properties, such as barrier, thermal and mechanical properties. Antimicrobial activity of the composite films was developed by using chitosan and MBES particles. The composite films were produced by a solution casting method. The composites showed a significantly good mechanical and thermal stability. The resulting chitosan/MBES films showed better oxygen permeation rate as compared to normal commercial films. It was concluded that the new prepared composite films have a high potential, mechanical, morphological, antimicrobial activity and improve long-lasting antimicrobial efficiency in food packaging applications (Yuvaraj and Rajeswari, 2018).

2.3.3 Phase Changing Material (PCM)

In a study, novel materials with heat management properties have been improved by means of encapsulation of a phase changing material (PCM) in a biopolymeric matrix using the electro-spinning technique. This study optimizes the methodology to achieve micro-, submicro- and nanoencapsulation structures based on zein (a maize protein) and dodecane (PCM paraffin which has a transition temperature at −10°C). It was concluded that these encapsulation technologies could be of interest in the food industry in order to develop new smart packaging materials with the ability to maintain temperature control so as to preserve the cold chain (Pérez-Masiá et al., 2013).

2.3.4 pH-Sensitive Dyes

In 2014, a colorimetric mixed-pH dye-based indicator was identified for the development of intelligent packaging, as a 'chemical barcode' for real-time monitoring of skinless chicken-breast spoilage. The relationship between the number of microorganisms and the number of volatile compounds was also investigated. It was declared that this indicator includes two groups of pH-sensitive dyes, where the first one is a mixture of bromothymol blue and methyl red, while the other is a mixture of bromothymol blue, bromocresol green and phenol red. Carbon dioxide (CO_2) was used as a spoilage metabolite because the degree of spoilage was correlated to the amount of increased CO_2, which was more than the level of total volatile basic nitrogen (TVB-N) during the storage period. Color changes, in terms of the total color difference of a mixed-pH dye-based indicator, correlated well with CO_2 levels of skinless chicken breast. Additionally, it was declared that the indicator response correlates with microbial growth patterns, thus enabling real-time monitoring of spoilage either at various constant temperatures or with temperature fluctuations (Rukchon et al., 2014).

2.3.5 Organic Acids

In 2013, a study on antimicrobial activity of low and medium molecular weight chitosan and organic acids (benzoic acid and sorbic acid and commercially available nanosized benzoic and sorbic-acid solubilisate equivalents) was examined and compared with commercial mixtures of organic acids used as meat coatings (Articoat DLP-02® and Sulac-01®). The results found open opportunities for the nano-sized solubilizes derived from food compatible sources, to be used in smart antimicrobial packaging applications, as less of the antimicrobial substances in question is required to deliver the same antimicrobial effect (Cruz-Romero et al., 2013).

2.3.6 Nanoparticles, Nanomaterials and Nanoemulsions

Semiconductors and metal-based nanoparticles can be used effectively in many areas, such as biosensors, sensor-related products and combining with packaging materials. Laser ablation is one of the most efficient physical methods for nanofabrication (Kabashin and Meunier, 2006). Laser ablation method can be defined as a method which was used to prepare nanometer-diameter catalyst clusters that define the size of wires produced by vapor-liquid-solid growth. This approach was used to prepare bulk quantities of uniform nanowires with very low diameters (as nanometers) (Morales and Lieber, 1998). Modification and fabrication of nanomaterials in liquid based on laser irradiation has become a rapidly growing field. Compared to other typically chemical methods, laser ablation/irradiation in liquid (LAL) is a simple and green technique that normally operates in water or organic liquids under ambient conditions. Nanomaterials can be easily developed with special morphologies, microstructures and phases by using the LAL method. Thus, a one-step formation of various functionalized nanostructures can be achieved for any application, like detection and packaging (Zeng et al., 2012).

In a study, nanoparticle was synthesized, which was based on silica-capped zinc sulfide (ZnS) for use as a stable and long-term antibacterial agent. Silica was used due to its very important properties in food packaging applications for moisture absorption in tune with its property of biocompatibility and water solubility. The results of the study denoted that pure ZnS represents excellent antibacterial action but it can last only a few days (Kumar et al., 2018).

A new eco-friendly method was studied for the treatment of poly (3-hydroxybutyrate) PHB as a candidate for food packaging applications. It was noted that PHB was modified by bacterial cellulose nanofibers (BC), using a melt-compounding technique and by plasma treatment or zinc oxide (ZnO) nanoparticle plasma coating for better properties and antibacterial activity. The results were noted that plasma treatment preserved the thermal stability, crystallinity and melting behavior of PHB–BC nanocomposites, regardless of the amount of BC nanofibers. Additionally, a remarkable increase in stiffness and strength and an increase in the antibacterial activity were also noted. Lastly, it was shown that plasma treatment also inhibits the growth of *Staphylococcus aureus* and *Escherichia coli* by 44 per cent and 63 per cent, respectively. Another significant benefit of the ZnO plasma coating was given as an important change in the thermal and mechanical behavior of PHB–BC nanocomposite as well as in the surface structure and morphology. Strong chemical bonding of the metal nanoparticles on PHB surface was achieved by ZnO plasma coating. Results also showed that ZnO plasma treatment completely inhibits the growth of *Staphylococcus aureus*, thanks to the presence of a continuous layer of self-aggregated ZnO nanoparticles. Finally, a plasma-treated PHB–BC nanocomposite can be proposed as a green solution for the food packaging industry (Panaitescu et al., 2018).

In a study, nanoemulsions can be used to improve the performance of sustainable food packaging devices, especially for the successful incorporation of new compounds and functionalities into conventional films and coatings. The nanoemulsion, featuring unique optical stability and rheological properties, has been improved to protect, encapsulate and deliver hydrophobic bioactive and functional compounds, including natural preservatives (such as essential oils from plants), nutraceuticals, vitamins, colors and flavors. The surfactants (including naturally-occurring proteins and carbohydrates), dispersants and oil-soluble functional compounds can be used to design food-grade nanoemulsions, intended for packaging applications. Bottom-up and top-down approaches can be used to fabricate nanoemulsions, which include high-energy methods (such as high-pressure homogenizers, microfluidics, ultrasound and high-speed devices) and low-energy methods (for instance, phase inversion and spontaneous emulsification). Finally, incorporation of nanoemulsions in biopolymer matrixes can be used for food packaging applications, especially their potential antimicrobial activity against food-borne pathogens (Espitia et al., 2019).

In a study, the isolation of cellulose nanofibers (CNF) using cellulase immobilized on cheap and easily formed polymeric gel disks was discussed. These gel disks are based on carrageenan gel coated with hyper-branched polyamidoamine that can be covalently bound to cellulase through glutaraldehyde spacer. It was stated that thermal and mechanical stability of the coated gel disks significantly improved. Free and immobilized cellulase exhibited maximum activities at 50°C and pH 5. In addition to this, immobilized cellulase exhibited broader temperature stability than in the free form. Additionally, it was also stated that immobilized cellulase gel disks can be easily separated and reused with great reusability capacity of about 85 per cent of the initial activity after six cycles. As a result, it has been concluded that enhanced thermal stability and reusability of immobilized cellulase paved the way for its use in industrial production of CNF in biomedical and food packaging applications (Yassin et al., 2019).

2.3.7 Gas Indicators

Generally, specific gas compounds (CO_2, O_2 and volatile organic compounds—VOCs) are emitted into the environment during food product spoilage. Main food spoilage indicator for packed food can be increase of CO_2 gas level. New packaging trends are starting to work on this point by defining CO_2 gas levels inside the packaging medium, immediately transmitting into a visible message for the consumer (like changing the color of an indicator) during storage period before or within the spoilage mechanism. There are still limitations on food packaging applications, such as high equipment cost, bulkiness and energy input requirement, including safety concerns (Puligundla et al., 2012).

2.3.8 Time Temperature Indicators

They are presented as smart packaging materials which implement on to the packaging and give information to the consumer about the temperature of food products. These indicators generally defined as a thermochromic ink dot represent the product at the correct serving temperature following refrigeration or microwave heating. Thermochromic-based indicators are designed to inform the consumer about the temperature of food products (Vaikousi et al., 2009; Kuswandi et al., 2011).

2.3.9 Films and Coatings

Films and coatings are samples of the new trends in the packaging technology. Films can be briefly defined as thin sheets formed in advance and subsequently applied on the product like wrapper or among the layers. Therefore, it can be used as covers, layer separation or packaging (Krochta, 2002; Ribeiro-Santos et al., 2017). In 2016, an electrospun polyvinyl alcohol/cinnamon essential oil/beta-cyclodextrin (PVA/CEO/β-CD) antimicrobial nanofibrous film (with 240 ± 40 nm average diameter) was successfully fabricated. It showed in this study that cinnamon essential oil (CEO) was encapsulated into the β-CD cavity and there existed molecular interaction among PVA, CEO, and β-CD, which enhanced the thermal stability of CEO. The addition of CEO/β-CD made the nanofibrous film more hydrophilic. The PVA/CEO/β-CD nanofibrous film showed excellent antimicrobial activity and its minimum inhibitory concentration (MIC) against *Escherichia coli* and *Staphylococcus aureus* was approximately 0.9–1 mg/mL (corresponding CEO concentration 8.9–9.9 µg/mL) and minimum bactericidal concentration (MBC) was approximately 7–8 mg/mL (corresponding CEO concentration 69.3–79.2 µg/mL) (Wen et al., 2016).

In a study, the antimicrobial properties of chitosan and chitosan-zinc oxide (ZnO) nanocomposite coatings on PE films were studied. Oxygen plasma pretreatment of PE films led to increased adhesion by 2 per cent of chitosan and the nanocomposite coating solutions to the packaging films. Uniform coatings on PE surfaces were obtained. Increasing insolubility and water-contact angle and decrease of swelling are the results of incorporation of ZnO nanoparticles into the chitosan matrix compared to the chitosan coating. Results showed that PE coated with chitosan-ZnO nanocomposite films completely inactivated and prevented the growth of food pathogens, while chitosan-coated films showed only 10-fold decline in the viable cell counts of *Salmonella enterica*, *Escherichia coli* and *Staphylococcus aureus* after 24 hours of incubation as compared to the control (Al-Naamani et al., 2016).

It was shown in a study that (polyethylene terephthalate) (PET) film, which was coated with chitosan (CS) layers can be easily used to decrease the oxygen permeability through the polymeric films for food packaging applications. Oxygen transmission rate (OTR) of the 130 µm PET films could be decreased from 11 to only 0.31 cm3/m² per day with a coated layer of 2 µm of CS. It was also declared that additional decrease can be obtained with the addition of vermiculite (VMT) to CS matrix in a high proportion (40 to 50 w/w per cent). Thus, the film showed high-barrier behavior due to the brick wall nanostructure, which produced an extremely tortuous path for oxygen molecules (Essabti et al., 2018).

A starch-based flexible coating for food packaging papers with excellent hydrophobicity and antimicrobial properties was fabricated. It was declared that the homogeneous dispersion of the ZnO nanoparticles (NPs) in the composite film within 5 per cent of ZnO NP dosage was revealed according to FTIR (Fourier transform infrared) and XRD (X-ray diffraction) spectra results. It was also confirmed that the roughness on the composite film increased with the increased dosages of ZnO NPs by using SEM (scanning electron microscope) and AFM (atomic force microscope) micrographs. Hydrophobic characteristics denoted that dramatic enhancement was achieved in the values and stabilities of DCAs (dynamic contact angles) in the resultant film and coated paper. TG (thermogravimetry) results established increase in the thermal stabilities of the composite films. A decreased water vapor transmission rate was significantly observed in the coated paper. It was concluded that a flexible coating with excellent antimicrobial activity towards *Escherichia coli* can be achieved when 20 per cent of guanidine-based starch and 2 per cent of CMC (carboxymethyl cellulose) is added. Besides, the migration of ZnO NPs into the food simulants was observed well below the overall migration legislative limit. The results showed that starch-based flexible composite film and coated paper can be an effective approach towards developing a green-based material for food packaging applications (Ni et al., 2018).

A biodegradable film was produced by using jackfruit waste flour (obtained from the rind part) and polyvinyl alcohol, which are expected to have potential food packaging applications. It was declared that jackfruit waste flour was prepared from the rind part, while polyvinyl alcohol composites with different concentrations of jackfruit waste flour were prepared by using a solution-casting method. The results showed that jackfruit waste flour as a low-cost material can be exploited in food packaging materials to produce biodegradable flexible films, which can lead to minimizing of synthetic polymer utilization in food packaging (Sarebanha and Farhan, 2018).

In a study, the effects of 1, 3 and 5 pr cent of zinc oxide (ZnO) nanoparticles on antimicrobial properties and permeability of polylactic acid film (a bio-composite film) were examined. Films that included all three percentages of ZnO nanoparticles showed inhibitory effects on *Escherichia coli* and *Staphylococcus aureus*, while pure polylactic acid film did not have any antimicrobial activity. According to the results, bio-composite films with 1 per cent and 3 per cent nano-zinc oxide showed a 19 per cent water vapor permeability enhancement as compared to pure poly-lactic acid. Furthermore, adding 3 per cent of nano-zinc oxide showed an impact on the reduction of permeability to oxygen. It was concluded that polylactic acid films containing nano-zinc oxide have a high potential for antimicrobial food packaging applications and can enhance the safety of food products (Nasab and Tabari, 2018).

A study indicates the role of multicarboxylic acids as crosslinkers to polyvinyl alcohol (PVA) films produced by solution-casting method for food packing applications. Effect of incorporating different carboxylic acids [such as oxalic acid (OA), succinic acid (SA) and citric acid (CA)] on physicochemical and bioactive properties of PVA was also examined. The PVA/CA films presented higher bacterial inhibition efficiency mainly due to the ability of CA to alter local pH and cause alteration in the permeability of microbial membrane by disrupting the bacterial substrate transport. Additionally, the microbial shelf-life of sliced carrots that were covered by PVA/CA film increased from one to five days as compared to the control and commercial (Suganthi et al., 2018).

In a work, a one-step co-continuous bilayer-coating process to generate a multilayer film, that rendered super hydrophobicity to a polyethylene terephthalate (PET) substrate, was reported. A continuous coating based on ultrathin polylactide (PLA) fibers was deposited on to the PET films by means of electrospinning, which increased the water contact angle of the substrate. Severally nanostructured silica (SiO_2) microparticles were electrosprayed on to the coated PET/PLA films to obtain super hydrophobic behavior. It was found that co-continuous deposition of PLA fibers and nanostructured SiO_2 microparticles onto PET films constituted a useful strategy to increase the surface hydrophobicity of the PET substrate, thereby obtaining an optimal apparent water contact angle of 170° and a sliding angle of 6°. Regrettably, a reduction in background transparency was observed as compared to the uncoated PET film, especially after electrospraying of the SiO_2 microparticles. But the films were seen to have good contact transparency. It was concluded that the materials developed showed significant potential in easy emptying of transparent food packaging applications (Pardo-Figuerez et al., 2018).

An eco-friendly nanocomposite film was developed by using polypropylene carbonate (PPC) as matrix and copper nanoparticles (CuNPs) modified tamarind-nut powder (TNP) as a new reinforcement. Firstly, PPC/10 wt. per cent TNP composites were fabricated using simple casting technique. Thereby, copper nanoparticles (CuNPs) were *in situ* generated in the PPC/TNP composite films, using different concentrations of copper sulfate source solutions. The crystallinity, thermal stability and the tensile properties of this eco-friendly nanocomposite film were developed with the addition of CuNPs. The film also presented excellent antibacterial activity against *Escherichia coli*, *Pseudomonas aeruginosa*, *Bacillus licheniformis* and *Staphylococcus aureus*. Thus, this film can be considered for biomedical and food-packaging applications (Devi et al., 2019).

In 2019, blend and bilayer bio-based active films were developed by solvent casting technique, using chitosan (CS) and gelatin (GL) as biopolymers, glycerol as a plasticizer and lauroyl arginate ethyl (LAE) as an antimicrobial compound. Blend films had higher tensile strength and elastic modulus and lower water vapor permeability than bilayer films (p<0.05). Bilayer films denoted as effective barriers against UV light and represented lower transparency values (p<0.05). It was shown that these bio-based active films incorporated with LAE (0.1 per cent, v/v) inhibited the growth of *Listeria monocytogenes, Escherichia coli, Salmonella typhimurium* and *Campylobacter jejuni*. It was highlighted that the development of blend and bilayer bio-based active films, based on CS and GL enriched with LAE for food packaging applications, showed improved physical, mechanical, barrier and antimicrobial properties (Haghighi et al., 2019).

In a study, chitosan (CH)- and polycaprolactone (PCL)-based films containing nanocellulose (NC) and grape seed extract (GSE) were prepared in monolayer form and bilayer form. The bilayer films with GSE kept the pH of the chicken breast fillets stable during storage. The results indicated that CH- and PCL-based bilayer films could be a promising material to transfer functional compounds as active packaging material layers in food packaging applications (Sogut and Seydim, 2019).

In a study, starch-kefiran-ZnO (SKZ) solution was prepared by UV irradiation in different time periods to show the effect of UV-C light on SKZ solution in different exposure times (1, 6, and 12 hours). Nano-ZnO (ZN) was used as a photo-initiator and reinforcement agent, simultaneously. UV treatments were affected by the mechanical properties of the films. The tensile strength increased because of the enhanced interaction between the biopolymer mixture and nanofiller but elongation at break decreased. Besides, UV absorption and water contact angle increased because of the better distribution of ZnS in the polymer matrix after the UV exposure. It was declared that the modified UV-SKZ could be an appropriate process to food packaging. Also, UV can be used as a nano-ZnO compatibilizer in food packaging materials according to the results (Shahabi-Ghahfarrokhi and Babaei-Ghazvini, 2019).

3. Industrial Application of Food Packaging

There are numerous studies which regard new trends in food packaging. Here are recent and interesting food packaging examples given with brief definitions (Table 1).

In 2009, time-temperature indicators were commercially present in the USA and UK markets as plastic containers for pouring syrup for pancakes to represent that the syrup is at the right temperature. Additionally, another example is orange juice pack labels, which include thermochromic-based designs to inform the consumer about the temperature of orange juice (Vaikousi et al., 2009; Kuswandi et al., 2011). Furthermore, time-temperature indicators can be used as freshness indicators for specific fish and meat products by a detection mechanism, which is based on pH change. Fresh-Check® and CheckPont® are commercial samples of these freshness indicators. The working principle of these freshness indicators can be defined through the following steps. Firstly, the fish product spoils and a pH increase occurs within the headspace of an enclosed food package. Latterly, this pH change represents fish spoilage and can be controlled by pH indicating sensor. This pH change can be easily controlled by using pH indicator dyes, which change color when placed in an acidic or basic environment. This is because of the fact that when fish spoils, it releases a variety of basic volatile amines which are detectable with appropriate pH-indicating sensors. The detection mechanism works by using a pH-sensitive dye (dye is entrapped within a polymer matrix on to the packaging material), which changes color visibly when in

Table 1. Examples of Industrial Application of Smart and Intelligent Packaging Systems.

Industrial name	Detection mechanism	Action mechanism	Applicable to	References
Fresh-Check® **CheckPont**®	Freshness (time-temperature) indicator	pH chance	Meat and fish product	Pacquit et al., 2006; Kuswandi et al., 2011
RipeSense ™	Ripeness sensor	Gas release	Pears, kiwifruit, melon, mango, avocado, and other stone fruit	Wallach, 1998; Kuswandi et al., 2011
Nanocomposite Films	Antibacterial activity of silver nanoparticle	antibacterial activity against *Staphylococcus aureus* and *Escherichia coli* cells	Not commercially applicable yet	Fortunati et al., 2012
Polyaniline (PANI) film	Chemical based indicator	A visible color change to a variety of basic volatile amines	Fish	Kuswandi et al., 2012
Phase change materials (PCMs)	Thermal protection	thermal protection to the packaged food	Refrigerated packaged food products	Chalco-Sandoval et al., 2015
Antimicrobial nanofibrous film	Active food packaging	Antimicrobial effect	Strawberry	Wen et al., 2016
Colorimetric sensor	Spoilage detection	increasing in signal intensity of the chemo-sensitive compounds	Fish	Morsy et al., 2016
Active/bioactive packaging systems	Sterilization process	Encapsulatean antioxidant, alpha-tocopherol	Not commercially applicable yet	Fabra et al., 2016
Biodegradable food packaging films	Wrapped with the composite film	Extend the shelf-life	Carrot pieces	Sarojini et al., 2019

direct contact with the volatile compounds that caused spoilage (Pacquit et al., 2006; Kuswandi et al., 2011).

RipeSense™ is a sensor label which reacts to the aromas released by a fruit as it ripens and represents ripeness stage of the fruit to the consumer, who wants to know if the fruit is ready to eat or not. The sensor is initially red and graduates to orange and finally turns to yellow. This sensor has already been applied for pears and can also be applied as a ripeness indicator for kiwifruit, melon, mango, avocado and other stone fruits, etc. (Wallach, 1998; Kuswandi et al., 2011).

Nanocomposite films were prepared by the addition of cellulose nanocrystals (CNCs) to eventually surfactant modified and silver nanoparticles in the polylactic acid matrix, using melt extrusion followed by a film formation process. This nanocomposite film showed antibacterial activity against *Staphylococcus aureus* and *Escherichia coli* cells. Thus, this novel nanocomposite might offer good perspectives for food packaging applications which require an antibacterial effect that remains constant over time (Fortunati et al., 2012).

In 2012, an antibacterial nanomaterial was developed as an additive for food packaging applications. This nanomaterial consisted of copper nanoparticles embedded in polylactic acid. Thus, a combination of antibacterial properties of copper nanoparticles and biodegradability of the polymer matrix was achieved. Metal nanoparticles have been synthesized by way of laser ablation, which is an easy route to prepare nanostructures without any capping agent in a liquid environment. Thus, nanoparticle suspensions have been easily mixed to a polymer solution. The resulting hybrid solutions were deposited by drop casting, thereby acquiring self-standing antibacterial packages. This experimental result is promising for possible use of copper nanoparticles embedded in a polylactic acid matrix to prevent the *Pseudomonas* spp. proliferation in smart food packages (Longano et al., 2012).

A smart packaging system was studied, which is based on a colorimetric method and included a polyaniline (PANI) film. In this smart packaging system, a chemical sensor is attached in the headspace of the packed fish and is used for real-time monitoring of the microbial breakdown products. This indicator, which is entrapped in package, contains PANI film which responds through a visible color change to a variety of basic volatile amines [specifically known as total volatile basic nitrogen (TVBN)] released during fish spoilage period. On the other hand, it was found that this indicator response also correlates well with microbial growth patterns in fish samples, especially the changing microbial populations [total viable count (TVC) and *Pseudomonas* spp.]. These also declared that the responses allowed the real-time monitoring of fish spoilage either at various constant temperatures or with temperature fluctuations. The PANI film could be recycled several times, using an acid solution to regenerate the PANI surface. Hereby, the PANI film can be considered as a low-cost sensor suitable for smart packaging applications (Kuswandi et al., 2012).

In 2015, phase change materials (PCMs) were studied to control the main factors (temperature variations during storage and distribution stages), which directly affect the quality of perishable food products. When PCMs are incorporated into the packaging structures, PCMs are able to absorb or release a great amount of energy during their melting/crystallization process and, thus, could provide thermal protection to the packaged food. Thus polystyrene (PS)-based multilayer heat storage structures with energy storage and hence temperature-buffering capacity were developed for application in refrigerated foods. For that reason, polycaprolactone (PCL) was used as the encapsulating matrix of a phase change material (PCM), called RT5 (a commercial blend of paraffin with a transition temperature at 5°C), by using high throughput electrohydrodynamic processing. As a conclusion, this work shows the potential of these materials for an efficient temperature-buffering effect of relevance in food packaging applications in order to preserve the quality of refrigerated packaged food products (Chalco-Sandoval et al., 2015).

Another sensor example of food spoilage monitoring can be colorimetric sensor array, which follows fish spoilage over time at room temperature for up to 24 hours as well as at 4°C for nine days. In this study, the increase in signal intensity of the chemo-sensitive compounds represented a trend similar to the increase in microbial growth during storage (Morsy et al., 2016).

In another work, the electro-hydrodynamic process was used to encapsulate an antioxidant, alpha-tocopherol, using different hydrocolloid matrices (soy protein isolate, SPI, whey protein isolate, WPI, and zein) as shell materials. These hybrid structures were directly electrospun/electrosprayed as a coating layer on to one side of a thermoplastic wheat gluten film, thereby giving rise to active/bioactive bilayer packaging structures. The water vapor-barrier efficiency of thermoplastic wheat gluten films was developed by the presence of the active coating layer. Furthermore, when delivering these structures to an industrial sterilization process, the alpha-tocopherol stability was preserved, especially when zein was used as the shell material. This work ensures a new method for improving active/bioactive packaging systems of interest in food applications (Fabra et al., 2016).

In 2017, a heat management PS foam tray, containing an ultrathin fiber-structured PS/PCM coating, was prepared by using high throughput electrohydrodynamic processing. For that reason, polystyrene (PS) was used as the encapsulating matrix of a commercial phase change material (PCM), called RT5 (a blend of paraffin with a transition temperature at 5°C), by using the electrospinning technique. The PS tray was coated with the PS/PCM ultrathin fiber mats and a soft heat treatment was applied to develop the adhesion between the layers with the aim of imparting heat-management capacity to the trays. Results of the study denoted that RT5 could be properly encapsulated inside the PS matrix, with a good encapsulation efficiency (ca. 78 per cent) and the developed PS fibers had a heat storage capacity equivalent to ~ 34 wt. per cent of the PCM. The effect of storage time and temperature was evaluated on the heat storage capacity of the developed PS-trays with the ultrathin fiber-structured PS/PCM layer. The heat storage capacity affected both the storage time and the temperature. This work presents a new insight into the development of heat-management polymeric materials of interest in food packaging applications, in order to preserve the quality of refrigerated packaged food products (Chalco-Sandoval et al., 2017).

In 2019, biodegradable food packaging films were manufactured from *mahua* oil-based polyurethane (PU) and chitosan (CS), incorporated with different rates of zinc oxide nanoparticles. PU/CS with 5 per cent nano-ZnO-composite film represented improved tensile strength and stiffness. It also showed that the incorporation of zinc oxide nanoparticles enhanced the antibacterial properties, barrier properties and hydrophobicity of the film. It also indicated that the shelf-life period of carrot pieces wrapped with the composite film was extended up to nine days. Additionally, it was also declared that the film containing zinc oxide nanoparticles was effective in reducing the bacterial contamination when compared to the commercial polyethylene film, thanks to the low environmental impact of the film (Sarojini et al., 2019).

References

Alexandratos, N. and J. Bruinsma (2012). Word Agriculture towards 2030/2050: The 2012 Revision, *ESA Working paper No. 12-03*, FAO, Rome.

Al-Naamani, L., S. Dobretsov and J. Dutta (2016). Chitosan-zinc oxide nanoparticle composite coating for active food packaging applications, *Innovative Food Science & Emerging Technologies*, 38 (Part A): 231–237.

Angione, M.D., R. Pilolli, S. Cotrone, M. Magliulo, A. Mallardi, G. Palazzo, L. Sabbatini, D. Fine, A. Dodabalapur, N. Cioffi and L. Torsi (2011). Carbon-based materials for electronic bio-sensing, *Mater. Today*, 14: 424–433.

Arfat, Y.A., S. Benjakul, T. Prodpran, P. Sumpavapol and P. Songtipya (2014). Properties and antimicrobial activity of fish protein isolate/fishskin gelatin film containing basil leaf essential oil and zinc oxide nanoparticles, *Food Hydrocolloids*, 41: 265–273.

Avella, M., J.J.D. Vlieger, M.E. Errico, S. Fischer, P. Vacca and M.G. Volpe (2005). Biodegradable starch/clay nanocomposite films for food packaging applications, *Food Chemistry*, 93: 467–474.

Azeredo, H.M.C.D. (2009). Nanocomposites for food packaging applications, *Food Research International*, 42: 1240–1253.

Azeredo, H.M.C.D., L.H.C. Mattoso and T.H. McHugh (2011). Nanocomposites in food packaging—A review. *In:* B. Reddy (Ed.). 2011. *Advances in Diverse Industrial Applications of Nanocomposites*, Intech., London, UK.

Batista, R.A., P.J.P. Espitia, J.S.S. Quintans, M.M. Freitas, M.Â. Cerqueira, J.A. Teixeira and J.C. Cardoso (2019). Hydrogel as an alternative structure for food packaging systems, *Carbohydrate Polymers*, 205: 106–116.

Briand, D., A. Oprea, J. Courbat and N. Bârsan (2011). Making environmental sensors on plastic foil, *Mater. Today*, 14: 416–423.

Cao, Q., H.-S. Kim, N. Pimparkar, J.P. Kulkarni, C. Wang, M. Shim, K. Roy, M.A. Alam and J.A. Rogers (2008). Medium-scale carbon nanotube thin-film integrated circuits on flexible plastic substrates, *Nature*, 454: 495–500.

Carbone, M., D.T. Donia, G. Sabbatella and R. Antiochia (2016). Silver nanoparticles in polymeric matrices for fresh-food packaging, *Journal of King Saud University, Science*, 28: 273–279.

Chalco-Sandoval, W., M.J. Fabra, A. López-Rubio and J.M. Lagaron (2015). Development of polystyrene-based films with temperature buffering capacity for smart food packaging, *Journal of Food Engineering*, 164: 55–62.

Chalco-Sandoval, W., M.J. Fabra, A. Lopez-Rubio and J.M. Lagaron (2017). Use of phase-change materials to develop electro-spun coatings of interest in food packaging applications, *Journal of Food Engineering*, 192: 122–128.

Cruz-Romero, M.C., T. Murphy, M. Morris, E. Cummins and J.P. Kerry (2013). Antimicrobial activity of chitosan, organic acids and nano-sized solubilisates for potential use in smart antimicrobially-active packaging for potential food applications, *Food Control*, 34: 393–397.

Dainelli, D., N. Gontard, D. Spyropoulos, E. Zondervan-van den Beuken and P. Tobback (2008). Active and intelligent food packaging: Legal aspects and safety concerns, *Trends Food Sci. Technol.*, 19: 103–112.

Day, B.P.F. (2001). Active packaging—A fresh approach, *J. Brand Technol.*, 1: 32–41.

Delen, D., B.C. Hardgrave and R. Sharda (2007). RFID for better supply-chain management through enhanced information visibility, *Production and Operations Management*, 16: 613–624.

Devi, M.P.I., N. Nallamuthu, N. Rajini, T.S.M. Kumar, S. Siengchin, A.V. Rajulu and N. Ayrilmis (2019). Biodegradable poly(propylene) carbonate using *in situ* generated CuNPs coated

Tamarindus indica filler for biomedical applications, *Materials Today Communications*, https://doi.org/10.1016/j.mtcomm.2019.01.007.

Dobrucka, R. and R. Cierpiszewski (2014). Active and intelligent packaging food—research and development: A review, *Pol. J. Food Nutr. Sci.*, 64: 7–15.

EC (2009). EU guidance to the commission regulation on active and intelligent materials and articles intended to come into contact with food, *European Commission*.

Espitia, P.J.P., J.A. Fuenmayor and C.G. Otoni (2019). Nanoemulsions: Synthesis, characterization, and application in bio-based active food packaging, *Comprehensive Reviews in Food Science and Food Safety*, 18: 264–285.

Essabti, F., A. Guinault, S. Roland and G. Régnier, S. Ettaqi and M. Gervais (2018). Preparation and characterization of polyethylene terephthalate films coated by chitosan and vermiculite nanoclay, *Carbohydrate Polymers*, 201: 392–401.

Fabra, M.J., A. Lopez-Rubio and J.M. Lagaron (2016). Use of the electrohydrodynamic process to develop active/bioactive bilayer films for food packaging applications, *Food Hydrocolloids*, 55: 11–18.

Farahi, R.H., A. Passian, L. Tetard and T. Thundat (2012). Critical issues in sensor science to aid food and water safety, *ACS Nano.*, 6: 4548–4556.

Fortunati, E., I. Armentano, Q. Zhou, A. Iannoni, E. Saino, L. Visai, L.A. Berglund and J.M. Kenny (2012). Multifunctional bionanocomposite films of poly(lactic acid), cellulose nanocrystals and silver nanoparticles, *Carbohydrate Polym.*, 87: 1596–1605.

Freedonia Group (2015). *Active and Intelligent Packaging: US Industry Study with Forecasts for 2019 & 2024, Study* #3338.

Gontard, N. (2000). *Panorama des emballages alimentaire actif*, Tech & Doc. Editions, Londres, ISBN-10: 2743003871. *In:* N. Gontard (Ed.). 2000. *Les Emballages Actifs*, Paris, France.

Haghighi, H., R. De Leo, E. Bedin, F. Pfeifer, H.W. Siesler and A. Pulvirenti (2019). Comparative analysis of blend and bilayer films based on chitosan and gelatin-enriched with LAE (lauroyl arginate ethyl) with antimicrobial activity for food packaging applications, *Food Packaging and Shelf-Life*, 19: 31–39.

Heising, J.K., M. Dekker, P.V. Bartels and M.A. Van Boekel (2014). Monitoring the quality of perishable foods: Opportunities for intelligent packaging, *Critical Reviews in Food Science and Nutrition*, 54: 645–654.

Hutton, T. (2003). *Food Packaging: An introduction, Key Topics in Food Science and Technology*, Chipping Campden, Gloucestershire, UK: Campden and Chorleywood Food Research Association, Group, 7: 108.

Jayaramudu, J., G.S.M. Reddy, K. Varaprasad, E.R. Sadiku, S.S. Ray and A.V. Rajulu (2013). Preparation and properties of biodegradable films from *Sterculia urens* short fiber/cellulose green composites, *Carbohydrate Polymers*, 93: 622–627.

Jiang, X., D. Valdeperez, M. Nazarenus, Z. Wang, F. Stellacci, W.J. Parak and P. del Pino (2015). Future perspectives towards the use of nanomaterials for smart food packaging and quality control, Part, *Syst. Charact.*, 32: 408–416.

Kabashin, A.V. and M. Meunier (2006). Laser Ablation-Based Synthesis of Nanomaterials. *In:* J. Perriere, E. Millon and E. Fogarassy (Eds.). *Recent Advances in Laser Progressing of Materials*, Elsevier, New York, USA.

Kerry, J.P., M.N. O'Grady and S.A. Hogan (2006). Past, current and potential utilization of active and intelligent packaging systems for meat and muscle-based products: A review, *Meat Science*, 74: 113–130.

Krochta, J.M. (2002). Proteins as raw materials for films and coatings: Definitions, current status, and opportunities. *In: GENNADIOS, A. Protein-based Films and Coatings*, pp. 21–61, Boca Raton.

Kumar, S., A. Jain, S. Panwar, I. Sharma, H.C. Jeon, T.W. Kang and R.K. Choubey (2018). Effect of silica on the ZnS nanoparticles for stable and sustainable antibacterial application, *International Journal of Applied Ceramic Technology*, 1–10.

Kuswandi, B., Y. Wicaksono, J.A. Abdullah, L.Y. Heng and M. Ahmad (2011). Smart packaging: Sensors for monitoring of food quality and safety, *Sensing and Instrumentation for Food Quality and Safety*, 5: 137–146.

Kuswandi, B., A.J. Restyana, A. Abdullah, L.Y. Heng and M. Ahmad (2012). A novel colorimetric food package label for fish spoilage based on polyaniline film, *Food Control*, 25: 184–189.

Longano, D., N. Ditaranto, N. Cioffi, F. Di Niso, T. Sibillano, A. Ancona, A. Conte, M.A. Del Nobile, L. Sabbatini and L. Torsi (2012). Analytical characterization of laser-generated copper nanoparticles for antibacterial composite food packaging, *Analytical and Bioanalytical Chemistry*, 403: 1179–1186.

Majeed, K., M. Jawaid, A. Hassan, A.A. Bakar, H.P.S.A. Khalil, A.A. Salema and I. Inuwa (2013). Potential materials for food packaging from nanoclay/natural fibres filled hybrid composites, *Materials and Design*, 46: 391–410.

Maksimović, M., V. Vujović and E. Omanović-Mikličanin (2015). Application of internet of things in food packaging and transportation, *Int. J. Sustainable Agricultural Management and Informatics*, 1: 333–350.

Malhotra, B., A. Keshwani and H. Kharkwal (2015). Antimicrobial food packaging: Potential and pitfalls, *Frontiers in Microbiology*, 6: 1–9.

McAlpine, M.C., H. Ahmad, D. Wang and J.R. Heath (2007). Highly ordered nanowire arrays on plastic substrates for ultrasensitive flexible chemical sensors, *Nat. Mater.*, 6: 379–384.

Miltz, J., N. Passy, C. and Mannheim (1995). Trends and applications of active packaging systems, 162: 201–210. *In:* M. Jagerstad and P. Ackerman (Eds.). 1995. *Food and Packaging Materials - Chemical Interaction*, The Royal Soc. of Chemistry, London, England.

Morales, A.M. and C.M. Lieber (1998). A laser ablation method for the synthesis of crystalline semiconductor nanowires, *Science*, 279: 208–211.

Morris, J.K. (2011). How safe is our food? *Emerg. Infect. Dis.*, 17: 126–128.

Morsy, M.K., K. Zor, N. Kostesha, T.S. Alstrøm, A. Heiskanen, H. El-Tanahi, A. Sharoba, D. Papkovsky, J. Larsen, H. Khalaf, M.H. Jakobsen and J. Emneus (2016). Development and validation of a colorimetric sensor array for fish spoilage monitoring, *Food Control.*, 60: 346–352.

Nasab, M.S. and M. Tabari (2018). Antimicrobial properties and permeability of poly lactic acid nanocomposite films containing zinc oxide, *Nanomedicine Research Journal*, 3: 125–132.

Ni, S., H. Zhang, H. Dai and H. Xiao (2018). Starch-based flexible coating for food packaging paper with exceptional hydrophobicity and antimicrobial activity, *Polymers*, 10: 1260.

Olafsdóttir, G., E. Martinsdóttir, J. Oehlenschläger, P. Dalgaard, B. Jensen, I. Undeland, I.M. Mackie, G. Henehan, J. Nielsen and H. Nilsen (1997). Method to evaluate fish freshness in research and industry, *Trends Food Sci. Tech.*, 8: 258–265.

Oliveira, J.P., G.P. Bruni, S.L.M. el Halal, F.C. Bertoldi, A.R.G. Dias and E.R. Zavareze (2019a). Cellulose nanocrystals from rice and oat husks and their application inaerogels for food packaging, *International Journal of Biological Macromolecules*, 124: 175–184.

Oliveira, J.P., G.P. Bruni, M.J. Fabra, E.R. Zavareze, A. López-Rubio and M. Martínez-Sanz (2019b). Development of food packaging bioactive aerogels through the valorization of gelidium sesquipedale seaweed, *Food Hydrocolloids*, 89: 337–350.

Othman, S.H. (2014). Bio-nanocomposite materials for food packaging applications: Types of biopolymer and nano-sized filler, *Agriculture and Agricultural Science Procedia*, 2: 296–303.

Pacquit, A., K.T. Lau, H. McLaughlin, J. Frisby, B. Quilty and D. Diamond (2006). Development of a volatile amine sensor for the monitoring of fish spoilage, *Talanta*, 69: 515–520.

Pacquit, A., J. Frisby, D. Diamond, K.T. Lau, A. Farrell, B. Quilty and D. Diamond (2007). Development of a smart packaging for the monitoring of fish spoilage, *Food Chemistry*, 102: 466–470.

Panaitescu, D.M., E.R. Ionita, C.A. Nicolae, A.R. Gabor, M.D. Ionita, R. Trusca, B.E. Lixandru, I. Codita and G. Dinescu (2018). Poly(3-hydroxybutyrate) Modified by nanocellulose and plasma treatment for packaging applications, *Polymers*, 10: 1249.

Pardo-Figuerez, M., A. López-Córdoba, S. Torres-Giner and J.M. Lagaron (2018). Superhydrophobic bio-coating made by co-continuous electrospinning and electrospraying on polyethylene terephthalate films proposed as easy emptying transparent food packaging, *Coatings*, 8: 364.

Pelaz, B., S. Jaber, D. Jimenez de Aberasturi, V. Wulf, J.M. de la Fuente, J. Feldmann, H.E. Gaub, L. Josephson, C.R. Kagan, N.A. Kotov, L. Liz-Marzán, H. Mattoussi, P. Mulvaney, C.B. Murray, A.L. Rogach, P.S. Weiss, I. Willner and W.J. Parak (2012). The state of nanoparticle-based nanoscience and biotechnology: Progress, promises, and challenges, *ACS Nano*, 6: 8468–8483.

Pelaz, B., G. Charron, C. Pfeiffer, Y. Zhao, J.M. de la Fuente, X.J. Liang, W.J. Parak and P. del Pino (2013). Interfacing engineered nanoparticles with biological systems: Anticipating adverse nano-bio interactions, *Small.*, 9: 1573–1584.

Pérez-Masiá, R., A. López-Rubio and J.M. Lagarón (2013). Development of zein-based heat-management structures for smart food packaging, *Food Hydrocolloids*, 30: 182–191.

Poyatos-Racionero, E., J.V. Ros-Lis, J.L. Vivancos and R. Martínez-Manez (2018). Recent advances on intelligent packaging as tools to reduce food waste, *Journal of Cleaner Production*, 172: 3398–3409.

Puligundla, P., J. Jung and S. Ko (2012). Carbon dioxide sensors for intelligent food packaging applications, *Food Control.*, 25: 328–333.

Rhim, J.-W., H.-M. Park and C.S. Ha (2013). Bio-nanocomposites for food packaging applications, *Progress in Polymer Science*, 38: 1629–1652.

Ribeiro-Santos, R., M. Andrade, N. Ramos de Melo and A. Sanches-Silva (2017). Use of essential oils in active food packaging: Recent advances and future trends, *Trends in Food Science & Technology*, 61: 132–140.

Rooney, M.L. (2005). Introduction to active food packaging technologies, ISBN: 978-0-12-31632-1, 63–69. *In:* J.H. Han (Ed.). *Innovations in Food Packaging*, Elsevier Ltd., London, UK.

Rukchon, C., A. Nopwinyuwong, S. Trevanich, T. Jinkarn and P. Suppakul (2014). Development of a food spoilage indicator for monitoring freshness of skinless chicken breast, *Talanta*, 130: 547–554.

Salari, M., M.S. Khiabani, R.R. Mokarram, B. Ghanbarzadeh and H.S. Kafil (2019). Preparation and characterization of cellulose nanocrystals from bacterial cellulose produced in sugar beet molasses and cheese whey media, *International Journal of Biological Macromolecules*, 122: 280–288.

Sarebanha, S. and A. Farhan (2018). Eco-friendly composite films based on polyvinyl alcohol and jackfruit waste flour, *Journal of Packaging Technology and Research*, 2: 181–190.

Sarojini, K.S., M.P. Indumathi and G.R. Rajarajeswari (2019). Mahua oil-based polyurethane/chitosan/nano ZnO composite films for biodegradable food packaging applications, *International Journal of Biological Macromolecules*, 124: 163–174.

Shahabi-Ghahfarrokhi, I. and A. Babaei-Ghazvini (2019). Using photo-modification to compatibilize nano-ZnO in development of starch-kefiran-ZnO green nanocomposite as food packaging material, *International Journal of Biological Macromolecules*, 124: 922–930.

Smits, E., J. Schram, M. Nagelkerke, R. Kusters, G. van Heck, V. van Acht, M. Koetse, J. van den Brand, G. Gelinck and H. Schoo (2012). Development of printed RFID sensor tags for smart food packaging, IMCS 2012—*The 14th International Meeting on Chemical Sensors*, DOI 10.5162/IMCS2012/4.5.2. 403–406.

Sogut, E. and A.C. Seydim (2019). The effects of chitosan- and polycaprolactone-based bilayer films incorporated with grape seed extract and nanocellulose on the quality of chicken breast fillets, *LWT–Food Sci. Technol.*, 101: 799–805.

Suganthi, S., S. Mohanapriya, V. Raj, S. Kanaga, R. Dhandapani, S. Vignesh and J.K. Sundar (2018). Tunable physicochemical and bactericidal activity of multicarboxylic-acids-crosslinked polyvinyl alcohol membrane for food packaging applications, *Chemistry Select*, 3: 11167–11176.

Summers, L. (1992). *Intelligent Packaging*, Centre for Exploitationof Science and Technology, London, UK.

Takei, K., T. Takahashi, J.C. Ho, H. Ko, A.G. Gillies, P.W. Leu, R.S. Fearing and A. Javey (2010). Nanowire active-matrix circuitry for low-voltage macroscale artificial skin, *Nat. Mater.*, 9: 821–826.

Tang, X.Z., P. Kumar, S. Alavi and K.P. Sandeep (2012). Recent advances in biopolymers and biopolymer-based nanocomposites for food packaging materials, *Critical Reviews in Food Science and Nutrition*, 52: 426–442.

Trovatti, E., S.C.M. Fernandes, L. Rubatat, C.S.R. Freire, A.J.D. Silvestre and C.P. Neto (2012). Sustainable nanocomposite films based on bacterial cellulose and pullulan, *Cellulose*, 19: 729–737.

Vaikousi, H., C.G. Biliaderis and K.P. Koutsoumanis (2009). Applicability of a microbial time–temperature indicator (TTI) for monitoring spoilage of modified atmosphere packed minced meat, *Int. J. Food Microbiol.*, 133: 272–278.

Valdés, M., A. Valdés González, J. García Calzón and M. Díaz-García (2009). Analytical nanotechnology for food analysis, *Microchim. Acta.*, 166: 1–19.

Vanderroost, M., P. Ragaert, F. Devlieghere and B. De Meulenaer (2014). Intelligent food packaging: The next generation, *Trends in Food Science & Technology*, 39: 47–62.

Wallach, D.F.H. and A. Novikov (1998). Methods and Devices for Detecting Spoilage in Food Products, WO 98/20337 (Biodetect Corporation, Nashua).

Wen, P., D.H. Zhu, H. Wu, M-H. Zong, Y.R. Jing and S.Y. Han (2016). Encapsulation of cinnamon essential oil in electrospun nanofibrous film for active food packaging, *Food Control.*, 59: 366–376.

Yam, K.L., P.T. Takhistov and J. Miltz (2005). Intelligent packaging: Concepts and applications, *Journal of Food Science*, 70: 1–10.

Yassin, M.A., A.A.M. Gad, A.F. Ghanem and M.H.A. Rehim (2019). Green synthesis of cellulose nanofibers using immobilized cellulose, *Carbohydrate Polymers*, 205: 255–260.

Yezza, I.A. (2009). Printed intelligence in packaging: Current and potential applications of nanotechnology, *Symposium on Nanomaterials for Flexible Packaging*, 2009, Columbus, April 30, Code 80863.

Yuvaraj, S. and N. Rajeswari (2018). Chitosan/modified banana epidermis starch composite films for food packaging applications, *Journal of Applied Packaging Research*, 10: 14–27.

Zeng, H., X.-W. Du, S.C. Singh, S.A. Kulinich, S. Yang, J. He and W. Cai (2012). Nanomaterials via laser ablation/irradiation in liquid: A review, *Advance Functional Materials*, 22: 1333–1353.

CHAPTER 4

Value-added Utilization of Fruit and Vegetable Pomace in Food Packaging

Jooyeoun Jung

1. Introduction

The sustainability of food packaging is becoming important in the food industry (Peelman et al., 2013). Plastics are widely used as food packaging materials and are currently consumed more than 200 million tons worldwide, with an annual growth of approximately 5 per cent (http://www.european-bioplastics.org/). These are generally non-biodegradable and compostable, thus mostly being landfilled. As a consequence, it causes an important environmental issue with waste generation, which leads to restriction of their use. In recent years, much effort has been put on developing environmentally-compatible products by incorporating biocomposite materials as a potential alternative to petroleum-based synthetic polymers (Avella et al., 2009; Gaikwad et al., 2016). As the alternative, biodegradable and/or compostable bioplastics, such as polylactide (PLA), starch, and PHA (polyhydroxyalkanoates) have been greatly utilized (Peelman et al., 2013). However, the packaging industry has been more interested in utilizing the bio-based materials from food processing by-products for resolving two issues simultaneously (Prochon et al., 2013; Valdés et al., 2014): (1) reducing food wastes generated by the food industry, and (2) enhancing sustainability of food packaging by adding value to those wastes and utilizing them as food packaging materials.

Fruit and vegetable processing produces significant amounts of by-products, which constitute about 25–30 per cent of a whole commodity group (Sagar et al., 2018). It approximately generates 1.81, 32.0 and 15.0 million tons in India, China and the United States of America, respectively (Wadhwa et al., 2013). These by-products

Department of Food Science and Technology, University of Nebraska-Lincoln, USA.
Email: jjung9@unl.edu

are mainly composed of pomace, seed, skin, rind or stem. Unfortunately, most of them are disposed in two different ways, such as landfill and incineration (Deng et al., 2012). In addition, because of the low content of nutrients and digestible energy, their use in animal feed is limited (Rohm et al., 2015).

Many studies report that fruit and vegetable pomaces still contain significant amounts of valuable bioactive compounds (e.g., anthocyanin, carotenoids, polyphenols, vitamins, enzymes, oils) and/or polymeric materials (e.g., pectin, lignocellulosic compounds) even after pressing, extracting, or filtering. Their quantities and/or qualities can be dependent on types of fruit and vegetable, cultivars, or processing methods. Many literatures have been published that say that incorporation of these bioactive compounds to food packaging system could provide high antioxidant or antimicrobial effects and extend the shelf-life of packaged food products (Ayala-Zavala et al., 2010; Deng and Zhao, 2011; Valdés et al., 2014). It was also reported that food applications with active packaging and antioxidant or antimicrobial packaging will be expanded to reach a value of approximately US$ 29.0 billion by 2020 (Anonymous, 2017; de la Caba et al., 2019). Hence, the value-added utilization of abundant and available pomace to develop biodegradable food packaging system with high potential has created great interest in the food packaging industry. These would not only reduce environmental concern and become great benefits for the food industry with sustainable waste management, but also improve the quality or shelf-life of packaged food products (Fig. 1).

This chapter reports chemical composition and natural bioactive compounds of various fruit and vegetable pomaces and introduces various utilizations of fruit and vegetable pomaces in food packaging as natural extracts, fillers, edible films and coatings. Innovative processing methods, including green extraction and encapsulation technologies for enhancing their extraction efficiency, stability and sustainability in food packaging and their challenges and future perspectives in terms of environmental, social and economic issues will be discussed in this chapter.

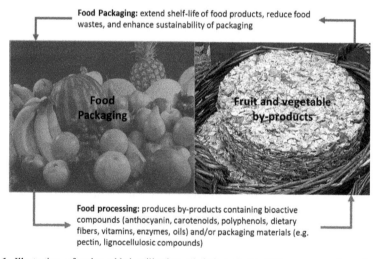

Fig. 1. Illustration of value-added utilization of fruit and vegetable pomaces for enhancing sustainability of food packaging.

2. Chemical Composition of Fruit and Vegetable Pomaces

Chemical composition of fruit and vegetable pomaces is varied, depending on the type of fruit and vegetable or processing methods. Chemical compositions of wine grape, fruit and vegetable pomaces are summarized in Table 1. Carbohydrate contents are relatively higher in *Muller Thurgau* (white wine grape) and *Morio Muscat* (white wine grape) than *Cabernet Sauvignon* (red wine grape), *Merlot* (red wine grape), and *Pinot Noir* (red wine grape). However, protein, lipid and ash contents of grape pomaces were relatively higher in red wine grape cultivars than in white wine grape cultivars. In fruit pomaces, raspberry pomace contains higher protein and lipid contents than other fruit pomaces. It could be probably due to the large amount of seeds in raspberry pomace. For vegetable pomace, their protein contents are relatively higher than fruit pomace. Carrot pomace contains less amounts of carbohydrates than other pomaces, but has highest ash content. Due to these variations among different fruit and vegetable pomaces or cultivars, their utilization can be selective, depending on the requirements of food packaging or target food products.

Table 1. Approximate Composition of Fruit and Vegetable Pomaces.

Content (%, w/w dry basis)						
Types of pomace	Cultivar	Carbohydrates*	Protein	Lipid	Ash	References
Wine grape	Muller Thurgau	88.29	6.54	2.64	2.53	Deng et al., 2012
	Morio Muscat	90.18	5.38	1.14	3.31	
	Cabernet Sauvignon	73.73	12.34	6.33	7.59	
	Merlot	77.60	11.26	3.35	7.19	
	Pinot Noir	76.96	12.13	4.74	6.17	
Fruit	Apple	80.02	4.18	5.08	2.03	Cho and Hwang, 2000
	Raspberry	76.7	10.0	11.1	2.2	Brodowska, 2017
	Pear	88.0	2.7	1.0	1.4	Grigelmo-Miguel and Martín-Belloso, 1999
	Orange	82.4	7.9	0.5	2.9	
	Peach	83.5	7.1	1.5	2.9	
Vegetable	Tomato	70.96	19.27	5.85	3.92	Del Valle et al., 2006
	Carrot	32.98	9.70	1.54	10.30	Chantaro et al., 2008
	Artichoke	71.1	15.6	1.0	7.5	Grigelmo-Miguel and Martín-Belloso, 1999
	Asparagus	70.3	16.2	2.0	5.3	

* Carbohydrate contents was determined by subtracting the sum of protein, lipid, and ash contents from 100%.

3. Natural Bioactive Compounds of Fruit and Vegetable Pomaces

Natural bioactive compounds, including total dietary fiber (TDF), total phenolic content (TPC) and radical scavenging activity (RSA) of fruit and vegetable pomaces are summarized in Table 2. For dietary fiber, all pomaces have higher insoluble dietary fiber (IDF) than soluble dietary fiber (SDF). It was seen that lemon, orange and carrot pomaces contain higher SDF (~ 9.2–~ 15.4 per cent, w/w dry matter, DM) than other pomaces. For wine grape pomaces, TPC and RSA were relatively higher in red wine grape pomace cultivars (*Cabernet Sauvignon, Merlot* and *Pinot Noir*) than white wine grape pomace cultivars (*Muller Thurgau* and *Morio Muscat*). Among fruits, raspberry and orange pomaces contain higher TPC and RSA than other pomaces. Many efforts have been made to avoid use of synthetic additives due to their potential toxic effects (Silva-Weiss et al., 2013). Hence, incorporation of natural extract from fruit and vegetable pomaces to food packaging, such as films or coatings, could diminish the concern regarding the use of artificial additives and meet the consumers' demand for natural products, as well as provide antioxidant or antimicrobial effects with extended shelf-life of food products (Cho and Hwang, 2000; Wang et al., 2007; Wang et al., 2013).

Table 2. Total Dietary Fiber (TDF) including Insoluble Dietary Fiber (IDF) and Soluble Dietary Fiber (SDF), Total Phenolic Content (TPC) and Radical Scavenging Activity of Wine, Fruit and Vegetable Pomaces.

TDF (%, w/w DM)				TPC (mg GAE/g DM)	RSA (mg AAE/g DM)	References
Types of pomace	Cultivar	IDF	SDF			
Wine grape	*Muller Thurgau*	29.6	0.7	15.8	25.6	Deng et al., 2012
	Morio Muscat	17.9	0.8	11.6	20.5	
	Cabernet Sauvignon	56.9	0.8	26.7	39.7	
	Merlot	53.2	1.5	25.0	40.2	
	Pinot Noir	59.9	1.7	21.4	32.2	
Fruit	Apple	29.1	2.5	7.4	3.9	Gouw et al., 2017a; Gouw et al., 2017b
	Blueberry	49.0	1.0	8.2	8.9	
	Cranberry	57.9	0.8	2.5	3.8	
	Raspberry	38.1	0.3	24.5	43.4	
	Grapefruit	56.0	4.6	18.7	7.0*	Figuerola et al., 2005; Sun et al., 2013; Xi et al., 2017
	Lemon	50.9	9.2	4.71	8.20[+]	
	Orange	54.0	10.3	43.5	14.2*	
Vegetables	Tomato	79.85	5.42	9.45	2.99[++]	García Herrera et al., 2010; Kalogeropoulos et al., 2012
	Carrot	57.62	15.38	8.53	5.0*	Pieszka et al., 2015; Turksoy et al., 2011; Yu et al., 2018

*mg Trolox equivalent/g dry matter (DM)
[+]µM Rutin equivalents (TE)/g wet matter (WM)
[++]µM Trolox equivalent/g dry matter (DM)

4. Pectin and Lignocellulosic Compositions of Fruit and Vegetable Pomaces

The use of fruit and vegetable pomaces containing high pectin and lignocellulosic compounds as fillers in biodegradable food packaging system or edible film/coating represents an attractive and sustainable way of valorization (Sánchez-Safont et al., 2018). In addition, fruit and vegetable pomaces are considered as the cheapest and most environmentally-virtuous lignocellulosic fibers. However, it should be noted that their performances in food packaging could vary depending on fruit and vegetable pomace as those could contain different amounts of pectin and lignocellulosic compounds and/or compositions, specifically when it comes to mechanical performance. Table 3 summarizes pectin and lignocellulosic composition for different fruit and vegetable pomaces. The pectin content is highest in sugar beet pomace, followed by olive, pear, cranberry and apple pomaces. Carrot and cherry pomaces contain relatively less amounts of pectin than other pomaces. Cellulose content is highest in blueberry and cranberry pomaces, followed by apple and carrot pomaces. Hemicellulose content is highest in apple or sugar beet pomaces, followed by olive, cranberry and blueberry pomaces. Lignin content is lowest in olive and apple pomaces, but cherry was observed to have the highest level than other pomaces. Hence, the potential of fruit and vegetable pomaces as filler or edible film/coating for food packaging should be evaluated by considering available amounts of pectin and lignocellulosic compounds and/or compositions in fruit and vegetable pomaces, as these could contribute functional or mechanical properties of food packaging and determine their effects on food products.

Table 3. Pectin and Lignocellulosic Composition of Wine, Fruit and Vegetable Pomaces.

Types of pomace	Cultivar	Pectin (%, w/w DM)	Lignocellulosic composition (%, w/w DM)			References
			Cellulose	Hemicellulose	Lignin	
Fruits	Apple	10.1	65.8	34.2	8.7	Gouw et al., 2017a; Gouw et al., 2017b
	Blueberry	7.4	76.2	23.8	36.8	
	Cranberry	11.4	73.8	26.2	43.5	
	Cherry	1.51	18.4	10.7	69.4	Nawirska and Kwaśniewska, 2005
	Pear	13.4	34.5	18.6	33.5	
Vegetables	Carrot	3.88	51.6	12.3	32.2	Nawirska and Kwaśniewska, 2005
	Tomato	7.55	19.0	12.0	36.0	Toushik et al., 2017
	Sugar beet	32.0	30.0	32.0	4.0	
	Olive	17.4	36.4	26.8	39.2	

5. Utilization of Fruit and Vegetable Pomaces in Food Packaging

Figure 2 illustrates various utilizations of fruit and vegetable pomaces in food packaging. Various forms (e.g., fresh, dried or powdered) of fruit and vegetable pomaces could be utilized into biodegradable boards, containers, films, edible coating,

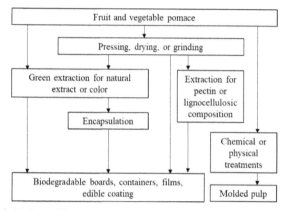

Fig. 2. Various utilizations of fruit and vegetable pomaces in food packaging.

or molded pulp, but these need to be differently processed, depending on the types of fruit and vegetable pomaces or target food products. The following sections introduce various food packaging applications, using fruit and vegetable pomaces and also discusses about processing methods and their effects on various food products.

5.1 Incorporation of Natural Extracts from Fruit and Vegetable Pomaces into Food Packaging

Natural extracts obtained from fruit and vegetables pomaces contain antioxidant, antibacterial and anti-inflammatory compounds, such as polyphenols, flavonoids, etc. (Cerruti et al., 2011; Yilmaz and Toledo, 2006). These natural extracts have been applied to various fields, such as pharmaceutical and cosmetic, but this chapter will mainly discuss food packaging. These natural extracts obtained from fruit and vegetable pomaces can be incorporated into food packaging with antioxidant or antimicrobial effects, and can be called 'active packaging' with enhanced functionalities. It could also control the release of active agents during an extended period of time to maintain the quality or shelf-life of products, without the need for direct addition of any substances to food products (Zhou et al., 2010). The active packaging systems and edible film/coating using natural extract have been studied and also applied to various food products (Table 4).

Park and Zhao (2006) incorporated cranberry pomace extract to low methoxyl pectin (LMP) or high methoxyl pectin (HMP) at a concentration of 0.50 per cent or 0.75 per cent (w/w wet basis) with 0.25 per cent (w/w wet basis) sorbitol or glycerol for improving film functionality (Park and Zhao, 2006). The higher (0.75 per cent) pectin concentration resulted in increased mechanical properties. It could create natural colorful and fruit-flavor edible films from fruit pomace water extracts. Wine grape pomace was also utilized as a natural additive for modified starch Mater-Bi (Bastioli, 1998; Cerruti et al., 2011). Starch is widely available and easily modified to get a thermoplastic polymer, but due to its hydrophilic nature, thermoplastic starch applications are limited (Rosa et al., 2009). To overcome this experimental drawback, starch is modified by blending with synthetic polymers, such as polyesters or vinyl alcohol copolymers and produce the Mater-Bi trademark by Novamont

(Bastioli, 1998). It has been seen that grape pomace extract is functioned as both plasticizer and antimicrobial agent, thus resulting in the increase of flexibility and decrease of biodegradability of films. Deng and Zhao (2011) also utilized wine grape pomace extract for creating films with plant-based polysaccharides, such as pectin, sodium alginate and Ticafilm® for enhancing mechanical, water barrier, nutritional and antibacterial properties. All films have relatively low WVP and mechanical properties were varied, depending on the types of polysaccharides. Derived films had antibacterial activity against both *Escherichia coli* (5-log reduction) and *Listeria innocua* (1.7- to 3.0-log reduction) in comparison to control. They suggested that films can be used as orally-dissolving films in pharmaceutic applications due to high water solubility and fast release rates of phenolic compounds. Ferreira et al. (2014) also created chitosan films incorporated with grape pomace extract. Chitosan has been extensively studied due to its biocompatibility, biodegradability and low toxicity (Busilacchi et al., 2013). Chitosan itself has antioxidant and antibacterial effect (Jung and Zhao, 2012, 2013), thus being suitable for food packaging usage. Hence, the incorporation of grape pomace extract with antioxidant or antimicrobial effectinto chitosan films is a feasible way to expand the potential applications of chitosan to biodegradable food packaging system with enhanced functional properties. Chitosan films were created with three different types of grape pomace extracts: 0.15 per cent of hot water extract (mainly polysaccharides), 0.15 and 0.3 per cent of chloroform extract (a grape skin wax extract), and 0.3 and 0.75 per cent of n-hexane extract (grape seed oil). Chitosan-based films with incorporation of grape pomace aqueous extract made the most hydrophilic and smoothest film with enhanced antioxidant properties. The incorporation of grape pomace wax extract into chitosan solution produced a film more heterogeneous and rough, but allowed improvement of the antioxidant properties with no significant changes in solubility. The wax extract also contributed to the increase of films' flexibility and decrease of stiffness. The chitosan-based films with grape pomace oil extract were the most hydrophobic ones, presenting higher antioxidant activities. It thus demonstrated that all these films can be an alternative to synthetic packaging materials and as a carrier for active compounds from natural fruit and vegetable pomace extracts, which could improve quality or shelf-life of food.

Biodegradable and antioxidant food packaging film was developed by combining wastes from the production of gelatin capsules and blueberry pomace ethanol extract (de Moraes Crizel et al., 2016). Oil nutraceutical capsules comprise mainly gelatin, which is generated in large quantities at a high waste treatment. These gelatin capsules of biodegradable and active biofilm are easy to obtain as they contain gelatin. Hence, food packaging films can be made with wastes from difference sources. Addition of blueberry pomace extract does not affect the macro-structure of films and increases the antioxidant effects against sunflower oil. Meanwhile, Licciardello et al. (2015) created corona-treated (812 W) polypropylene film (final surface energy > 44 dyne/cm) coated with lacquers by an automated coating machine (Licciardello et al., 2015). The lacquers were Superol-Barrier varnish, referred to as Shellac and Cellax–Barrier–Varnish or as cellulose nitrate. The lacquers were coated on corona-treated polypropylene film with and without the addition of grape pomace extract and olive leaf extract. The antioxidant capacity of film samples coated with Shellac and cellulose nitrate, containing grape pomace extract and olive leaf extract, was assessed and the antioxidant levels of films containing grape pomace extract were

0.272, 0.483 and 0.728 Troloxmeq/L at 5 per cent concentration and 0.705, 0.786 and 0.893 Troloxmeq/L at 10 per cent concentration for water, 10 per cent ethanol and 50 per cent ethanol food simulants, respectively.

Synthetic additives have been used in the food industry to extend the shelf-life of food products. Because of growing concern about the potential health hazards associated with synthetic additives and increased consumer demand for natural products, there is a growing interest in the use of natural additives for food processing and packaging (Shahidi and Zhong, 2010). The use of active packaging films containing natural extracts can improve the quality and/or shelf-life of food products and also fulfill the consumers' demand for natural products.

5.2 Incorporation of Fruit and Vegetable Pomaces as Fillers in Food Packaging

For reducing food waste and harmful effects of petrochemical-derived plastic residues on the environment, biodegradable food packaging has been considered as an alternative to overcome these issues (Coussy et al., 2013). As the alternative, biodegradable and/or compostable bioplastics, such as polylactide (PLA), starch and PHA (polyhydroxyalkanoates), have been greatly utilized (Peelman et al., 2013), but its price is higher than conventional plastics. It also has the limit on highly respiring food products, such as fruit and vegetables, as its barrier properties are too high to control O_2/CO_2 exchanges during storage (Berthet et al., 2016). Hence, one strategy to reduce the final cost of materials and enhancing functional properties and biodegradability of food packaging is to apply low-cost lignocellulosic fillers obtained from fruit and vegetable pomaces (Berthet et al., 2016).

Table 4 shows utilization of fruit and vegetable pomaces as filler for biodegradable composites or films. In the study of Gaikwad et al. (2016), apple pomace powder was created and directly reinforced to the polyvinyl alcohol (PVA) at 1, 5, 10, and 30 per cent on the weight basis of the PVA. The created film has antioxidant effect, thus delaying the lipid oxidation. PVA with apple pomace can be thus utilized as active food packaging. Wheat straw, brewing spent grains and olive mills can be utilized as filler for poly (3-hydroxybutyrate-co-valerate) (PHVB) (Berthet et al., 2015). They extracted the lignocellulosic fibers from each material by following this procedure: (1) cut milling and grinding for wheat straw, sieving 500 μm), caustic treatment, (2) grinding for brewing spent grains and cryo-centrifugal milling, phenolic extraction, residual oil removal, protein chemical removal, and (3) grinding for olive pomace. The particle size of wheat straw, brewing spent grains and olive pomace was 109 μm, 148 μm, and 46 μm respectively. Then, the composite material was prepared by melt extrusion on a lab-scale twin-screw extruder for PHVB pellet and lignocellulosic fibers. The temperature from the polymer feeding to the die varied from 180°C to 140°C. The resulting material was pelletized with a 2–3 mm length and dried at 60°C for at least 8 hours. Composite films were then produced by using a heated hydraulic press at 150 bar and 170°C for 5 minutes. It was obvious that the mechanical properties were degraded by adding fibers to PHBV due to poor fiber/matrix adhesion, degradation of PHBV polymer chains and decrease of PHBV's crystallinity. However, PHBV/lignocellulosic fibers could be still applicable for food packaging, especially PHBV/wheat straw fiber composites for respiring food products

Table 4. Applications of Wine, Vegetable and Fruit Pomaces in Food Packaging.

Types of pomace	Application types	Factors	Other components	References
Apple	Filler	Concentration at the range of 1–30% (w/w)	Polyvinyl alcohol (PVA)	Gaikwad et al., 2016
Wheat straw, brewing spent grains and olive mills	Filler	Particle size at 46, 109, and 148 μm	Poly(3-hydroxybutyrate-co-valerate)(PHVB)	Berthet et al., 2015
Blueberry	Filler	Concentration at the range of 4, 8 and 12 % (w/w, based on starch)	Cassava starch	Luchese et al., 2018
Blueberry	Filler and extract	Concentration of fiber and extract	Gelatin	de Moraes Crizel et al., 2016
Olive pomace	Filler	Concentration of pomace flour or microparticles	Chitosan	de Moraes Crizel et al., 2018a
Wine grape	Extract	Types of polysaccharides	Pectin, sodium alginate, and Ticafilm®	Deng and Zhao, 2011
Wine grape	Extract	Types of extraction method and concentration of extract	Chitosan	Ferreira et al., 2014
Cranberry	Extract	Concentrations of extract	Pectin	Park and Zhao, 2006
Wine grape	Coated to film	Types of coating matrix	Shellac and cellulose nitrate	Licciardello et al., 2015
Apple, blueberry and cranberry	Pulp	Concentration of each pomace	Recycled newspaper pulp and cellulose nanofiber	Gouw et al., 2017a

(e.g., fruit or vegetable), whereas PHBV/olive mills composites for water sensitive products (e.g., baked goods). There are also studies utilizing blueberry pomace as the filler for food packaging. In the study of Luchese et al. (2018), freeze-dried, milled and sieved (100 mesh) blueberry pomace powder was incorporated into cassava starch at the level of 4, 8 and 12 per cent (w/w based on cassava starch). It was seen that blueberry pomace powder-incorporated films showed good light barrier due to the presence of aromatic compounds in blueberry pomace. So, it can be considered as active packaging, preventing food deterioration caused by UV radiation. In another study, blueberry pomace was utilized as the packaging material in the form of both extract and filler and incorporated into gelatin capsule waste (mainly composed of gelatin) (de Moraes Crizel et al., 2016). The blueberry pomace extract was obtained by using a centrifugal extractor. For blueberry pomace fiber, fresh blueberry pomace was dried in a forced-air dry oven at 55°C for 24 hours. Dried blueberry pomace was milled and screened to make the particles smaller than 125 mm. Similar to the result of blueberry pomace powder-incorporated cassava starch film, the addition of both fiber and extract increased the function as the UV light barrier, thus reducing lipid oxidation of sunflower oil. The same research group also investigated the incorporation of olive pomace powder into chitosan (de Moraes Crizel et al., 2018a). Olive pomace was freeze-dried and ground in a mill for making the particle size smaller than 500 nm. This flour was further homogenized with water and dried by using the spray dryer with the inlet drying temperature of 160°C. Both olive pomace flour and microparticles were incorporated into chitosan solutions at the level of 10 per cent, 20 per cent and 30 per cent in relation to the mass of chitosan and cast on to polystyrene petri dishes. In comparison with olive pomace-flour-incorporated chitosan film, the addition of 10 per cent olive microparticles enhanced the tensile strength. Both flour- and micorparticles-incorporated chitosan films had the protective packaging properties for nuts against oxidation during 31 days. Park and Zhao (2006) derived cranberry pomace-extract-incorporated pectin films. The created films could retain natural color and fruit flavor from fruit pomace water extracts. There are also several studies utilizing grape pomace as the packaging material. Deng and Zhao (2011) made wine grape pomace-extract-incorporated hydrocolloid films (e.g., pectin, sodium alginate and Ticafilm®). Grape pomace extract was obtained by hot water extraction. Depending on the type of hydrocolloid matrix, physicochemical, mechanical and water barrier properties of derived films varied. They could also measure antibacterial activity against both *Escherichia coli* and *Listeria innocua.* Pomace-extract-incorporated films could have 5 log reductions in *E. coli* and 1.7 to 3.0 log reductions in *L. innocua* in comparison with control. Ferreira et al. (2014) also created grape pomace-extract-incorporated chitosan film. Different from studies mentioned above, this study used three different types of extract using hot water, wax and oil. For hot water extract, dried pomace was mixed with preheated water at 70°C at a ratio of 1:10 (w/v) and placed into a 50°C water bath for 45 minutes. The mixture was centrifuged, filtered and freeze-dried. For wax extraction, the residue obtained from the hot water extraction was dried, homogenized with chloroform at a ratio of 1:10 (w/v) and placed into 70°C water bath for 30 minutes. The mixture was filtered and the solvent was evaporated. They could extract waxes at a yield of 6.5 per cent and was confirmed by the FT-IR. For oil extraction, seeds were collected, washed and dried at 40°C. The milled seeds (< 0.5 mm mesh) were extracted in a 200 mL Soxhlet extractor with n-hexane at 60°C for 5 hours. The solvent was then evaporated and oil was obtained with the extraction yield of 16.7 per cent. The prepared extracts were incorporated into 1.5 per cent chitosan solution at different levels (0.15, 0.3, or 0.75 per cent w/v). This solution (31 g) was transferred into a plexiglass plate with 144 cm^2 and 3 mm height and dried at 35°C for 18 hours. It aimed at deriving films with high antioxidant and low solubility. It was seen that the incorporation of oil extract could decrease 75 per cent solubility in water due to improved hydrophobicity. The antioxidant mechanism of chitosan films was altered, depending on the type of extract. The hydrophobic films incorporated with wax or oil showed higher antioxidant capacity in organic medium (ABTS and DPPH assays), whereas most hydrophilic films with hot water extract showed an improvement in FRAP and in reducing power assays. In the study of Licciardello et al. (2015), grape pomace was also utilized to create antioxidant packaging, but differently applied to films from the studies mentioned above. Grape-pomace extract was extracted by using a mixture of ethanol, water and acetic acid (50:49:1; v/v/v, 50°C) for 60 minutes at a liquid/solid ratio of 1:20. Prepared grape pomace extract was incorporated into the shellac-based lacquer with the level of 5 and 10 per cent (w/w). Then the polyolefin film was coated with the lacquers by an automated coating machine set at 25 mm/s speed, dried at 50°C for one minute. The derived film was evaluated for antioxidant capacity, using suitably modified ABTS test. It showed that antioxidant effect of films could be increased with increasing level of incorporated grape pomace extract in coating.

As shown above, fruit and vegetable pomaces have been utilized in food packaging, but these needed to be processed before applying to other polymeric materials or compounds, such as extraction or chemical or physical treatments, which could produce significant amounts of residue or energy cost. Hence, this would not be the best strategy for utilizing food wastes for demanding various benefits from these approaches, thus requiring the method that utilizes the generated wastes directly with less residue, processing and cost. For example, Gouw et al. (2017) created molded pulp-packaging boards combined with recycled newspaper and fruit pomace slurry. This method was intended to utilize fruit pomace directly as the packaging material with less processing steps. For producing pomace slurry, about 200 g of fruit pomace (apple, cranberry and blueberry pomaces) was blended with 1 liter of tap water for 20 minutes. It was aimed to create good mechanical and water barrier properties of fruit pomace-pulp-molded board combined with newspaper slurry, plasticizer and cellulose nanofiber. The results suggest that fruit pomace can partially substitute newspaper to create fruit pomace boards with better or similar properties to 100 per cent newspaper board, but the compatibility between newspaper and fruit pomace fibers depends on the type of fruit pomace.

6. Edible Films and Coatings

Edible films and coatings can protect food products from physical, chemical and biological deterioration, thus resulting in shelf-life extension and safety improvement (Kester and Fennema, 1986). The greatest benefits of edible films and coatings is their edibility (Debeaufort et al., 1998). Since fruit and vegetable pomaces contain the greatest amount of bioactive compounds or dietary fibers, their incorporation into edible packaging could enhance the value of packaged food without directly adding substances to food products. As shown in Fig. 3, edible coating could protect food from physical damage or UV light, but control gas exchange or release of active compounds

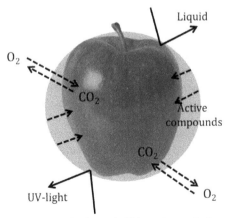

Fig. 3. Protective mechanisms of edible coating on food products.

from coating over the extended storage period (Fig. 3). It would be especially beneficial to food products respiring during storage, such as fruit or vegetable.

Edible coatings or films have been applied to minimally-processed fruit and vegetable. These are mostly consumed without washing process so applied that the protective coating should be edible and better to be natural. Ayala-Zavala et al. (2010) published the hypothesis that the antioxidant and antimicrobial protection of fresh-cut fruits could be possibly obtained from the fruit's own byproducts (Ayala-Zavala et al., 2010). In the study of Moreira et al. (2015), apple fiber obtained from apple pomace was incorporated into polysaccharide-based coating formulation and its effect was investigated on fresh-cut apples (Moreira et al., 2015). Sodium alginate, gellan gum, or pectin solutions were mixed with apple fiber obtained from apple pomace. Glycerol was added as the plasticizer. The apple pieces were dipped into the chilled polysaccharide solutions for 2 minutes and then allowed to drip off for 1 minute. Samples were submerged in the calcium chloride solution with ascorbic acid for 2 minutes for crosslinking with applied polysaccharides on to the fruit surface. Coated samples were placed in polypropylene trays sealed with a polypropylene film and stored in darkness at $4 \pm 1°C$. The coated apples retained their initial firmness and color during storage. It was seen that gellan-gum-edible coating could reduce mesophilic and psychrophilic counts in coated apples during the storage period. The incorporation of apple fiber could enhance antioxidant activity of the coating. In sensory evaluation, the coating could prolong storage periods of apple samples with sensory scores above the sensory acceptability threshold (3) for any attributes.

The biodegradable coatings and films were also prepared by using the fruit and vegetable pomace flour (Fai et al., 2016). The fruit and vegetable pomace flour consisted of the following species: Selecta orange (*Citrus sinensis*), passion fruit (*Passiflora edulis*), watermelon (*Citrullus lanatus*), lettuce (*Lactuca sativa*), courgette (*Cucurbita pepo*), carrot (*Daucus carota*), spinach (*Spinacea oleracea*), mint (*Mentha* sp.), taro (*Colocasia esculenta*), cucumber (*Cucumis sativus*) and rocket (*Eruca sativa*). Coating solutions were prepared by using 8 g of fruit and vegetable pomace flour per 0.1 liter of different solutions as extractors: ammonium hydroxide and methaphosphoric acid at pH 7.0 and pH 9.0 and ammonium hydroxide and orthophosphoric acid at pH 9.0. Coating solutions were heated at 70°C for 45 minutes, cooled at room temperature and filtered through polyester cloth meshes and centrifugation. Carrots were prepared in two different ways, including slices or shredded by using a vegetable processor and coated by prepared coating solutions by either immersion for 5 minutes or spraying. Coated samples were stored at 5°C for 15 days in a dark cooling chamber. The result showed that immersion and spray-coated shredded carrots had 12 and 25 per cent less mass loss than the control during storage, respectively. However, results showed that cutting methods for carrots had a significant effect than coating method. Coated shredded carrots performed better than those that were sliced and it would be because of the larger surface area with a higher incorporation of edible coating solution. More coating adsorbed to a tissue of peeled vegetable surface could allow shredded carrots to absorb more bioactive compounds, thus providing an additional functional benefit (i.e., antioxidants and functional barriers).

In the study of Khalifa (2016), both olive leaves and pomace extracts were incorporated into chitosan coating, which was applied to apple and strawberry. For testing antimicrobial effect of coating, the apple and strawberry fruits were disinfected with sodium hypochlorite. Subsequently, cross-shaped wounds were made on the apple and strawberry, using sterilized puncher and inoculated by *P. expansum* and *R. stolonifer* spores suspension, respectively. Results showed that olive-pomace-extract-incorporated chitosan coating could retain the quality of apple and strawberry during storage. This coating also enhanced its antifungal effect against *P. expansum* and *R. stolonifer* counted and decay area of fruit.

The use of edible films and coatings incorporated with natural extracts obtained from fruit and vegetable pomaces has been increasing for various food products. This approach could offer environmental advantages over plastic materials due to biodegradability and also reduce both food and packaging wastes by extending shelf-life of food and directly consuming food without the necessity of disposing packaging.

7. Green Extraction Technology for Fruit and Vegetable Pomaces

Food industries have been challenged to find green extraction technologies to displace conventional solvent extraction techniques for not only reducing energy use, but also enhancing valorization of food wastes for economic sustainability (Rombaut et al., 2014). Several alternative techniques significantly reduce solvent consumption and increase the speed of the extraction as well as ensuring a safe and high quality of extract (Otero-Pareja et al., 2015). Several green extraction tools, such as microwave, ultrasound and supercritical fluid extraction are highlighted in this section.

The most common equipment used for extraction is ultrasonic treatment, including bath and probe. In comparison with the ultrasonic baths with low efficiency, the ultrasonic probe is more efficient and powerful. Ultrasounds are characterized by their frequency range (from 20 kHz to 10 MHz) that generate mechanical deformation, cell wall destruction and further release of the cellular content in a solid, liquid or gaseous media (Rombaut et al., 2014). Some examples of ultrasound-assisted extraction for fruit and vegetable pomaces are introduced in Table 5. Minjares-Fuentes et al. (2014) extracted the pectin from grape pomace by using the ultrasound-assisted acid extraction. They could find that the treatment condition using citric acid at pH 2 at 75°C for 60 minutes had the highest pectin yield (~ 32.3 per cent). To obtain extract rich in antioxidants (e.g., polyphenols), ultrasound-assisted water extraction was applied to apple pomace. The optimal condition was found at 40°C, 40 minutes and 0.764 W/cm^2, which showed higher total phenolic and antioxidant activity than at conventional extraction. Total phenolics and anthocyanin were extracted from blueberry wine pomace by using ultrasound-assisted extraction. The optimized condition was found at 61.03°C, a liquid–solid ratio of 21.70 mL/g, 23.67 min. Under this condition, the yield of anthocyanin and total pheholics was higher than those at a conventional solvent extraction.

The use of microwaves for extraction of biological compounds was first reported in 1986 (Ganzler et al., 1986). Microwave can directly affect the glandular and vascular system of a biological sample, thus liberating active compounds to the solvent. Similar to ultrasound-assisted extraction, microwave-assisted extraction was applied to various fruit and vegetable pomaces and it showed similar or better yield than the conventional extraction with reduced extraction time. It was seen that the microwave treatment cleaved and liberated phenolic compounds from citrus

Table 5. Green Extraction Technology Applied for Fruit and Vegetable Pomaces.

Types of extraction technology	Types of pomaces	Target compounds	Treatment conditions	References
Ultrasound	Grape pomace	Pectin	Citric acid at pH 2.0, 75°C, 60 min.	Minjares-Fuentes et al., 2014
Ultrasound	Apple pomace	Polyphenol	40°C for 40 min. at 0.764 W/cm²	Pingret et al., 2012a
Ultrasound	Blueberry wine pomace	Phenolic and anthocyanin	61.03°C, a liquid-solid ratio of 21.70 mL/g, 23.67 min	He et al., 2016
Microwave	Citrus mandarin pomace	Phenolic	250 or 500 W	Hayat et al., 2010
Microwave	Sour cherry pomace	Phenolic	700 W, ethanol-water, 12 min, 20 mL solvent/g solid	Simsek et al., 2012
Ultrasound and microwave	Red grape pomace	Polyphenol	25 kHz at 300 W, 20°C for 60 min for ultrasound 200 W, 50°C, and 60 min for microwave	Drosou et al., 2015
Supercritical carbon dioxide extraction	Apricot pomace	β-carotene	30.4–50.7 MPa and 313–333 K	Şanal et al., 2004
Supercritical carbon dioxide extraction	Industrial tomato waste	Carotenoids, tocopherols fatty acids, sitosterols	460 bar and 80°C	Vági et al., 2007
Supercritical carbon dioxide and ethanol extraction	Sour cherry pomace	Phenolics	20–60 MPa and 40–60°C	Yener and Bayındırlı, 2008
Supercritical carbon dioxide and methanol consecutive extractions	Grape pomace	Fatty acids and polyphenols	103–250 bar and 35°C	Aizpurua-Olaizola et al., 2015

pomace, thus resulting in an increase of free phenolic compounds and enhancement of antioxidant capacity of the extracts (Hayat et al., 2010). However, it was important to provide a suitable microwave power and extraction time since higher power and longer treatment time cause degradation of some flavanones and flavonol compounds. In the study of Simsek et al. (2012), the optimum condition of microwave-assisted extraction was found at 700 W, ethanol-water, 12 minutes and 20 mL solvent/g solid for sour cherry pomace. This resulted in higher antiradical efficiency and concentration of phenolics, but shortened the extraction time. The efficiency of microwave-assisted extraction for red grape pomace was also determined in regard to the extraction yield, antiradical activity and phenolic contents, and compared with that of ultrasound-assisted extraction (Drosou et al., 2015).

The supercritical fluid extraction uses the unique properties of supercritical fluids to facilitate the extraction of target compounds from solid samples. A supercritical fluid can have a good solvating power, high diffusivity, low viscosity and marginal surface tension when it is above its critical temperature and pressure (Pingret et al., 2012b). Thus, it achieves fast and efficient extraction by permitting a rapid mass transfer and penetrating the pores in the sample matrix (Rombaut et al., 2014). In a study of Şanal et al. (2004), extraction of β-carotene was carried out from apricot pomace, using supercritical carbon dioxide extraction. It was seen that the highest extraction yield of β-carotene was obtained at 40.5 MPa and 328 K at the end of 90-minute extraction time. The tomato pomace extract at the optimal condition of supercritical extraction (460 bar and 80°C) could contain 90.1 per cent of lycopene rich in tocopherols and phytosterol (Vági et al., 2007). For sour cherry pomace, the extract obtained at the optimal condition (54.8–59 MPa, 50.6–54.4°C, 20 per cent ethanol, 40 minutes) of supercritical extraction contained 0.60 mg garlic acid equivalent/g sample of total phenolic contents and 2.30 mg DPPH/g sample of antioxidant effect, respectively (Yener and Bayındırlı, 2008). In the study of Aizpurua-Olaizola et al. (2015), the sequential supercritical fluid extraction was applied to extract fatty acids and polyphenols from wine pomace. For the first fraction, supercritical fluid pressure was maintained at 250 bar at 35°C for the extraction of fatty acids. For the second fraction, the supercritical fluid pressure was set at 103 bar at 35°C with the use of 40 per cent methanol as co-solvent for the extraction of polyphenols. This consecutive extraction method could obtain the extract with high yields of fatty acids and polyphenols in very short time. Overall, the supercritical fluid extraction method could enhance the extraction efficiency on the target compounds by using non-toxic and less solvent at shorter time.

8. Encapsulation

Active compounds present in fruit and vegetable pomaces undergo oxidation and/ or degradation when exposed to extreme environmental conditions in terms of temperature, oxygen and light (Hasbay and Galanakis, 2018). Hence, stabilization is important to retain and/or enhance their functional effects in food packaging applications during the extended storage period of food products.

Encapsulation has been widely studied as a route to stabilize and protect bioactive compounds extracted from fruit and vegetable pomaces, allowing them to be protected against degradation and/or oxidation for a longer period of time and encouraging their application to various food products. Wall materials used in both encapsulation and packaging materials include gums, polysaccharides, lipids, proteins, fibers and mixtures of these materials (Davidov-Pardo et al., 2013). The selection of an adequate wall material is important for the stability of the compounds in packaging. Since food packaging, such as film or coating, should contain these wall materials forming matrix, natural extracts within the packaging system could be protected and made relatively stable than in other applications. This section, however, discusse the incorporation of encapsulated natural extract from fruit and vegetable pomaces into food packaging for enhancing their stability and homogenous properties over film or coating and also controlling their release during the extended storage of food products (Fig. 4).

Active biodegradable film incorporated with encapsulated anthocyanin was developed and its effect was evaluated on the quality of extra-virgin olive oil during storage (Stoll et al., 2017). Natural extract was extracted by using an acidified alcoholic solvent (70 per cent ethanol, HCl 0.1 per cent) at 40°C, in a proportion of 1:80 (grape pomace/solvent) under protection against light. The extract was then concentrated and 30 per cent (w/v) maltodextrin (degree of dextrose equivalent, DE=20) was mixed with

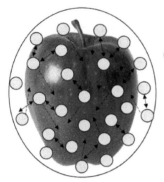

◯ Encapsulated natural extract
↗ Repulsion between droplets
    for enhancing homogenous properties
↙ Release of active compounds from
    encapsulated natural extract

Fig. 4. Incorporation of encapsulated natural extracts into food packaging film or coating.

the extract. The samples were freeze-dried for 72 hours and produced encapsulated anthocyanin from grape pomace extract. Film matrix was composed of 5 per cent cassava starch and incorporated with encapsulated anthocyanin (microcapsules) in a proportion of 8 per cent by stirring. The film was created by a casting method and dried at 40°C for 9 hours and formed into pouches to pack the oil. In comparison with polypropylene pouches carrying oil (four days), the starch film with encapsulated anthocyanin could maintain the quality of oil by showing under the limits established by Codex Alimentarius for over eight days (13.6 meq O_2/kg of peroxides).

Biodegradable films with gelatin and papaya peel microparticles were developed for enhancing antioxidant effect and its effect on preventing oxidation of lard (de Moraes Crizel et al., 2018b). First, the papaya peels were made into powder. For the preparation of microparticles with gelatin, the papaya peel powder was mixed with gelatin and Tween 80 and the solution was homogenized for 30 seconds at 5000 rpm. The formulation wase spray-dried for producing microparticles. For the film matrix, the gelatin capsule residue obtained from nutraceutical industry of chia oil nutraceutical capsules was mixed with microparticles and made into biodegradable packaging films with antioxidant activity using the casting method. It was seen that the films with microparticles presented a higher antioxidant effect than those without microparticles. For food application (i.e., lard), the samples packaged in these films presented a significantly lower amount of oxidation compounds after 22 days under high temperature (around 40°C) and light. When compared with samples packaged with polyethylene, gelatin films with microparticles did not affect the profile of fatty acids in the lard samples.

9. Challenges and Future Perspectives

Food packaging is an important part of the food industry. There are clear trends in food packaging in regard to addressing sustainability related to both resource efficiency and waste utilization by adding the value to food processing wastes. The development of biodegradable materials from food processing wastes would result in great business potential for the food packaging industry as well as meeting the consumers' demand for natural and sustainable food packaging. For decades, there has been a lot of research utilizing fruit and vegetable pomaces to food packaging system, which has been also perceived to have great potential and commercial markets (Verghese et al., 2015), but it has been rarely reached in large-scale production (Guillard et al., 2018). It could be discussed for the following reasons:

1) The technical limits related to the functional properties of food packaging materials from fruit and vegetable pomaces are different from conventional plastics.
2) The resistance of the food industry and consumers to adopt unknown technologies.
3) The costs of the new implementation, the inefficiency and lack of competitiveness of the new technologies and regulatory barriers.
4) The lack of collaboration and exchange between stakeholders of the food chain (R&D centers, food and packaging manufacturers, legislators, consumers).
5) The lack of large-scale production or trials for the scale-up and commercialization.
6) The lack of communications between industries and end-users, thus not knowing or understanding new technologies (Aday and Yener, 2015).

Food packaging must protect food and also provide convenience to consumers. Meanwhile, it should follow legal and environmental perspectives and meet consumer expectations for the quality of packaged food products (Kim and Seo, 2018; Wilson et al., 2018). This chapter demonstrates that food packaging incorporated with fruit and vegetable pomaces could meet those principles of food packaging by extending shelf-life and improving the quality of packaged food products.

The development of food packaging incorporated with various fruit and vegetable pomaces suggests the use of environment-friendly and cost-effective processes and add value to fruit and vegetable pomaces contributing to sustainable development of biodegradable materials. The significant amount of waste generated during fruit and vegetable processing should be properly processed so that it could be converted to natural extract or colors with high antioxidant or antimicrobial effects and renewable and biodegradable packaging materials. The use of fruit and vegetable pomaces in food packaging is promising and has much potential, but the existing research is still scarce and more in-depth research to increase their utilizations in bulk and to demonstrate its safety and commercial applications as an alternative to conventional plastics is needed.

References

Aday, M.S. and U. Yener (2015). Assessing consumers' adoption of active and intelligent packaging, *Brit. Food J.*, 117: 157–177.

Aizpurua-Olaizola, O., M. Ormazabal, A. Vallejo, M. Olivares, P. Navarro, N. Etxebarria and A. Usobiaga (2015). Optimization of supercritical fluid consecutive extractions of fatty acids and polyphenols from *Vitis vinifera* grape wastes, *J. Food Sci.*, 80: E101–E107.

Anonymous (2017). Protective packaging market: Global industry analysis and opportunity assessment, *Future Market Insights*, 2015–2025.

Avella, M., A. Buzarovska, M. Errico, G. Gentile and A. Grozdanov (2009). Eco-Challenges of Bio-Based Polymer Composites, *Materials*, 2: 911.

Ayala-Zavala, J.F., C. Rosas-Domínguez, V. Vega-Vega and G.A. González-Aguilar (2010). Antioxidant enrichment and antimicrobial protection of fresh-cut fruits using their own byproducts: looking for integral exploitation, *J. Food Sci.*, 75: R175–R181.

Bastioli, C. (1998). Properties and applications of Mater-Bi starch-based materials. *Polym. Degrad. Stabil.*, 59: 263–272.

Berthet, M.A., H. Angellier-Coussy, D. Machado, L. Hilliou, A. Staebler, A. Vicente and N. Gontard (2015). Exploring the potentialities of using lignocellulosic fibers derived from three food by-products as constituents of biocomposites for food packaging, *Ind. Crop Prod.*, 69: 110–122.

Berthet, M.A., H. Angellier-Coussy, V. Guillard and N. Gontard (2016). Vegetal fiber-based biocomposites: Which stakes for food packaging applications? *J. Appl. Polym. Sci.*, 133.

Brodowska, A.J. (2017). Raspberry pomace—Composition, properties and application, *Eur. J. Biol. Res.*, 7(2): 86–96.

Busilacchi, A., A. Gigante, M. Mattioli-Belmonte, S. Manzotti and R.A.A. Muzzarelli (2013). Chitosan stabilizes platelet growth factors and modulates stem cell differentiation toward tissue regeneration, *Carbohyd. Polym.*, 98: 665–676.

Cerruti, P., G. Santagata, G. Gomez d'Ayala, V. Ambrogi, C. Carfagna, M. Malinconico and P. Persico (2011). Effect of a natural polyphenolic extract on the properties of a biodegradable starch-based polymer, *Polym. Degrad. Stabil.*, 96: 839–846.

Chantaro, P., S. Devahastin and N. Chiewchan (2008). Production of antioxidant high dietary fiber powder from carrot peels, *LWT–Food Sci. Technol.*, 41: 1987–1994.

Cho, Y.J. and J.-K. Hwang (2000). Modeling the yield and intrinsic viscosity of pectin in acidic solubilization of apple pomace, *J. Food Eng.*, 44: 85–89.

Coussy, H., V. Guillard, C. Guillaume and N. Gontard (2013). Role of packaging in the smorgasbord of action for sustainable food consumption, *Agro-food-Industry Hi Tech.*, 24: 15–19.

Davidov-Pardo, G., I. Arozarena and M.R. Marín-Arroyo (2013). Optimization of a wall material formulation to microencapsulate a grape seed extract using a mixture design of experiments, *Food Bioprocess Tech.*, 6: 941–951.

de la Caba, K., P. Guerrero, T.S. Trung, M. Cruz-Romero, J.P. Kerry, J. Fluhr, M. Maurer, F. Kruijssen, A. Albalat, S. Bunting, S. Burt, D. Little and R. Newton (2019). From seafood waste to active seafood packaging: An emerging opportunity of the circular economy, *J. Clean. Prod.*, 208: 86–98.

de Moraes Crizel, T., T.M. Haas Costa, A. de Oliveira Rios and S. Hickmann Flôres (2016). Valorization of food-grade industrial waste in the obtaining active biodegradable films for packaging, *Ind. Crops Prod.*, 87: 218–228.

de Moraes Crizel, T., A. de Oliveira Rios, V.D. Alves, N. Bandarra, M. Moldão-Martins and S. Hickmann Flôres (2018a). Active food packaging prepared with chitosan and olive pomace, *Food Hydrocolloid.*, 74: 139–150.

de Moraes Crizel, T., A. de Oliveira Rios, V.D. Alves, N. Bandarra, M. Moldão-Martins and S. Hickmann Flôres (2018b). Biodegradable films based on gelatin and papaya peel microparticles with antioxidant properties, *Food Bioprocess Tech.*, 11: 536–550.

Debeaufort, F., J.A. Quezada-Gallo and A. Voilley (1998). Edible films and coatings: Tomorrow's packagings: A review, *Criti. Rev. Food Sci. Nutr.*, 38: 299–313.

Del Valle, M., M. Cámara and M.E. Torija (2006). Chemical characterization of tomato pomace, *J. Sci. Food Agr.*, 86: 1232–1236.

Deng, G.F., C. Shen, X.R. Xu, R.D. Kuang, Y.-J. Guo, L.S. Zeng, L.L. Gao, X. Lin, J.F. Xie, E.Q. Xia, S. Li, S. Wu, F. Chen, W.-H. Ling and H.B. Li (2012). Potential of fruit wastes as natural resources of bioactive compounds, *Int. J. Mol. Sci.*, 13: 8308–8323.

Deng, Q. and Y. Zhao (2011). Physicochemical, nutritional and antimicrobial properties of wine grape (cv. Merlot) pomace extract-based films, *J. Food Sci.*, 76: E309–E317.

Drosou, C., K. Kyriakopoulou, A. Bimpilas, D. Tsimogiannis and M. Krokida (2015). A comparative study on different extraction techniques to recover red grape pomace polyphenols from vinification byproducts, *Ind. Crop. Prod.*, 75: 141–149.

Fai, A.E.C., M.R. Alves de Souza, S.T. de Barros, N.V. Bruno, M.S.L. Ferreira and É.C.B.D.A. Gonçalves (2016). Development and evaluation of biodegradable films and coatings obtained from fruit and vegetable residues applied to fresh-cut carrot (*Daucus carota* L.), *Postharvest Biol. Tec.*, 112: 194–204.

Ferreira, A.S., C. Nunes, A. Castro, P. Ferreira and M.A. Coimbra (2014). Influence of grape pomace extract incorporation on chitosan films properties, *Carbohyd. Polym.*, 113: 490–499.

Figuerola, F., M.A.L. Hurtado, A.M.A. Estévez, I. Chiffelle and F. Asenjo (2005). Fiber concentrates from apple pomace and citrus peel as potential fiber sources for food enrichment, *Food Chem.*, 91: 395–401.

Gaikwad, K.K., J.Y. Lee and Y.S. Lee (2016). Development of polyvinyl alcohol and apple pomace bio-composite film with antioxidant properties for active food packaging application, *J. Food Sci. Tec.*, 53: 1608–1619.

Ganzler, K., A. Salgó and K. Valkó (1986). Microwave extraction: A novel sample preparation method for chromatography, *J. Chromatogr. A*, 371: 299–306.

García Herrera, P., M.C. Sánchez-Mata and M. Cámara (2010). Nutritional characterization of tomato fiber as a useful ingredient for food industry, *Innov. Food Sci. Emerg. Technologies*, 11: 707–711.

Gouw, V.P., J. Jung, J. Simonsen and Y. Zhao (2017a). Fruit pomace as a source of alternative fibers and cellulose nanofiber as reinforcement agent to create molded pulp packaging boards, *Compos. Part A-Appl. S.*, 99: 48–57.

Gouw, V.P., J. Jung and Y. Zhao (2017b). Functional properties, bioactive compounds and in vitro gastrointestinal digestion study of dried fruit pomace powders as functional food ingredients, *LWT - Food Sci. Technol.*, 80: 136–144.

Grigelmo-Miguel, N. and O. Martín-Belloso (1999). Comparison of dietary fiber from by-products of processing fruits and greens and from cereals, *LWT - Food Sci. Technol.*, 32: 503–508.

Guillard, V., S. Gaucel, C. Fornaciari, H. Angellier-Coussy, P. Buche and N. Gontard (2018). The next generation of sustainable food packaging to preserve our environment in a circular economy context, *Fronti. Nutr.*, 5.

Hasbay, I. and C.M. Galanakis (2018). 7 - Recovery technologies and encapsulation techniques, pp. 233–264. *In*: C.M. Galanakis (Ed.). *Polyphenols: Properties, Recovery and Applications*, Woodhead Publishing.

Hayat, K., X. Zhang, U. Farooq, S. Abbas, S. Xia, C. Jia, F. Zhong and J. Zhang (2010). Effect of microwave treatment on phenolic content and antioxidant activity of citrus mandarin pomace, *Food Chem.*, 123: 423–429.

He, B., L.L. Zhang, X.Y. Yue, J. Liang, J. Jiang, X.L. Gao and P.-X. Yue (2016). Optimization of ultrasound-assisted extraction of phenolic compounds and anthocyanins from blueberry (*Vaccinium ashei*) wine pomace, *Food Chem.*, 204: 70–76.

Jung, J. and Y. Zhao (2012). Comparison in antioxidant action between α-chitosan and β-chitosan at a wide range of molecular weight and chitosan concentration, *Bioorgan. Med. Chem.*, 20: 2905–2911.

Jung, J. and Y. Zhao (2013). Impact of the structural differences between α- and β-chitosan on their depolymerizing reaction and antibacterial activity, *J. Agr. Food Chem.*, 61: 8783–8789.

Kalogeropoulos, N., A. Chiou, V. Pyriochou, A. Peristeraki and V.T. Karathanos (2012). Bioactive phytochemicals in industrial tomatoes and their processing byproducts, *LWT–Food Sci. Technol.*, 49: 213–216.

Kester, J.J. and O.R. Fennema (1986). Edible films and coatings: A review, *Food Technol.*, Chicago, 48: 47–59.

Khalifa, I., H. Barakat, H.A. El-Mansy and S.A. Soliman (2016). Improving the shelf-life stability of apple and strawberry fruits applying chitosan-incorporated olive oil processing residues coating, *Food Packaging Shelf*, 9: 10–19.

Kim, D. and J. Seo (2018). A review: Breathable films for packaging applications, *Trends Food Sci. Tech.*, 76: 15–27.

Licciardello, F., J. Wittenauer, S. Saengerlaub, M. Reinelt and C. Stramm (2015). Rapid assessment of the effectiveness of antioxidant active packaging—Study with grape pomace and olive leaf extracts, *Food Packaging and Shelf-Life*, 6: 1–6.

Luchese, C.L., T. Garrido, J.C. Spada, I.C. Tessaro and K. de la Caba (2018). Development and characterization of cassava starch films incorporated with blueberry pomace, *International J. Biol. Macromolecules*, 106: 834–839.

Minjares-Fuentes, R., A. Femenia, M.C. Garau, J.A. Meza-Velázquez, S. Simal and C. Rosselló (2014). Ultrasound-assisted extraction of pectins from grape pomace using citric acid: A response surface methodology approach, *Carbohydr. Polym.*, 106: 179–189.

Moreira, M.R., L. Cassani, O. Martín-Belloso and R. Soliva-Fortuny (2015). Effects of polysaccharide-based edible coatings enriched with dietary fiber on quality attributes of fresh-cut apples, *J. Food Sci. Tech.*, 52: 7795–7805.

Nawirska, A. and M. Kwaśniewska (2005). Dietary fiber fractions from fruit and vegetable processing waste, *Food Chem.*, 91: 221–225.

Otero-Pareja, M.J., L. Casas, M.T. Fernández-Ponce, C. Mantell and E.J.M.D.I. Ossa (2015). Green extraction of antioxidants from different varieties of red grape pomace, *Molecules*, 20: 9686.

Park, S.I. and Y. Zhao (2006). Development and characterization of edible films from cranberry pomace extracts, *J. Food Sci.*, 71: E95–E101.

Peelman, N., P. Ragaert, B. De Meulenaer, D. Adons, R. Peeters, L. Cardon, F. Van Impe and F. Devlieghere (2013). Application of bioplastics for food packaging, *Trends Food Sci. Tech.*, 32: 128–141.

Pieszka, M., P. Gogol, M. Pietras and M. Pieszka (2015). Valuable components of dried pomaces of chokeberry, black currant, strawberry, apple and carrot as a source of natural antioxidants and nutraceuticals in the animal diet, *Am. Anim. Sci.*, 15: 475–491.

Pingret, D., A.S. Fabiano-Tixier, C.L. Bourvellec, C.M.G.C. Renard and F. Chemat (2012a). Lab and pilot-scale ultrasound-assisted water extraction of polyphenols from apple pomace, *J. Food Eng.*, 111: 73–81.

Pingret, D., A. Fabiano Tixier and F. Chemat (Eds.) (2012b). *Comprehensive Sampling and Sample Preparation*, Academic Press, Oxford.

Prochoń, M. and A. Przepiórkowska (2013). Innovative application of biopolymer keratin as a filler of synthetic acrylonitrile-butadiene rubber NBR, *J. Chem.*, 2013: 8.

Rohm, H., C. Brennan, C. Turner, E. Günther, G. Campbell, I. Hernando, S. Struck and V. Kontogiorgos (2015). Adding Value to Fruit Processing Waste: Innovative Ways to Incorporate Fibers from Berry Pomace in Baked and Extruded Cereal-based Foods – A SUSFOOD Project, *Foods* (Basel, Switzerland), 4: 690–697.

Rombaut, N., A.S. Tixier, A. Bily and F. Chemat (2014). Green extraction processes of natural products as tools for biorefinery, *Biofuel. Bioprod. Bior.*, 8: 530–544.

Roza, M.F., B.S. Chiou, E.S. Medeiros, D.F. Wood, T.G. Williams, L.H.C. Mattoso, W.J. Orts and S.H. Imam (2009). Effect of fiber treatments on tensile and thermal properties of starch/ethylene vinyl alcohol copolymers/coir biocomposites, *Bior. Techn.*, 100: 5196–5202.

Sagar, N.A., S. Pareek, S. Sharma, E.M. Yahia and M.G. Lobo (2018). Fruit and vegetable waste: bioactive compounds, their extraction, and possible utilization, *Compr. Rev. Food Sci. F.*, 17: 512–531.

Şanal, İ.S., A. Güvenç, U. Salgın, Ü. Mehmetoğlu and A. Çalımlı (2004). Recycling of apricot pomace by supercritical CO_2 extraction, *J. Supercrit. Fluid.*, 32: 221–230.

Sánchez-Safont, E.L., A. Aldureid, J.M. Lagarón, J. Gámez-Pérez and L. Cabedo (2018). Biocomposites of different lignocellulosic wastes for sustainable food packaging applications, *Compos.*, Part B, *Eng.*, 145: 215–225.

Shahidi, F. and Y. Zhong (2010). Novel antioxidants in food quality preservation and health promotion, *Eur. J. Lipid Sci. Tech.*, 112: 930–940.

Silva-Weiss, A., M. Ihl, P.J.A. Sobral, M.C. Gómez-Guillén and V. Bifani (2013). Natural additives in bioactive edible films and coatings: functionality and applications in foods, *Food Eng. Rev.*, 5: 200–216.

Simsek, M., G. Sumnu and S. Sahin (2012). Microwave assisted extraction of phenolic compounds from sour cherry pomace, *Sep. Sci. Technol.*, 47: 1248–1254.

Stoll, L., A.M.D. Silva, A.O.E.S. Iahnke, T.M.H. Costa, S.H. Flôres and A.D.O. Rios (2017). Active biodegradable film with encapsulated anthocyanins: Effect on the quality attributes of extra-virgin olive oil during storage, *J. Food Process. Pres.*, 41: e13218.

Sun, Y., L. Qiao, Y. Shen, P. Jiang, J. Chen and X. Ye (2013). Phytochemical profile and antioxidant activity of physiological drop of citrus fruits, *J. Food Sci.*, 78: C37–C42.

Toushik, S.H., K.T. Lee, J.S. Lee and K.S. Kim (2017). Functional applications of lignocellulolytic enzymes in the fruit and vegetable processing industries, *J. Food Sci.*, 82: 585–593.

Turksoy, S., Ouml and B. Zkaya (2011). Pumpkin and carrot pomace powders as a source of dietary fiber and their effects on the mixing properties of wheat flour dough and cookie quality, *Food Sci. Technol. Res.*, 17: 545–553.

Vági, E., B. Simándi, K.P. Vásárhelyiné, H. Daood, Á. Kéry, F. Doleschall and B. Nagy (2007). Supercritical carbon dioxide extraction of carotenoids, tocopherols and sitosterols from industrial tomato by-products, *J. Supercrit. Fluid*, 40: 218–226.

Valdés, A., A.C. Mellinas, M. Ramos, M.C. Garrigós and A. Jiménez (2014). Natural additives and agricultural wastes in biopolymer formulations for food packaging, *Front. Chem.*, 2.

Verghese, K., H. Lewis, S. Lockrey and H. Williams (2015). Packaging's role in minimizing food loss and waste across the supply chain, *Packag. Technol. Sci.*, 28: 603–620.

Wadhwa, M., M.P.S. Bakshi and H.P.S.M. (Eds.) (2013). Utilization of fruit and vegetable wastes as livestock feed and as substrates for generation of other value-added products. *In: FAO* (Ed.). Bangkok (Thailand), Regional Office for Asia and the Pacific.

Wang, J., B. Wang, W. Jiang and Y. Zhao (2007). Quality and shelf-life of mango (*Mangifera indica* L. cv. 'Tainong') coated by using chitosan and polyphenols, *Food Sci. Technol. Int.*, 13: 317–322.

Wang, L., Y. Dong, H. Men, J. Tong and J. Zhou (2013). Preparation and characterization of active films based on chitosan incorporated tea polyphenols, *Food Hydrocolloid.*, 32: 35–41.

Wilson, C.T., J. Harte and E. Almenar (2018). Effects of sachet presence on consumer product perception and active packaging acceptability—A study of fresh-cut cantaloupe, *LWT –Food Sci. Technol.*, 92: 531–539.

Xi, W., J. Lu, J. Qun and B. Jiao (2017). Characterization of phenolic profile and antioxidant capacity of different fruit part from lemon (*Citrus limon* Burm.) cultivars, *J. Food Sci. Tech.*, 54: 1108–1118.

Yener, M.E. and A. Bayındırlı (2008). Extraction of total phenolics of sour cherry pomace by high pressure solvent and subcritical fluid and determination of the antioxidant activities of the extracts AU - Adil, İncinur Hasbay, Sep., *Sci. Tech.*, 43: 1091–1110.

Yilmaz, Y. and R.T. Toledo (2006). Oxygen radical absorbance capacities of grape/wine industry byproducts and effect of solvent type on extraction of grape seed polyphenols, *J. Food Compos. Anal.*, 19: 41–48.

Yu, G., J. Bei, J. Zhao, Q. Li and C. Cheng (2018). Modification of carrot (*Daucus carota* Linn. var. Sativa Hoffm.) pomace insoluble dietary fiber with complex enzyme method, ultrafine comminution, and high hydrostatic pressure, *Food Chem.*, 257: 333–340.

Zhou, G.H., X.L. Xu and Y. Liu (2010). Preservation technologies for fresh meat—A review, *Meat Sci.*, 86: 119–128.

CHAPTER 5

Bio-based Materials Inspired from Brazilian Cerrado Wastes

Ricardo Stefani,[1,*] *Lidyane Mendes Bento*[1,2] and
Genilza da Silva Mello[1,3]

1. Introduction

Brazil is known for its great biodiversity as it is a rich source for renewable natural resources. Due to its territorial extension, this biodiversity ranges through the Amazon rainforest, Savanna formation (Cerrado) and semi-desert areas (Caatinga) among others. Thus, its flora is considered of great scientific and industrial interest— being a source for products of pharmaceutical (Luzia and Jorge, 2011; Maria et al., 2008), nutritional (Dias et al., 2012; Wu et al., 2013) and industrial importance (Genovese et al., 2008; Ramos and Wilhelm, 2005). Nevertheless, its full potential is still underexplored.

Some species native to Brazilian Cerrado are edible and have been studied in academy and industry as smart food packaging material for being antioxidant (Bonilla and Sobral, 2016; Pertuzatti et al., 2014) and quality indicator. Among the Brazilian ecosystems, the most explored is the Cerrado. The Cerrado in Brazil is represented by about 23 per cent of its territorial area in a heterogeneous and discontinuous ecosystem that ranges from the State of São Paulo to Goiàs, Minas Gerais, Mato Grosso and Tocantins.

Brazil owns 13.1 per cent of the global biota, 34,916 listed species of plants, including bryophytes, ferns, lyophytes, gymnosperms and angiosperms, with 200 new species catalogued per year of the latter (Stehmann and Sobral, 2017). In spite of the great diversity, the large number of species threatened with extinction of the Brazilian

[1] Laboratory of Materials Studies (LEMAT), Institute of Exact and Earth Sciences, Araguaia University Campus, Federal University of Mato Grosso (UFMT), Barra do Garças, Brazil.
[2] Faculty of Engineering, University of Cuiabá (UNIC), Cuiabá, Brazil.
[3] Faculty of Applied Agricultural Biological Social Sciences (UNEMAT), State University of Mato Grosso, Nova Xavantina, Brazil.
* Correspondig author: rstefani@ufmt.br

fauna and flora brings with it economic implications, genetic impoverishment and human health (Clark et al., 2014).

The Brazilian Cerrado is one of the most extensive phyto-geographical domains (> 200 million ha) threatened in the world. It occurs in the central part of Brazil and had until 2008 about 43 per cent of the biome converted into pasture, housing and planting areas (Machado et al., 2008). In addition to its territorial extension, the flora found in this biome is a rich source of bioactive compounds; however, the taxonomic classification of its flora/fauna is still insufficient, causing several groups to be still unknown. Hence, this leads to an unestimated lost in both bio- and chemical diversity, as unknown plant species could also be a source of unknown bioactive compounds. Besides bioactive micromolecules, the Cerrado is also a source of biopolymers, mainly carbohydrates, proteins and oils.

Many species have already been found in the world as a source of bioactive compounds and biopolymers, with tropical ecosystems being further studied for their diversity (Pereira et al., 2012; Reis et al., 2015; Valdés et al., 2015). Despite the consolidated knowledge regarding the pharmacological and functional properties of bioactive compounds from known species from Cerrado (Dias et al., 2012; Genovese et al., 2008; Oliveira et al., 2012; Wu et al., 2013), studies on the application of these compounds in industry are still rare and superficial. Some crops from Cerrado, such as pequi, babaçu, cashew, açaí, carnaúba and jabuticaba are in a mature stage from the point of view of agricultural and industrial exploration and are very important for income improvement and social development of certain poor Brazilian regions. The industrial processing of such species, just like any other agro-industrial process, generates residues. These residues are rich in oils, starch, proteins and antioxidant compounds, which are an inexpensive source of renewable biopolymer materials. Moreover, there has been great interest in employing these residues to obtain biodegradable polymers and composites with new and improved properties, including applications in the agro- and food industry.

It is clear that the Brazilian Cerrado flora is a rich and unexplored source for bio-based materials with potential industrial application, mainly food packaging, since the economically explored species from Cerrado are explored by the food industry to be used as liquors, in ice-cream and juices and it is natural that the food industry searches for means for the reuse of its residues and by-products. Thus, this chapter will review the literature and discuss the perspectives on the subject.

2. General Overview of Bio-based Materials for Food Packaging and Cerrado as Source of Biopolymers

Food packages are wrappers that maintain the quality of food during transportation and storage. In this way, the packaging increases the half-life of foods, avoiding contact with oxidizing substances, microorganisms, contaminating chemicals, light and other conditions that interfere with their shelf-life. The various materials used in the manufacture of food packaging provide physical protection as well as the essential physicochemical conditions for the maintenance of food quality. Beside the protection of food, it is increasing the interest of the industry for the so-called intelligent and active food packaging (Dudnyk et al., 2018; Nopwinyuwong et al., 2010; Souza et al., 2017), which that can monitor and/or change the environmental conditions of packaging in real-time. Intelligent and active food packaging can be developed using biodegradable or non-biodegradable polymers or a combination of both. Among the biodegradable polymers, most such as carbohydrates (starch, chitosan and alginate) and gelatins (proteins) are obtained from renewable sources and since they are biodegradable, they have a clear environmental advantage. Moreover, most of these biopolymers can produce uniform films and polymeric matrices, turning the biodegradable polymeric film into a low-cost active or smart polymer.

Thus, bio-packaging for food can be a versatile and low-cost approach to control and preserve the quality of a food product.

By combining a polymeric support and other substances, such as essential oil from spices (Hosseini et al., 2016; Rodríguez et al., 2008), lactic acid produced by bacteria (Kim et al., 2013) and plant extracts (Gorrasi et al., 2016; Mousavi et al., 2018; Singh et al., 2008), active packaging is being developed to reduce, inhibit or cease the development of superficial microorganisms in food. Several studies are carried out, focusing on carbohydrate materials that are biocompatible, non-toxic and biodegradable and have good film-forming property, such as starch and chitosan, which also have the characteristic of having antimicrobial properties (Hosseini et al., 2009).

Due to its antimicrobial properties, chitosan has been extensively investigated as a potential antimicrobial food packaging (Bonilla et al., 2014; Bonilla and Sobral, 2016; Lagaron et al., 2007; No et al., 2007). Besides chitosan, other natural polymers and natural extracts have been used together to produce antimicrobial active films. These include *Aloe vera* extracts (Asefa 2012) PET coatings (Gorrasi et al., 2016) and cassava starch combined with essential oils (Souza et al., 2013). Therefore, several studies are being carried out to evaluate the antimicrobial activity of smart biodegradable films combined with natural extracts (Abreu et al., 2015; Atarés and Chiralt, 2016; Dutta et al., 2009; Zhang et al., 2015).

Antioxidant food packaging is another type of active food packaging with growing interest, both in academy and in industry. The investigation of natural antioxidants in food packaging is in the form of films or micro- or nanoparticles (Barbosa-Pereira et al., 2014; Lorenzo et al., 2014; Medina-Jaramillo et al., 2017; Reis et al., 2015; Rocha et al., 2017).

Smart biodegradable films are produced mainly from carbohydrates, such as starch. The most common starch used to produce biodegradable food packaging are corn, tapioca and potato starch. Starch-based films have mechanical properties with tensile strength and elongation at break as compared to films made with synthetic polymers. Despite the advantage of such polymers being biodegradable with good tensile strength and elongation at break, starch-based films have drawbacks over synthetic polymer properties. For example, carbohydrate films are less flexible, have high wettability and gas permeability, which limits the application in food with high water activity. To overcome these limitations, starch has been either modified or combined with other compounds, forming composites. It is much simpler to produce starch composites and is the preferred approach to overcome these limitations. Starch composites with polyvynil alcohol (Zhai et al., 2017), carbon nanotubes (Castrejón-Parga et al., 2014), gums (Kim et al., 2015) and other compounds have been developed. These compounds interact both chemically and physically with starch, improving its mechanical and physicochemical properties.

Although bio-based smart packaging investigation is now mainstream in the academy, few have become commercially available (Janjarasskul and Suppakul, 2018; Pennane et al., 2014) and despite the industry interest, large-scale production of bio-based packaging is restricted due to their limitations, since most of them are based on starch or its modifications. Therefore, it is clear that further research is necessary to solve these issues before biobased food packaging can be used commercially. As these issues are difficult to solve in the short-term, there is a trend to develop biobased polymers as edible active packaging, since edible polymers do not require improved mechanical and physicochemical properties. Moreover, each starch has its own features, since the proportion of amylose and amylopectin determines both physical and chemical properties of starch. Hence, some plant species from Cerrado have been investigated as a rich and alternative starch source.

Echinolaena inflexa, which is a native Poaceae species from Cerrado, has been investigated as a starch source (Souza et al., 2010). The authors investigated the annual variation of starch content to determine the best season for high starch-extraction yield. Starch contents were evaluated by standard methods and it was present in much higher amounts than glucose, fructose and sucrose, ranging from 11 to 55 mg.g^{-1}, being higher during April and July.

Lobeira (*Solanum lycocarpum*) is a native and resistant plant from Cerrado. It produces non-edible and starch rich fruits, and the search and selection of starch for industrial uses have made it very attractive as a starch source. Its starch composition was investigated and the results demonstrate that Lobeira starch has an intrinsic-viscosity of 3515 mPa.s, at estimated molecular weight of 645.69 kDa and a crystallinity degree of 38 per cent, higher than that observed for a typical A-polymorph and lower than B-type starches, evidencing the mixed character of Lobeira Starch (Pascoal et al., 2013). Its amylose content, varying from 33–40 per cent, is higher than those of tapioca, corn and potato starches. These results shows that Lobeira starch has improved properties over traditional starch and open perspectives to advanced applications of this starch.

Five native species (*Cipura paludosa, Cipura xanthomelas, Sisyrinchium vaginatum, Trimezia cathartica* and *Trimezia juncifolia*) of the Iridaceae family have also been investigated for its starch contents (Carneiro et al., 2014). Starch was extracted from bulbs, rhizomes and corms by standard methods and the amylose contents were determined by colorimetric methods. Starch content may vary according to each species. In *S. vaginatum*, the starch content was nearly undetectable, while *C. xanthomelas* it was as high as 800 mg.g^{-1} in the bulbs. Thus, plants members of Iridaceae family, mainly those of Cipura genus, can be a rich and alternative source of starch.

Tucum-do-cerrado (*Bactris setosa*) has also been investigated for its starch content, but with no further characterization (Fustinoni-Reis et al., 2016). The tucum-do-cerrado is an edible fruit with starch contents around 27.8 mg.g^{-1}, which is not as high as those of Lobeira, but it is similar to starch contents of other fruits.

Despite the great interest in Cerrado plants as source of starch, unfortunately large extensions of Cerrado have been transformed into pastures and plantations to meet other economic interests linked to the production of bioenergy, food industry and agriculture. The Cerrado is large and many species are of restricted occurrence, rare and difficult to be cultivated and explored. Thus, the search for new sources of carbohydrates in plants of Cerrado should focus on grassy shrubs and widespread plants, such as those from Poaceae family and not in the fruitful ones. Moreover, despite the ecological and economic importance of starch, relatively little is yet known about its grasses. There is a particularly urgent need to extend the study of grass carbohydrate to regions not yet accessed, such as the rupestrian fields of Brazil (De Moraes et al., 2016).

3. Edible and Antimicrobial Packaging and Coatings from Brazilian Cerrado

Over the past decades, edible films and coatings have gained attention as a alternative to protect foods, such as fruits and other vegetables from degradation. Edible films are easy to prepare, cheap and can be incorporated with active compounds that help in food conservation. One main advantage of edible films and coatings is that they can be prepared from biodegradable and non-toxic natural polymers, such as polysaccharides and proteins. Thus, there is an increasing interest in the application of edible films and coatings based on these non-toxic and biodegradable polymers, extending the shelf-life of perishable products (Santos et al., 2017; Souza et al., 2015). Moreover, such films can be prepared with polymers extracted from plants or biomass from agro- and food industrial residues, wastes or by-products, aggregating value to biomass. Edible films and coating promote a barrier to gases and water vapor, and can also be added as antifungal, antibacterial and antioxidants agents. Therefore, they improve both the product shelf-life and appearance (Silva-Weiss et al., 2013). Edible films do not need to be mechanically improved and are prepared as hydrogels (Ganiari et al., 2017; Wu et al., 2018) or by extrusion/casting (Noori et al., 2018; Vásconez et al., 2009) because they are directly applied on the food.

Edible films and coatings must contain a polymer with a structural matrix with strong bonds between its molecular chains, such as polysaccharides (Fakhoury et al., 2012), proteins (Bonilla and Sobral, 2016), lipids and natural resins, used in an isolated or combined form, can carry antimicrobial (Arismendi et al., 2013; Rojas-Graü et al., 2007) or antioxidant substances (Atarés et al., 2010; Piñeros-Hernandez et al., 2017; Teixeira et al., 2014). As many natural products can improve active edible films and coatings, the Brazilian Cerrado has the potential to be a renewable source for new active substances for food packaging.

The Cerrado biome of Brazil is a type of savanna and has several species in its flora used by the population as foods, mostly the fruitful species (Dias et al., 2012; Rufino et al., 2010). These fruits have great nutritional value and peculiar sensory attributes, such as color, flavor and aromas, as well as great antioxidant and antimicrobial activity (Oliveira et al., 2012; Roesler et al., 2007). The Cerrado flora has about 3,000 fruitful species of shrubs, 1,000 arboreal species and 500 creepers, most of them edible (Bizzo et al., 2009). These plants have great antimicrobial and antioxidant potential and can be inserted into packaging or coatings in order to prolong the shelf-life of perishable products. Some of these fruits, such as Araticum, Baru, Babassu, Jabuticaba and Pequi have been economically explored.

Araticum (*Annona crassiflora* Mart) is a fruit found in the Brazilian Cerrado, distributed widely in the central part of Brazil. They consist of vitamins, iron, phosphorus and calcium. Roesler found in this fruit a high antioxidant activity compared to other vegetables (Roesler et al., 2007). This fruit is very common in the Brazilian middle-west and is consumed in natural and as ice-cream, juices and desserts, but unfortunately it degrades rapidly, in two or three days, mostly by browning. This blocks Araticum commerce with other regions from Brazil and abroad. Thus, efforts have been conducted to improve Araticum shelf-life. Films enriched with ascorbic acid have been developed to avoid Araticum browning, but unsuccessfully (Soares et al., 2007). Thus, edible films to preserve and extend shelf-life of Araticum are still to be developed.

Pequi (*Caryocar brasiliense*) is another important fruit obtained from Brazilian Cerrado. Besides being a natural source of antioxidants (Leão et al., 2017), this plant also presents antimicrobial activity. The extracts from its leaves showed antimicrobial activity against *P. aeruginosa, B. cereus* and *S. aureus* (Moreira et al., 2019; Pereira et al., 2019). Considering all these studies in favor of the efficacy of the flora of the Brazilian Cerrado in relation to its antimicrobial activity, films and coatings with extract made from these plants have been developed. An edible coating of chitosan, enriched with ethanolic extract of pequi peels and bark, to conserve post-harvest tomatoes have been developed and preserved physical and chemical parameters during storage (Breda et al., 2017). As pequi peels are rich in polyphenol compounds that are known for possessing antioxidant, antimicrobial and antifungal properties, this fruit has a potential use as a replacement for artificial food additives. Natural compounds have an advantage over artificial compounds due to biocompatibility, low toxicity and low production cost, since most of them can be extracted from agro-industrial residues. The edible chitosan/pequi peel extract coating was prepared by ethanolic extraction of pequi peels and chitosan films were prepared by classical casting technique. The peel extract was further added to chitosan hydrogel in a ratio of 4:1 (chitosan/extract) to form the active coating and tomatoes were individually dipped into the coating formulation. The coating demonstrated favorable effects and successfully delayed physical changes in tomatoes during storage, mainly color changes and weight loss. The coating also demonstrated efficacy when stabilizing the counting of molds and yeasts during the evaluation period of 16 days, having successfully shown antimicrobial and antifungal activity.

Carnaúba waxes (*Copernicia prunifera*) were described as edible coating to prevent the growth of microorganisms in cashew fruit, promoting a barrier to the growth of fungi and yeasts and increase of shelf-life of several different food products (Santos et al., 2017; Tavares et al., 2017). Santos et al. (2017) developed a chitosan film enriched with carnaúba wax, thereby improving the water barrier of chitosan films. It is known that the addition of lipid compounds, such as waxes, reduces the water vapor permeability of films or coatings, but decreases film transparency and affects the mechanical properties. Some vegetable oils, such as soybean, palm and sunflower are used as a source of lipids for improving the carbohydrate films' water barrier. Besides vegetable oils, wax, such as carnauba and bee are employed and these demonstrate better gas and moisture barrier properties (Tavares et al., 2017). Santos et al. (2017) produced films based on chitosan and different concentrations of carnauba wax to modify the optical and barrier properties of chitosan films. The films were developed with chitosan solution of 1 per cent (w/v) concentration, followed by addition of carnauba wax at 0, 15, 30, 40 and 50 per cent (w/w) concentrations. The aim was to modify the hydrophilic behavior of chitosan films without modification of optical properties, like transparency, since it is very common that the addition of oil and wax affects the film transparency. Although the aim was to avoid modification of optical properties, the films showed a direct relation of opacity growth with wax concentration. This result is similar to that found in related work (Saucedo-Pompa et al., 2009; Zhang et al., 2018). The results also show that water vapor transmission rate is affected by concentrations of wax from 30 to 40 per cent and the concentration of 30 per cent wax is sufficient to change the film from hydrophilic to hydrophobic. The results from contact angle measurements indicate that the hydrophilicity of films decreases with the addition of wax and the films with 40 and 50 per cent carnauba wax concentration showed hydrophobic behavior. In general, these results show that the addition of carnauba wax can improve barrier properties of biodegradable polymeric films. Since water and gas barrier is mandatory in food packaging, the addition of carnauba wax in natural biopolymer films is a promising approach to develop new and improved biodegradable food packaging.

Cashew (*Anacardium occidentale* L.) is a native plant from northeastern Brazil and is also economically important. Its production is estimated at 50,000 tons per year. It exudes a gum that is a by-product of cashew industry with no commercial value. The gum is defined as a complex hetero polysaccharide composed by galactanpiranoses and galacturonic acid, which provide anionic groups that can interact with cationic groups. Thus, cashew gum can easily form blends with other polymers. The potential of cashew gum has attracted the attention of many researchers interested in the development of biodegradable polymers to replace non-degradable polymers. Oliveira et al. (2018) have investigated casting-blend films from cashew gum and gelatin for food packaging application. The produced films are uniform, with great transparency and homogeneous, with direct relation of gelatin concentration with homogeneity and compactness. Indeed, other important properties, such as mechanical properties and thermal stability are directly related with gelatin concentration. In this work, authors found that gelatin addition to the cashew gum at 2.5:5 provided an elongation at break of the film above 100 per cent, which is characteristic of a ductile material. The elongation and tensile strength of the films were higher than the values reported in the literature using other polysaccharides. The films also have diminished water vapor permeability. As water vapor permeability, mechanical strength and thermal stability depend upon the cashew gum/gelatin ratio, films and biopolymers with a wide range of applications in food packaging can be developed, using cashew gum and gelatin. For example, the cashew gum/gelatin proportion can be changed to produce more or less hydrophobic films or more or less flexible, depending on the desired application in food packaging. Thus, cashew gum is a naturally occurring biopolymer that can be used to develop a wide range of new and improved packaging.

The search for less hydrophilic and resistant biopolymers has inspired the investigation of babassu (*Attalea speciosa* L.) as a source of new carbohydrate polymers. Babassu oil is of great industrial importance, with applications in food, cleaning and cosmetic products. Babassu mesocarp is a by-product of this important industry and is a rich source of starch and displays antioxidant properties. In order to investigate the potential application of babassu starch to develop biodegradable films, Maniglia et al. (2019) have developed films made from babassu starch and some plasticizers. The authors developed films with glycerol, sorbitol, glucose, or urea as plasticizers and the film's physical and functional properties were evaluated.

Among the chosen plasticizers, glycerol and sorbitol are widely applied as plasticizers for biopolymers, while the use of urea and glucose is not very common. It is well known that plasticizers affect physical and chemical properties of biopolymers (Li et al., 2015; Medina-Jaramillo et al., 2017). Thus, it is worth the investigation to determine which plasticizer is suitable to produce films based on babassu starch. The study of Maniglia et al. (2019) has achieved some interesting results. The choice of plasticizer does not affect the film optical properties, i.e., transparency and color, but affect mechanical properties and water vapor transmission rate. The most resistant and rigid films were prepared with sorbitol as plasticizer, but when glycerol was used as plasticizer, films became less resistant and less rigid. In other words, it is better to use glycerol as plasticizer when a more flexible film is desired. These results agree with the results for other starch types, such as the widely investigated cassava starch (Piñeros-Hernandez et al., 2017). Starch films plasticized with glycerol or urea present greater solubility in water. Starch films plasticized with sorbitol or glucose presented lower solubility in water and lower water vapor transmission rate and contact angle. In general, glycerol and urea increased babassu starch film's hydrophilicity, while sorbitol and glucose lowered it. At a first glance, the observation that glucose lowers the hydrophilicity could be surprising, but it is a cyclic and a larger molecule when compared to others plasticizers, with great molecular weight and size. This makes the water diffusion through the films more difficult and similar results were found with other starches (Moreno et al., 2017). Moreover, it is clear that film properties depend upon the concentration and type of plasticizer, so the properties of the produced film can be predicted by choosing the plasticizer type and concentration.

4. Antioxidant Packaging using Cerrado Species

Lipid oxidation is, after deterioration caused by microorganisms, the main cause of food spoilage. This oxidation is more common in foods rich in lipids, especially polyunsaturated ones and leads to the formation of unpleasant taste, rancidity, toxic aldehydes, cytotoxic compounds and polycyclic aromatic hydrocarbons, making the food unsuitable for human consumption. The stability of easily oxidable foods can be maintained by direct addition of antioxidants in the food or by packaging in vacuum, avoiding contact with O_2. However, fresh foods, such as fish and meat lose their best taste and smell when stored in vacuum packages. To reduce this adversity, there is a growing interest in bioactive substances with antioxidant capacity. Bioactive substances with antioxidant capacity have the ability to prevent the deterioration of fats and other constituents in foods and have the advantage of low toxicity when compared to synthetic antioxidants (Gómez-Estaca et al., 2014; López-de-Dicastillo et al., 2012). However, the direct incorporation of antioxidant substances into foods is effective only as long as there is an antioxidant to be consumed in the reaction and the food becomes unprotected when the reaction ceases. Thus, the incorporation of antioxidant substances directly in fresh food is a very limited approach to prevent food oxidation. To overcome this limitation, the encapsulation and the production of biopolymer-based film with natural and even artificial antioxidants have been investigated (Hosseini et al., 2013; Reis et al., 2015; Souza et al., 2014).

Currently, packaging and micro-/nanoparticles with antioxidant activity have been developed to release the antioxidant substance in a controlled manner. These systems are produced by direct incorporation of antioxidant substances into the polymeric matrix. Natural extracts or pure natural polyphenols, such as flavonoids and tannins, are generally used as antioxidant agents in such films and particles. These films and particles can act by releasing the antioxidant substance in the food or by sequestration of oxidizing molecules, such as oxygen, reactive oxygen species or metal ions (Leão et al., 2017; Medina-Jaramillo et al., 2017). Although the investigations of antioxidant natural substances as well as their incorporation in active films and particles are now widespread and routine, the investigation about the use of antioxidant substances and natural extracts from Brazilian Cerrado species is still scarce. However, knowledge of the antioxidant potential of these species is already widespread and well described in the current literature (Genovese et al., 2008; Wu et al., 2013). Thus, there is a potential for the use of such species as a source for natural antioxidants comparable to standard antioxidants, such as BHT and BHA. This potential is due to the abundance of antioxidant compounds in these species, such as ascorbic acid, polyphenols, carotenes and tocopherols (Genovese et al., 2008; Rufino et al., 2010).

Among the investigated active packaging inspired from Brazilian Cerrado, chitosan films containing buriti (*Mauritia flexuosa* L.f.) and 'macaúba' (*Acrocomia aculeata*) vegetable oils, which are known for their antioxidant potential, were produced (Amaral et al., 2017). *Mauritica flexuosa* is a palm tree native to Cerrado and amazon rainforest and its vegetable oil is rich in tocopherol and polyphenols. Due to the presence of polyphenols, its oil has attracted interest in the pharmaceutical and food industry. *Acronomia acuelata* is also a palm tree native to Cerrado and is native to Brazilian Cerrado, with occurrence also in Paraguay and Bolivia. Its almond is non-edible, but it is a rich source of oil and there is a growing interest for its use in biodiesel production (da Silva César et al., 2015). The films were produced using chitosan and buriti or macaúba vegetable oil in the proportion of biopolymer/oil of 1:1 (w/w). Both films with macaúba and buriti oils are homogeneous and with color ranging from transparent to yellowish. Regarding water vapor permeability, the incorporation of macaúba in chitosan films increased water vapor permeability due to the formation of a polymeric matrix with a great number of micropores. On the other hand, the incorporation of macaúba oil led to humidity diffusion. This means that for applications where less humidity diffusion is required, films made from macaúba oil and chitosan could be a choice. However, others parameters, such as crystallinity, morphology, mechanical resistance and antioxidant capacity should be investigated before practical application of these films.

Polyphenols from cagaita (*Eugenia dysintyerica* DC) having a great antioxidant potential were encapsulated and tested for their antioxidant and antimicrobial properties (Ribeiro et al., 2008). They showed great antioxidant activity and inhibitory activity against *Staphylococcus aureus* and *Listerya monocytogenes* similar to vancomycin, erythromycin and ceftriaxone used as controls (Daza et al., 2017). Cagaita active compounds were extracted, dried and encapsulated without previous purification and isolation. The extract was encapsulated with gum arabic or inulin, as the carrier using different spray-drying conditions. Spray-drying technique is an alternative to prevent deterioration of the bioactive compounds, ensuring that such compounds can be used in food conservation.

Jabuticaba, also spelled as *jaboticaba*, is a native fruit from Brazil with growing nutritional, commercial and industrial interest. Jabuticaba is consumed in ice-creams, juices, licors and desserts and also *in natura*. Among jabuticaba species, two are industrial and commercially viable: *Myrciaria cauliflora* and *Myrciaria jabuticaba*, as they are the last species that are most cultivated for industrial proposes. The fruits of jabuticaba are rich in anthocyanidins and other antioxidant compounds with total antioxidant capacity comparable to those of BHT (Wu et al., 2013). This make this fruit a source of natural antioxidants that can replace the artificial ones. Hence, this fruit has attracted interest from industry and academy for the production of antioxidant films and encapsulation for active food packaging. Silva et al. (2013) encapsulated extracts from jabuticaba peels, an industrial waste from jabuticaba processing. Although jabuticaba peels are considered as industrial waste, most of the identified active compounds in jabuticaba are located in its peels. So, there is a growing interest in the application of such peels as a source for food additives. The active compounds were extracted from peels using 70 per cent ethanol at pH 2 and further concentrated in a rotatory evaporator. The obtained extract was incorporated into microcapsules prepared from maltodextrin, starch and arabic gum as carrier agents at the concentration of 1:3 (v/v) extract/carrier solution. The final solution was homogenized and spray-dried. According to the results, the use of maltodextrin at 30 per cent, a spray-drying temperature of 180°C and the combination of arabic gum allowed the formation of homogeneous and stable particles and although the authors did not evaluate the remaining antioxidant capacity, it can be inferred that the microcapsules possess antioxidant capacity, due the presence of jabuticaba peels' extract.

Baldin et al. (2016) have encapsulated jabuticaba extract and added it to fresh sausages as antioxidant and antimicrobial agent. Aqueous jabuticaba extract was obtained under mechanical agitation, protection of light and in a ratio of fruit:water of 1:3, concentrated in rotatory evaporator and microencapsulated in a spray-drier, using maltodextrin as carrier agent. The microcapsules were further incorporated into sausages in the proportion of 2 and 4 per cent weight, sausages with 0 per cent microcapsules (control) were also produced; both control and sausages enriched with microcapsules were monitored during 15 days. The results demonstrate that the encapsulation protected the anthocyanins from degradation, since the contents of anthocyanins in microcapsule were higher than the values found in non-encapsulated extract. Thus it is clear that the encapsulation protects the extract compounds from degradation. Moreover, microcapsules exhibited good antioxidant activity and antimicrobial activity against *S. aureus* and *E. coli*. Due to microcapsules' antioxidant and antimicrobial activity, the lipid oxidation in sausages gets significantly reduced. Hence, jabuticaba extract encapsulated in maltodextrin has the potential as a natural dye, antioxidant and antimicrobial agent in meat processing industry.

Guaraná (*Paulina cupana* L.) is a native plant from Brazil and although it is generally assigned to Amazon rainforest biome, its occurrence starts in the border between Cerrado and Amazon biomass. Its fruits are rich in caffeine; antioxidant compounds and its antimicrobial activity have also been reported (Majhenič et al., 2007). Bonilla and Sobral (2016) developed a hybrid chitosan/gelatin film enriched with guarana seeds' ethanolic extract. The developed films presented good antioxidant capacity and good inhibitory activity against *E. coli* and *S. aureus*. The results suggest that these films could provide an alternative as biodegradable active biopolymers.

As seems, the food industry is most interested in finding new applications for wastes and by-products from the young industry of Cerrado crops. Although most of the research regarding reuse of wastes and by-products is conducted by academia, the collaboration with the industry is of most importance. Since finding new applications for wastes and by-products add value to these materials, the costs of production and profit could be optimized.

5. Other Technological Applications

Natural compounds from Cerrado biome have also been investigated in technological applications. De Lima et al. (2010) developed a biosensor based on chitosan as immobilizer and pequi polyphenol oxidase to detect thiodicarb. The biosensor was used for the determination of thiodicarb in fresh fruit and vegetable samples and the results were compared with HPLC results. Pequi polyphenol oxidase can be immobilized in chitosan in an excellent manner. Thus, taking advantage of this excellent immobilization, a new and effective biosensor for thiodicarb quantification was developed. The reported biosensor is of simple construction, has low detection limit, good repeatability and reproducibility. The results in detecting thiocarb in fruits and other vegetables were satisfactory when compared with those obtained using HPLC.

Another biosensor based on polyphenol oxidase was developed by Antunes et al. (2018). This biosensor is based on pure polyphenol oxidase extracted from jurubeba. Jurubeba (*Solanum paniculatum* L.) is a shrub-like solanaceae native to the Brazilian Cerrado. Its fruits are rich in phytochemicals as steroids, saponins, alkaloids and glycosides. Hence, this plant is very common in popular medicine and nutrition. As this plant is also rich in polyphenol oxidase, this enzyme has been extracted from the plant to develop a biosensor for the detection and quantification of paracetamol with reliable and reproducible results.

Jenipapo (*Genipa americana* L.) fruit extract has been also employed to develop a biosensor to determine the content of phenolic compounds in textile industrial waste (Lobón et al., 2018). The biosensor was developed to overcome the limitations of chromatographic methods, since they are generally unselective, require previous sample calibration and are expensive. As the previously discussed biosensors, this biosensor is based on polyphenol oxidase enzyme as Jenipapo is a rich source of this enzyme, due to the high concentration of phenolic compounds in this plant. The use of crude vegetal extracts on biosensors has an extremely low cost; so it is an alternative to isolated enzymes. The results of the designed biosensor were compared with those of spectrophotometric methods, using real samples from a textile industry. The results showed that there are no significant statistical differences between the biosensor and standard methods.

6. Conclusion and Perspectives

The demand for new compounds for the production of edible coatings, mainly in fruits, has been increasing in a great way because they are materials that have enormous potential to prolong the useful life of handling sensitive foods that have critical conservation, especially in microbiological control during storage. Usually, they are based on biodegradable materials, such as polysaccharides, lipids and some proteins. The use of coatings is demonstrably effective in maintaining the natural coloration of the fruits, reducing the respiratory rate and loss of mass, as well as acts in the compounds by helping to preserve the nutritional and functional values and combating the microbial activity in the food. The practice of applying edible coatings is economically viable, thanks to the materials used for production, which mostly have a low relative commercial value and a broad biocompatibility.

Despite the vast abundance of Cerrado species, few studies have applied this potential for the manufacture of food packaging. As active and intelligent food packaging has received a lot of attention in the last decades, the incorporation of bioactive substances obtained from Cerrado biodiversity, though still incipient, can drive the discovery of new and improved smart materials for the food industry, with the advantage of being abundant, renewable and low toxic. As many Cerrado species are a rich source of antioxidant compounds, these species can be used to extract antioxidant compounds for the investigation and preparation of antioxidant materials with potential applications, not only in the food industry but also in other fields, such as in dye, pharmaceutical and civil construction industry. There is a trend in the use of extracts over the use of pure substances. The purification of natural substances is laborious, expensive and the substances tend to degradate and the use of extracts over pure substances is cheaper and more simple.

Besides food industry, which is the industry of choice for economically and environmentally-sustainable Cerrado exploration, the investigation of new and improved technological applications of biopolymers and substances found in plants from Cerrado is growing. There is a need for attention and action to the sustainable exploration and development of Brazilian Cerrado. It is clear that the Brazilian Cerrado is a great and un-estimated source for new biopolymers and active substances with potential for food, pharmaceutical and technological applications. But there is still so much to do concerning the knowledge, research, exploration and development of this biome as a source for new and improved materials.

References

Abreu, A.S., M. Oliveira, A. De Sá, R.M. Rodrigues, M.A. Cerqueira, A.A. Vicente and A.V. Machado (2015). Antimicrobial nanostructured starch-based films for packaging, *Carbohydrate Polymers*, 129: 127–134.

Amaral, I.B.C., S.R. Arrudas, J.R. de Meira and A.B. Reis (2017). Análise do processo difusivo de filmes de quitosana contendo óleo de palmeiras (Aracaceae) do cerrado brasileiro, *Unimontes Científica*, 19: 13–26.

Antunes, R.S., L.F. Garcia, V.S. Somerset, E.D.S. Gil and F.M. Lopes (2018). The use of a polyphenoloxidase biosensor obtained from the fruit of jurubeba (*Solanum paniculatum* L.) in the determination of paracetamol and other phenolic drugs, *Biosensors*, 8.

Arismendi, C., S. Chillo, A. Conte, M.A. Del Nobile, S. Flores and L.N. Gerschenson (2013). Optimization of physical properties of xanthan gum/tapioca starch edible matrices containing potassium sorbate and evaluation of its antimicrobial effectiveness, *LWT – Food Science and Technology*, 53: 290–296.

Asefa, G. (2012). *Development and Evaluation of Antimicrobial Aloe-based Packaging Films*, Addis Ababa Institute of Technology (AAiT), School of Graduate Studies, Department of Chemical Engineering.

Atarés, L., J. Bonilla and A. Chiralt (2010). Characterization of sodium caseinate-based edible films incorporated with cinnamon or ginger essential oils, *Journal of Food Engineering*, 100: 678–687.

Atarés, L. and A. Chiralt (2016). Essential oils as additives in biodegradable films and coatings for active food packaging, *Trends in Food Science & Technology*, 48: 61–62.

Baldin, J.C., E.C. Michelin, Y.J. Polizer, I. Rodrigues, S.H.S. de Godoy, R.P. Fregonesi, M.A. Pires, L.T. Carvalho, C.S. Fávaro-Trindade, C.G. de Lima, A.M. Fernandes and M.A. Trindade (2016). Microencapsulated jabuticaba (*Myrciaria cauliflora*) extract added to fresh sausage as natural dye with antioxidant and antimicrobial activity, *Meat Science*, 118: 15–21.

Barbosa-Pereira, L., J.M. Cruz, R. Sendón, A.R.B. Quirós, A. Ares, M. Castro-López, M.J. Abad, J. Maroto and P. Paseiro-Losada (2014). Development of antioxidant active films containing tocopherols to extend the shelf life of fish, *Food Control*, 31: 236–246.

Bizzo, H.R., C.H. Ana Maria and C.M. Rezende (2009). Óleos essenciais no Brasil: Aspectos gerais, desenvolvimento e perspectivas. *Quimica Nova*, 32: 588–594.

Bonilla, J., E. Fortunati, L. Atarés, A. Chiralt and J.M. Kenny (2014). Physical, structural and antimicrobial properties of poly vinyl alcohol-chitosan biodegradable films, *Food Hydrocolloids*, 35: 463–470.

Bonilla, J. and P.J.A. Sobral (2016). Investigation of the physicochemical, antimicrobial and antioxidant properties of gelatin-chitosan-edible film mixed with plant ethanolic extracts, *Food Bioscience*, 16: 17–25.

Breda, C.A., D.L. Morgado, O.B.G. de Assis and M.C.T. Duarte (2017). Effect of chitosan coating enriched with pequi (*Caryocar brasiliense* Camb.) peel extract on quality and safety of tomatoes (*Lycopersicon esculentum* Mill.) during storage, *Journal of Food Processing and Preservation*, 41: e13268.

Carneiro, R.V., R.C.L. Figueiredo-Ribeiro, M.G. Moraes, M.A.M. Carvalho and V.O. Almeida (2014). Diversity of non-structural carbohydrates in the underground organs of five Iridaceae species from the Cerrado (Brazil), *South African Journal of Botany*, 96: 105–111.

Castrejón-Parga, K.Y., H. Camacho-Montes, C.A. Rodríguez-González, C. Velasco-Santos, A.L. Martínez-Hernández, D. Bueno-Jaquez, J.L. Rivera-Armenta, C.R. Ambrosio, C.C. Conzalez, M.E. Mendoza-Duarte and P.E. García-Casillas (2014). Chitosan-starch film reinforced with magnetite-decorated carbon nanotubes, *Journal of Alloys and Compounds*, 615: S505–S510.

Clark, N.E., R. Lovell, B.W. Wheeler, S.L. Higgins, M.H. Depledge and K. Norris (2014). Biodiversity, cultural pathways, and human health: A framework, *Trends in Ecology & Evolution*, 29: 198–204.

Daza, L.D., A. Fujita, D. Granato, C.S. Fávaro-Trinade and M.I. Genovese (2017). Functional properties of encapsulated Cagaita (*Eugenia dysenterica* DC.) fruit extract, *Food Bioscience*, 18: 15–21.

De Lima, F., B.G. Lucca, A.M.J. Barbosa, V.S. Ferreira, S.K. Moccelini, A.C. Franzoi and I.C. Vieira (2010). Biosensor based on pequi polyphenol oxidase immobilized on chitosan crosslinked with cyanuric chloride for thiodicarb determination, *Enzyme and Microbial Technology*, 47: 153–158.

De Moraes, M.G., M.A.M. De Carvalho, A.C. Franco, C.J. Pollock and R.D.C.L. Figueiredo-Ribeiro (2016). Fire and drought: Soluble carbohydrate storage and survival mechanisms in herbaceous plants from the Cerrado, *BioScience*, 66: 107–117.

Dias, F., B. Abadio, I. Galv and R. Botelho (2012). Physicochemical characteristics and antioxidant activity of three native fruits from Brazilian savannah (Cerrado), *Alimentos e Nutrição*, 23: 1–6.

Dudnyk, I., E.R. Janeček, J. Vaucher-Joset and F. Stellacci (2018). Edible sensors for meat and seafood freshness, *Sensors and Actuators B: Chemical*, 259: 1108–1112.

Dutta, P.K., S. Tripathi, G.K. Mehrotra and J. Dutta (2009). Perspectives for chitosan-based antimicrobial films in food applications, *Food Chemistry*, 114: 1173–1182.

Fakhoury, F.M., S. Maria Martelli, L. Canhadas Bertan, F. Yamashita, L.H. Innocentini Mei and F.P. Collares Queiroz (2012). Edible films made from blends of manioc starch and gelatin – Influence of different types of plasticizer and different levels of macromolecules on their properties, *LWT – Food Science and Technology*, 49: 149–154.

Fustinoni-Reis, A.M., S.F. Arruda, L.P.S. Dourado, M.S.B. da Cunha and E.M.A. Siqueira (2016). Tucum-do-cerrado (*Bactris setosa* mart.) consumption modulates iron homeostasis and prevents iron-induced oxidative stress in the rat liver, *Nutrients*, 8: 38.

Ganiari, S., E. Choulitoudi and V. Oreopoulou (2017). Edible and active films and coatings as carriers of natural antioxidants for lipid food, *Trends in Food Science and Technology*, 68: 70–82.

Genovese, M.I., M. Da Silva Pinto, A.E. De Souza Schmidt Goncalves and F.M. Lajolo (2008). bioactive compounds and antioxidant capacity of exotic fruits and commercial frozen pulps from Brazil, *Food Science and Technology International*, 14: 207–214.

Gómez-Estaca, J., C. López-de-Dicastillo, P. Hernández-Muñoz, R. Catalá and R. Gavara (2014). Advances in antioxidant active food packaging, *Trends in Food Science & Technology*, 35: 42–51.

Gorrasi, G., V. Bugatti, L. Tammaro, L. Vertuccio, G. Vigliotta and V. Vittoria (2016). Active coating for storage of Mozzarella cheese packaged under thermal abuse, *Food Control*, 64: 10–16.

Hosseini, M.H., S.H. Razavi and M.A. Mousavi (2009). Antimicrobial, physical and mechanical properties of chitosan-based films incorporated with thyme, clove and cinnamon essential oils, *Journal of Food Processing and Preservation*, 33: 727–743.

Hosseini, S.F., M. Rezaei, M. Zandi and F. Farahmandghavi (2016). Development of bioactive fish gelatin/chitosan nanoparticles composite films with antimicrobial properties, *Food Chemistry*, 194: 1266–1274.

Hosseini, S.F., M. Zandi, M. Rezaei and F. Farahmandghavi (2013). Two-step method for encapsulation of oregano essential oil in chitosan nanoparticles: Preparation, characterization and *in vitro* release study, *Carbohydrate Polymers*, 95: 50–56.

Janjarasskul, T. and P. Suppakul (2018). Active and intelligent packaging: The indication of quality and safety, *Critical Reviews in Food Science and Nutrition*, 58.

Kim, E., D.Y. Choi, H.C. Kim, K. Kim and S.J. Lee (2013). Calibrations between the variables of microbial TTI response and ground pork qualities, *Meat Science*, 95: 362–7.

Kim, S.R.B., Y.G. Choi, J.Y. Kim and S.T. Lim (2015). Improvement of water solubility and humidity stability of tapioca starch film by incorporating various gums, *LWT – Food Science and Technology*, 64: 475–482.

Lagaron, J.M., P. Fernandez-Saiz and M.J. Ocio (2007). Using ATR-FTIR spectroscopy to design active antimicrobial food packaging structures based on high molecular weight chitosan polysaccharide, *Journal of Agricultural and Food Chemistry*, 55: 2554–62.

Leão, D.P., A.S. Franca, L.S. Oliveira, R. Bastos and M.A. Coimbra (2017). Physicochemical characterization, antioxidant capacity, total phenolic and proanthocyanidin content of flours prepared from pequi (*Caryocar brasilense* Camb.) fruit by-products, *Food Chemistry*, 225: 146–153.

Li, J., F. Ye, J. Liu and G. Zhao (2015). Effects of octenylsuccination on physical, mechanical and moisture-proof properties of stretchable sweet potato starch film, *Food Hydrocolloids*, 46: 226–232.

Lobón, G., L. Garcia, D. Ferraz, E. Gil, D. Thomaz, R. Luque, F. Lopes and R. Antunes (2018). Development of a polyphenol oxidase biosensor from jenipapo fruit extract (*Genipa americana* L.) and determination of phenolic compounds in textile industrial effluents, *Biosensors*, 8: 47.

López-de-Dicastillo, C., J. Gómez-Estaca, R. Catalá, R. Gavara and P. Hernández-Muñoz (2012). Active antioxidant packaging films: Development and effect on lipid stability of brined sardines, *Food Chemistry*, 131: 1376–1384.

Lorenzo, J.M., R. Batlle and M. Gómez (2014). Extension of the shelf-life of foal meat with two antioxidant active packaging systems, *LWT – Food Science and Technology*, 59: 181–188.

Luzia, D.M.M. and N. Jorge (2011). Study of antioxidant activity of non-conventional Brazilian fruits, *Journal of Food Science and Technology*, 51: 1167–1172.

Machado, R.B., L.M.S. Aguiar, A. Castro, C.C. Nogueira and M.B. Ramos-Neto (2008). Caracterização da fauna e flora do Cerrado. *Palestras do XI Simpósio Nacional sobre o Cerrado e II Simpósio Internacional sobre, Savanas Tropicais*, 12–17.

Majhenič, L., M. Škerget and Ž. Knez (2007). Antioxidant and antimicrobial activity of guarana seed extracts, *Food Chemistry*, 104: 1258–1268.

Maniglia, B.C., L. Tessaro, A.P. Ramos and D.R. Tapia-Blácido (2019). Which plasticizer is suitable for films based on babassu starch isolated by different methods? *Food Hydrocolloids*, 89: 143–152.

Medina-Jaramillo, C., O. Ochoa-Yepes, C. Bernal and L. Famá (2017). Active and smart biodegradable packaging based on starch and natural extracts, *Carbohydrate Polymers*, 176: 187–194.

Moreira, R.V., M.P. Costa, V.S. Castro, C.E. Paes, Y.S. Mutz, B.S. Frasao, S.B. Mano and C.A. Conte-Junior (2019). Antimicrobial activity of pequi (*Caryocar brasiliense*) waste extract on goat Minas Frescal cheese presenting sodium reduction, *Journal of Dairy Science*.

Moreno, O., J. Cárdenas, L. Atarés and A. Chiralt (2017). Influence of starch oxidation on the functionality of starch-gelatin based active films, *Carbohydrate Polymers*, 178: 147–158.

Mousavi Khaneghah, A., S.M.B. Hashemi and S. Limbo (2018). Antimicrobial agents and packaging systems in antimicrobial active food packaging: An overview of approaches and interactions, *Food and Bioproducts Processing*, 111: 1–19.

No, H.K., S.P. Meyers, W. Prinyawiwatkul and Z. Xu (2007). Applications of chitosan for improvement of quality and shelf life of foods: A review, *Journal of Food Science*, 72: R87–100.

Noori, S., F. Zeynali and H. Almasi (2018). Antimicrobial and antioxidant efficiency of nanoemulsion-based edible coating containing ginger (*Zingiber officinale*) essential oil and its effect on safety and quality attributes of chicken breast fillets, *Food Control*, 84: 312–320.

Nopwinyuwong, A., S. Trevanich and P. Suppakul (2010). Development of a novel colorimetric indicator label for monitoring freshness of intermediate-moisture dessert spoilage, *Talanta* 81: 1126–32.

Oliveira, M.A., R.F. Furtado, M.S.R. Bastos, R.C. Leitão, S.D. Benevides, C.R. Muniz, H.N. Cheng and A. Biswas (2018). Performance evaluation of cashew gum and gelatin blend for food packaging, *Food Packaging and Shelf-Life*, 17: 57–64.

Oliveira, V.B., L.T. Yamada, C.W. Fagg and M.G.L. Brandão (2012). Native foods from Brazilian biodiversity as a source of bioactive compounds, *Food Research International*, 48: 170–179.

Pascoal, A.M., M.C.B. Di-Medeiros, K.A. Batista and K.F. Fernandes (2013). Physical-Chemical Characterization of Lobeira (*Solanum lycocarpum*) Starch, pp. 1–1 XLII Annual Meeting of SBBQ, Foz do Iguaçu, Brazil.

Pennane, K., C. Focas, V. Kumpusalo-Sanna, K. Keskitalo-Vuokko, I. Matullat, M. Ellouze, S. Pentikäinen, M. Smolader, V. Korhonen and M. Ollila (2014). European consumers' perception of time-temperature indicators in food packaging, *Packaging Technology and Science*, 28: 303–323.

Pereira, F.F.G., M.K.S.B. Feitosa, M. de S. Costa, S.R. Tintino, F.F.G. Rodrigues, I.R.A. Menezes, H.D.M. Coutinho, J.G.M. da Costa and E.O. de Sousa (2019). Characterization, antibacterial activity and antibiotic modifying action of the *Caryocar coriaceum* Wittm. pulp and almond fixed oil, *Natural Product Research*, 1–5.

Pereira, M.C., R.S. Steffens, A. Jablonski, P.F. Hertz, A.D.O. Rios, M. Vizzotto and S.H. Flôres (2012). Characterization and antioxidant potential of Brazilian fruits from the Myrtaceae family, *Journal of Agricultural and Food Chemistry*, 60: 3061–7.

Pertuzatti, P.B., M.T. Barcia, D. Rodrigues, P.N. Da Cruz, I. Hermosín-Gutiérrez, R. Smith and H.T. Godoy (2014). Antioxidant activity of hydrophilic and lipophilic extracts of Brazilian blueberries, *Food Chemistry*, 164: 81–88.

Piñeros-Hernandez, D., C. Medina-Jaramillo, A. López-Córdoba and S. Goyanes (2017). Edible cassava starch films carrying rosemary antioxidant extracts for potential use as active food packaging, *Food Hydrocolloids*, 63: 488–495.

Ramos, L.P. and H.M. Wilhelm (2005). Current status of biodiesel development in Brazil, *Applied Biochemistry and Biotechnology*, 121-124: 807–19.

Reis, L.C.B., C.O. de Souza, J.B.A. da Silva, A.C. Martins, I.L. Nunes and J.I. Druzian (2015). Active biocomposites of cassava starch: The effect of yerba mate extract and mango pulp as antioxidant additives on the properties and the stability of a packaged product, *Food and Bioproducts Processing*, 94: 382–391.

Ribeiro, E.M.G., L.M.J. De Carvalho, G.M.D. Ortiz, F.S.N. Cardoso, D.S. Viana, J.L.V. Carvalho, P.B. Gomes and N.M. Tebaldi (2008). An overview on cagaita (*Eugenia dysenterica* DC) macro and micro components and a technological approach, *Food Industry*, pp. 3–22.

Rocha, M.A.M., M.A. Coimbra and C. Nunes (2017). Applications of chitosan and their derivatives in beverages: A critical review, *Current Opinion in Food Science*, 15: 61–69.

Rodríguez, A., C. Nerín and R. Batlle (2008). New cinnamon-based active paper packaging against Rhizopusstolonifer food spoilage, *Journal of Agricultural and Food Chemistry*, 56: 6364–9.

Roesler, R., L.C. Carrasco, R.B. Holanda, C. Alves, S. Sousa and G.M. Pastore (2007). *Antioxidant activity of Cerrado fruits, Ciência e Tecnologia de Alimentos*, 27: 53–60.

Rojas-Graü, M.A., R.J. Avena-Bustillos, C. Olsen, M. Friedman, P.R. Henika, O. Martín-Belloso, Z. Pan, and T.H. McHugh (2007). Effects of plant essential oils and oil compounds on mechanical, barrier and antimicrobial properties of alginate-apple puree edible films, *Journal of Food Engineering*, 81: 634–641.

Rufino, M.D.S.M., R.E. Alves, E.S. de Brito, J. Pérez-Jiménez, F. Saura-Calixto and J. Mancini-Filho (2010). Bioactive compounds and antioxidant capacities of 18 non-traditional tropical fruits from Brazil, *Food Chemistry*, 121: 996–1002.

Santos, F.K.G. dos, K.N. de O. Silva, T.D.N. Xavier, R.H. de L. Leite and E.M.M. Aroucha (2017). Effect of the addition of carnauba wax on physicochemical properties of chitosan films, *Materials Research*, 20: 479–484.

Saucedo-Pompa, S., R. Rojas-Molina, A.F. Aguilera-Carbó, A. Saenz-Galindo, H. de La Garza, D. Jasso-Cantú and C.N. Aguilar (2009). Edible film based on candelilla wax to improve the shelf life and quality of avocado, *Food Research International*, 42: 511–515.

Silva, P.I., P.C. Stringheta, R.F. Teofilo and I.R.N. De Oliveira (2013). Parameter optimization for spray-drying microencapsulation of jaboticaba (*Myrciaria jaboticaba*) peel extracts using simultaneous analysis of responses, *Journal of Food Engineering*, 117: 538–544.

Silva-Weiss, A., V. Bifani, M. Ihl, P.J.A. Sobral and M.C. Gómez-Guillén (2013). Structural properties of films and rheology of film-forming solutions based on chitosan and chitosan-starch blend enriched with murta leaf extract, *Food Hydrocolloids*, 31: 458–466.

da Silva César, A., F. de Azedias Almeida, R.P. De Souza, G.C. Silva and A.E. Atabani (2015). The prospects of using *Acrocomia aculeata* (macaúba) a non-edible biodiesel feedstock in Brazil, *Renewable and Sustainable Energy Reviews*, 49: 1213–1220.

Singh, G., I.P.S. Kapoor, P. Singh, C.S. de Heluani, M.P. de Lampasona and C.A.N. Catalan (2008). Chemistry, antioxidant and antimicrobial investigations on essential oil and oleoresins of *Zingiber officinale*, *Food and Chemical Toxicology*, 46: 3295–3302.

Soares Soares, M., M. Caliari, R. Vera and C. Silveira Melo (2007). Filmes plásticos e ácido ascórbico na qualidade de araticum minimamente processado, *Ciência Rural*, 37.

Souza, A., C. Sandrin, M. Calió, S. Meirelles, V. Pivello and R. Figueiredo-Ribeiro (2010). Seasonal variation of soluble carbohydrates and starch in *Echinolaena inflexa*, a native grass species from the Brazilian savanna and in the invasive grass *Melinis minutiflora*, *Brazilian Journal of Biology*, 70: 395–404.

Souza, A.C., G.E.O. Goto, J.A. Mainardi, A.C.V. Coelho and C.C. Tadini (2013). Cassava starch composite films incorporated with cinnamon essential oil: Antimicrobial activity, microstructure, mechanical and barrier properties, *LWT – Food Science and Technology*, 54: 346–352.

Souza, M.P., A.F.M. Vaz, M.T.S. Correia, M.A. Cerqueira, A.A. Vicente and M.G. Carneiro-da-Cunha (2014). Quercetin-loaded lecithin/chitosan nanoparticles for functional food applications, *Food and Bioprocess Technology*, 7: 1149–1159.

Souza, M.P., A.F.M. Vaz, H.D. Silva, M.A. Cerqueira, A.A. Vicente and M.G. Carneiro-da-Cunha (2015). Development and characterization of an active chitosan-based film containing quercetin, *Food and Bioprocess Technology*, 8: 2183–2191.

Souza, V.G.L., A.L. Fernando, J.R.A. Pires, P.F. Rodrigues, A.A.S. Lopes and F.M.B. Fernandes (2017). Physical properties of chitosan films incorporated with natural antioxidants, *Industrial Crops and Products*, 107: 565–572.

Tavares, R., E.J. de Jesus, E.N.F. Santos, B.B. Amorim, C.A.A. Gonçalves, L.L. Costa and F.B.B. Jardim (2017). Avaliação da cera de carnaúba *(Copernicia prunifera)* como revestimento de mandioquinha-salsa (*Arracacia xanthorrhiza* Bancroft.), *Global Science and Technology*, 9.

Teixeira, B., A. Marques, C. Pires, C. Ramos, I. Batista, J.A. Saraiva and M.L. Nunes (2014). Characterization of fish protein films incorporated with essential oils of clove, garlic and origanum: Physical, antioxidant and antibacterial properties, *LWT – Food Science and Technology*, 59: 533–539.

Valdés, A., A.C. Mellinas, M. Ramos, N. Burgos, A. Jiménez and M.C. Garrigós (2015). Use of herbs, spices and their bioactive compounds in active food packaging, *RSC Adv.*, 5: 40324–40335.

Vásconez, M.B., S.K. Flores, C.A. Campos, J. Alvarado and L.N. Gerschenson (2009). Antimicrobial activity and physical properties of chitosan-tapioca starch-based edible films and coatings, *Food Research International*, 42: 762–769.

Wu, S.B., C. Long and E.J. Kennelly (2013). Phytochemistry and health benefits of jaboticaba, an emerging fruit crop from Brazil, *Food Research International*, 54: 148–159.

Wu, S., W. Wang, K. Yan, F. Ding, X. Shi, H. Deng and Y. Du (2018). Electrochemical writing on edible polysaccharide films for intelligent food packaging, *Carbohydrate Polymers*, 186: 236–242.

Zhai, X., J. Shi, X. Zou, S. Wang, C. Jiang, J. Zhang, X. Huang, W. Zhang and M. Holmes (2017). Novel colorimetric films based on starch/polyvinyl alcohol incorporated with roselle anthocyanins for fish freshness monitoring, *Food Hydrocolloids*, 69: 308–317.

Zhang, L., R. Li, F. Dong, A. Tian, Z. Li and Y. Dai (2015). Physical, mechanical and antimicrobial properties of starch films incorporated with ε-poly-L-lysine, *Food Chemistry*, 166: 107–14.

Zhang, Y., B.K. Simpson and M.J. Dumont (2018). Effect of beeswax and carnauba wax addition on properties of gelatin films: A comparative study, *Food Bioscience*, 26: 88–95.

CHAPTER 6

Fish-gelatin and Carob Seed Peel By-product for Developing Novel Edible Films

Irene Albertos Muñoz,* Ana Belén Martin-Diana
and *Daniel Rico*

1. Introduction

The Bioeconomy Strategy of European Union was launched in 2012 and revised in 2018. This aimed to accelerate the deployment of a sustainable European bioeconomy in order to maximize its contribution towards the 2030 agenda and its Sustainable Development Goals (SDGs) as well as the Paris Agreement. The main objectives of this strategy is to ensure food and nutrition security, manage natural resources sustainably, reduce dependence on non-renewable resources, adapt the climate-change creation of jobs and maintain the European competitiveness. In term of environmental benefits, the bioeconomy can contribute to the defossilisation of plastic industry. Plastic would be replaced by bio-based, recyclable and marine-biodegradable materials to contribute to plastic-reduction, circular economy and maintenance of the health of seas and oceans (EC, 2018a).

Nowadays, most of the food packaging is made of plastic, with the global production in 2015 being more than 300 million metric tons (Worm et al., 2017). Furthermore, packaging is currently the largest industrial sector in Europe—40 per cent of plastic is allocated to packaging (Association of Plastic Manufacturers, 2016). In packaging, plastics help ensure food safety and reduce food wastage. However, plastic

Agrarian Technological Institute of Castilla and León (ITACyL). Department of Research and Technology. Government of Castilla and León, Valladolid, Spain.
* Corresponding author: albmunir@itacyl.es

disposal currently poses an important environmental problem. In 2017, the Commission highlighted this issue on plastic production and use, and commited to work towards the goal of ensuring that all plastic packaging will be recyclable by 2030 (EC Communication from the Commission to the European Parliament, the Council, the European Economic and Social Committee and the Committee of the Regions, 2018b).

Bio-based plastic can substitute plastic from fossil sources. Also it plastic can be produced within multi-product biorefinery approaches—biorefineries can improve the efficiency of biomass utilization by increasing parallel exploitation of sideflows, reducing and/or recovering waste and residues. As a consequence, the resource efficiency and waste prevention as well as recycling and circularity are boosted. The transformation of waste and residues into beneficial products, such as bioplastics, will further use different sources of biomass (agriculture/forestry/fishing-based residues, food production and household 'biodegradable waste', aquatic resources). These sources for bio-based plastic meet the waste management standards, due to ease of recycling. In the Commission's legislative proposal of single use plastics, a report should be submitted, in a certain timeline, indicating whether sufficient scientific and technical progress has been made and criteria or a standard for biodegradability in the marine environment applicable to single and their single-use substitutes have been developed (EC Proposal for a Directive of the European Parliament and of the council on the reduction of the impact of certain plastic products on the environment, 2018c). Research and innovation activities across the whole innovation chain will be essential in achieving this objective. Activities will be underpinned by the 6rs (reinvent/rethink, refuse, reduce, reuse/repair, recycle and replace/rebuy) framework: redesign-reduce-remove-reuse-recycle-recover (EC, 2018a).

The aim of this chapter is presented edible films as a feasible way of reducing plastic consumption. In this work, carob seed peel by-product has been used in the formulation into fish-gelatin film.

2. Edible Films and Coatings as Environment-friendly Strategy

The development of edible films and coatings is an environment-friendly practice that allows reduction on the impact and disposal costs associated with synthetic polymeric films (Silva-Weiss et al., 2013). Edible film is a stand-alone wrapping material, whereas a coating is formed and applied directly on the food surface itself. Edible films are promising packaging systems due to their non-polluting nature. Currently, edible films cannot entirely replace traditional packaging due to handling and hygiene reasons, although edible films can reduce the use of conventional packaging (Cordeiro de Azeredo, 2012), extend the shelf-life and improve the quality of any food system. They can reduce the exchange of moisture, oxygen, carbon dioxide, flavor and aroma transfer between the food components and the surrounding atmosphere (Krochta and Mulder-Johnston, 1997) (Fig. 1). In addition, edible films are excellent carriers for the gradual release of bioactive compounds over storage (Campos et al., 2011).

Furthermore, some of the biomaterials used for edible films and coatings are extracted from renewable agricultural resources or industrial by-products due to greater availability, lower cost in comparison with fossil resources and as an alternative to provide more value to these by-products. Some examples are whey protein of dairy

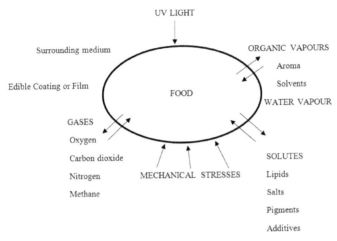

Fig. 1. External agents that can be controlled by edible films and coatings (adapted from: Debeaufort and Voilley, 2009).

industry, gelatin from slaughterhouses, sunflower protein from oilcakes, starch and cellulose from vegetables, chitosan from crustacean and carrageenan from seaweeds, among others.

There are no universal formulas for edible films and coatings; specific barrier requirement and food product specifications will determine the type of layer that is best for each situation. However, ideal edible films and coatings should have the following characteristics (Pavlath and Orts, 2009) (Fig. 2).

Edible films and coatings are defined by two essential elements—firstly, being edible implies that the film or coating and all its components must be safe to eat; secondly, edible films must be composed of bilipids, proteins and/or polysaccharides. These film- and coating- forming materials can be hydrophilic and hydrophobic, but in order to maintain edibility, only water or etanol can be used as a solvent during the processing. Other components can also be added into the material matrix to enhance the functionality, especially in films. These additives include plasticizers

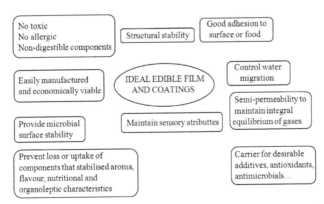

Fig. 2. Characteristics of an ideal edible film and/or coating (adapted from Pavlath and Orts, 2009).

(glycerol, sorbitol, propylene glycol), cross-linking agents (such as transglutaminase for proteins and citric or tannic acid for polysaccharides), reinforcements (fibers and nanoreinforcements) and emulsifiers (tweens, fatty acid salts and lecithins) (Salgado et al., 2015) with all of them enhancing the final properties of the edible film.

2.1 Edible Films and Coatings Classification According to Their Polymeric Base

Lipids: Lipids (oils, fats, waxes, resins, essential oils and emulsifiers) are used in films formulations and coatings for limiting moisture migration. Polar compounds, such as resins are good barriers for O_2, CO_2 and ethylene (Šuput et al., 2015). However, lipids do not have the ability to form cohesive and independent films. Therefore, they are either used as a coating or incorporated into other biopolymers to make composite films (Dehghani et al., 2018; Debeaufort and Voilley, 2009). Wax coatings have been used for centuries to prevent the loss of moisture and to create a shiny fruit surface. The use of coatings in food dates back to China in the 12th century, where they were used for citric fruits (Dehghani et al., 2018).

Proteins: Research and development on films and coatings made from proteins has been conducted over the past 20 years. Part of the interest in proteins has been driven by environmental reasons. Edible protein films and coatings have been produced from renewable sources of by-products and waste streams (Dangaran et al., 2009).

Proteins exhibit excellent gas barrier properties, especially at low relative humidity (Popović et al., 2012). These inherent properties make them excellent film and coating materials. Films are formed and stabilized through electrostatic interactions, hydrogen bonding, van der Waals forces, covalent bonding and disulfide bridges (Krochta et al., 1994). Firstly, proteins must be denatured by heating, using solvent and/or changes in pH (Dehghani et al., 2018). The main limitations of protein films are low water-vapor barrier properties and limited mechanical strength. In comparison with conventional polymeric packaging material, proteins films have greater loss of moisture, approximately two to four orders of magnitude. The addition of plasticisers, such as glycerin and sorbitol, can improve the flexibility and limit the resistance to water-vapor transmission of this type of films (Sánchez-Ortega et al., 2014). Crops (cereal proteins, wheat) and fishery industry account for the major proteins source from co-products. Cereal proteins have relatively good oxygen barrier properties. The main disadvantage of myofibrillar proteins and milk is their low mechanical properties in comparison with other proteins (Murrieta-Martínez et al., 2018).

Polysaccharides: Polysaccharides are generally hydrophilic molecules, showing poor moisture barrier properties, with a selective permeability for O_2 and CO_2 and resistance to lipid migration (Deghani et al., 2018). The application of starch on edible films is promising because of having environmental appeal, low cost, flexibility, transparency and being tasteless and odorless (Bilbao-Sainz et al., 2010; Chiumarelli and Hubinger, 2012). Gums (agar, methylcellulose, hydroxypropylmethylcellulose, konjak gum, pullulan, inulin) are water-soluble hydrocolloids of considerable molecular weight. Hydrogen bonds between solvent and the polymer allow gums to dissolve in water. Gums have the ability to thicken and/or become gel aqueous solutions because of both hydrogen bonding between polymer chains and intermolecular friction when subjected to shear (Nieto, 2009). Chitosan is the second most abundant polysaccharide found in the nature, after cellulose. It is obtained by deacetylation of chitin, which is extracted from the exoskeleton of crustaceans and/or fungal cell walls (Dehghani et al., 2018). It has been widely used as an hydrocolloid to constitute edible films due to its biodegradability, biocompatibility, antimicrobial and antioxidant activities and non-toxicity (Ruiz-Navajas et al., 2013).

2.2 Bioactive Properties of Edible Films and Coatings

Edible coatings and films have a high potential to carry active ingredients, such as antioxidants, antimicrobials, flavor, colorants, nutraceuticals and probotics that can extend product shelf-life, and even enhance organoleptic properties and nutritional value of the food. Antioxidants and antimicrobials have been most studied due their properties and interest for food applications.

Antioxidants: Natural antioxidants are promising additives for films and coatings, in order to delay oxidation of food lipids (Ganiari et al., 2017). Antioxidants also are used to protect against oxidative degradation and enzymatic browning in minimally-processed fruit and vegetables, as for example cysteine, glutathione, ascorbic and citric acids. Most of these anti-browning agents are hydrophilic compounds, increasing the water-vapor transmission rate and water loss (Martin-Belloso et al., 2009). Natural antioxidants, such as essential oils, extracts from herbs and spices and agricultural by-products, have been added to films and coatings to delay oxidation of lipid foods. Natural antioxidants increase the phenolic content and anti-radical activity of the film. The interactions of specific natural ingredients and the distribution of the additive in the polymer matrix might affect the characteristics and properties of the film. Another factor that must be taken into account is the stability and release of the antioxidants on storage. For instance, the release of antioxidant additives is better controlled when using a blend of polymers instead of a single polymeric matrix (Ganiari et al., 2017).

Antimicrobial: Antimicrobial edible films and coatings increase food safety by inhibiting the growth of pathogenic and spoilage microorganisms. Food-borne pathogen outbreaks are a major concern for consumers and industry. Besides pathogens, the growth of spoilage microorganism is the main factor limiting the shelf-life in food products. Thus, it is essential to implement appropiate antimicrobial preservation technologies, such as edible films and coatings.

In the case of edible films and coatings, microbial control is achieved by using antimicrobial biopolymers and/or adding active ingredients. Chitosan has been widely studied for its antimicrobial properties. The antimicrobial mechanism of chitosan is based on the fact that its positive charges may compete with Ca^{2+} for the negatively charged bacterial membrane (Coma et al., 2002). The antimicrobial effectiveness of chitosan films *in vitro* agar assay against different food spoilage and pathogenic bacteria was negligible due to the lack of chitosan diffusion. However, the application of chitosan films on food products reduced the microbial load over storage (Albertos et al., 2015a). The antimicrobial ingredients can act by direct contact or indirect contact after a preliminary step of diffusion. The addition of these compounds in the packaging material concentrates them at the product surface where the protection is required. Furthermore, only very small amounts of additives are required, as compared to additives formulated through direct addition in the matrix of the food product (Guillard et al., 2009). The more commonly antimicrobial ingredients used are organic acids, bacteriocins, lactoperoxidase and essential oils (Campos et al., 2011).

2.3 Key Antioxidant and Antimicrobial Compounds in Edible Films and Coatings

Plant and phenolic extracts: Natural antioxidants and antimicrobials from plants have received attention. Green tea (Giménez et al., 2013), grape seed (Moradi et al., 2012) and olive leaves (Albertos et al., 2017) among others has been successfully applied on edible films. Polyphenols are the major plant compounds with antioxidant, antimicrobial and positive effects on human health.

Other authors proposed the addition directly of phenolic compounds, such as ferulic acid (Mathew and Abraham, 2008) or rosmaric acid (Frankel et al., 1996). However, the bioactive properties in plant extracts used to be higher than individual phenolics. Frankel et al. (1996) demonstrated that rosemary extract was more active than rosmarinic acid. Similarly, Lee and Lee (2010) studied the antioxidant activity of individual phenolic compounds in the olive leaf extract and demonstrated that the antioxidant capacity of the combined compounds was higher than those of the individual phenolic compounds, such as oleuropein.

Vitamins: Edible films and coatings can deliver vitamins. Tocopherol and ascorbic acid have been the most widely incorporated. The antioxidant activity of α-tocopherol has been well documented. Antioxidant capacity of films was enhanced when α-tocopherol was presented in the film matrix (Martins et al., 2012). Besides, an important reduction of oxygen permeability of the films was observed when α-tocopherol was incorporated to the film. Thus, α-tocopherol can improve the quality of foods sensible to oxidation. Furthermore, films enriched with α-tocopherol improve the water vapour permeability of the films due to its hydrophobic nature (Fabra et al., 2011).

L-ascorbic acid is another antioxidant that has been incorpotated in edible films and coatings. L-ascorbic acid is a reducing agent and a water-soluble antioxidant for food preservation. Several authors have explored the possibility of incorporating L-ascorbic acid as anti-browning agent into edible films and coatings in minimally processed fruits (Wong et al., 1994; Baldwin et al., 1996; Pérez-Gago et al., 2006; Martín-Belloso et al., 2009).

Essential oils (EOs): Essential oils (EOs) have been widely used as flavoring agents in food since the earliest recorded history. However, nowadays they attract increasing interest due to their antimicrobial and antioxidant properties (Holley and Patel, 2005). The antimicrobial and the antioxidant activity of plant EOs is usually associated to phenolic profile (Du et al., 2009). The antimicrobial mechanism may relate to the ability of phenolic compounds to alter microbial cell permeability, damage cytoplasmic membranes, interfere with cellular energy and disrupt the proton motive force (Burt, 2004).

The application of EOs through their formulation into film allows to reduce the amount of EOs needed for obtaining the desired results. In this way, disadvantages derived of their use, such as intense aroma and even their high cost, would be reduced (Sánchez-González et al., 2011).

Organic acids: Organic acids include lactic, acetic, malic and citric acid. The antimicrobial activity is based on pH reduction, disruption of substrate transport and reduction of proton motive force (Campos et al., 2011). Other weak lipophilic acids used in edible films and coatings are sorbic acid and potassium salt, commonly name as sorbates. They have the ability to penetrate the cell membrane when they are in the undissociated form, acididyfing the cytoplasm and inhibiting the growth (Samelis and Sofos, 2003).

Bacteriocins: Bacteriocins are antibacterial peptides produced by lactic acid bacteria. Bacteriocins have distinct mechanisms of action and can be divided into those that promote a bactericidal effect, with or without cell lysis, or bacteriostatic, inhibiting cell growth (Da Silva Sabo et al., 2014). One of the main advantages of these compounds is easy digestion by the human tract (Mills et al., 2011). Nisin is the most characterised bacteriocin, being recognised asa food-grade preservative (E-234). It exhibits antimicrobial activity towards a wide range of Gram-positive bacteria and is particularly effective against spores. In contrast, it shows little or no activity against Gram-negative bacteria, yeast and molds (Delves-Broughton, 2005).

Lactoperoxidase (LPO): Lactoperoxidase (LPO) is a heme-containing antimicrobial enzyme present in milk and in mammalians' saliva and tears. The antimicrobial activity of LPO is due to production of potent oxidising and bactericidal compounds (hypothiocyanite OSCN⁻ and hypothiocyanous acid HOSCN) in the presence of hydrogen peroxide (H_2O_2) (Sheikh et al., 2018).

3. Fish Gelatin

Gelatin is a soluble protein obtained by partial hydrolysis of collagen, the main constituent in bones, cartilages and skins. It is one of the most widely used biopolymersin edible film elaboration. Fish gelatin has received considerable attention in recent years. Specifically, fish gelatin has gained interest over mammalian gelatins, as it prevents the risk of Bovine Spongiform Encephalopathy, the potential Kosher and Halal market and the possibility of obtaining it from by-products, for example skin of Alaska pollock or pink salmon (Avena-Bustillos et al., 2011). The properties (gelatin, thermal, mechanical, water vapour and oxygen permeability) of mammalian, warm- and cold-water fish gelatin solutions and films were compared by Avena-Bustillos et al. (2011). Mammalian gelatins yielded stronger films, whereas fish gelatins produced more deformable films. The high water-vapour permeability (WVP) is one of the main drawbacks of gelatin. In cold-water fish gelatin, WVP was reduced in comparison with mammalian and warm-water fish gelatin and oxygen permeability of cold-water fish gelatin films was significantly lower than that of mammalian gelatin films (Avena-Bustillos et al., 2011).

Traditionally, fish gelatins are obtained from skin and bones from fish processing industries, containing 27–49 per cent of protein on a dry weight basis (Lowell et al., 1978). This allows transforming this by-product stream into a commercially valuable product. Gelatin properties change depending on fish origin (species and habitat). The molecular weight distribution and amino acid composition are the main factors influencing the physical and structural properties of gelatin (Gómez-Guillén et al., 2009). Furthermore, the solid waste from surimi processing, which represents 50–70 per cent of the original material, could be used for obtaining gelatin (Morrissey et al., 2005).

Fish gelatin has been extensively studied due to its film-forming ability and its usefulness as an outer film to protect food for drying and exposure to light and oxygen (Arvanitoyannis, 2002). Edible fish gelatin films were prepared by solvent casting and plasticizers should be required. The high hygroscopic nature and poor mechanical properties are the main limitations of fish gelatin edible films as protective barriers. Consequently, different strategies have been used to improve the permeability and mechanical attributes of these edible films (Hosseini and Gómez-Guillén, 2018) (Fig. 3).

One possibility is blending with biopolymers through extruding, laminating, coating, etc. Protein-polysaccharide interactions have gained growing attention, especially with chitosan. Chitosan is compatible with gelatin because it is positively charged, while gelatin is negatively charged under working conditions of pH (Hosseini and Gómez-Guillén, 2018). Other fish gelatin films were formulated by incorporating fibers, mainly lignin (Nuñez-Flores et al., 2013) and lignosulphonate (Nuñez-Flores et al., 2012) to improve their physical properties.

A further possibility to increase the mechanical properties of fish gelatin films is the incorporation of hydrophobic compounds, such as fatty acids, lipid alcohols, waxes and vegetable oils, which have resulted in diminishing the water-vapour permeability and water solubility (Gómez-Guillén et al., 2009).

One of the most promising strategies to overcome the poor properties of these films is the use of chemical, enzymatic and physical compounds to induce both intermolecular and intramolecular chemical bonding (BenBettaïeb et al., 2015). Different chemical cross-linking agents, such as glutaraldehyde, glyceraldehyde,

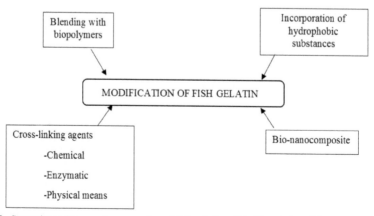

Fig. 3. Strategies to improve the properties of fish gelatin edible films (adapted from Hosseini and Gómez-Guillén, 2018).

formaldehyde and glyoxal, have been used to improve the attributes of protein films. However, most of these chemical cross-linking agents can induce toxicity or lead to other undesirable effects (Hosseini and Gómez-Guillén, 2018). The *enzyme* transglutaminase (TGase) is considered safe and induces enzymatic cross-linking by catalysing an acyl transfer reaction between the γ-carboxamide group of glutamine residues and the ε-amine group of lysine residues of peptide chains (Folk, 1983). The mechanical properties are enhanced in fish gelatin by using TGase (Gómez-Guillén et al., 2011). The addition of TGase has proven effective for producing more stable gels of Baltic cod skins at roo m temperature (Kołodziejska et al., 2004). Physical treatments, such as thermal, UV, γ and electron beam irradiation have created cross-linking between polysaccharides and proteins to improve barrier and mechanical properties in fish gelatin films (Hosseini and Gómez-Guillén 2018).

The term bio-nanocomposites means a biopolymer matrix reinforced with nanoparticles (< 100 nm) and included in low fractions (1–10 per cent in mass). These materials showed better barrier properties, mechanical strength and heat resistance than biopolymers alone or conventional micro- or macro-scale composites, due to their high aspect ratio and high surface area (Rhim et al., 2013). In the fabrication of bio-nanocomposites gelatin films, various types of nanofillers have been used, such as nanoclays (montmorillonite, sepiolite, etc.), polysaccharide nanofillers (nanofibers/nanowhiskers), metal ions (silver, copper) and metal oxide nanoparticles (zinc oxide, titanium dioxide) (Hosseini and Gómez-Guillén, 2018). Gelatin nanofibers have also been developed. Nanofibers are used for controlling release of bioactive compounds due to their large surface area to volume ratio, high encapsulation efficiency and controlled-release properties. Fish gelatin nanofibers are promising systems as carriers of essential oils, such as cinnamaldehyde and carvacrol (Liu et al., 2018a,b). The application of these essential oils in food preservation is limited by its particular flavor, volatility and lipophilic nature. This limitation can be avoided or lessened by encapsulation in nanofibers. Nanofibers can mask flavor, improve aqueous solubility and stability. Furthermore, recently nanofibers have attracted increased attention because they have improved bioactive activities as compared to traditional methods (Liu et al., 2018a,b). Among the various approaches to preparing fish gelatin nanofibers, electrospinning (Chiou et al., 2013) and blow spinning (Liu et al., 2017; Liu et al., 2018a,b) have been studied. Electrospinning uses electric force with the aim of melting up polymers to fiber diameters in the order of some hundred nanometers. Fish gelatin fibers can be produced at room temperature as different from mammalian sources, which require gelation temperature above 35°C. Consequently, modification in electrospinning apparatus should be done to work with mammalian gelatin (Chiou et al., 2013). Blow spinning combines electrospinning, solution and melt spinning. Blow spinning subsists on a pressurized air gas source for delivering the carrier gas and a syringe pump for polymer solution.

3.1 Fish Gelatin Properties

Fish gelatin is a source of biologically active peptides with promising health benefits. These peptides are inactive in the parent protein sequence but can be liberated during gastrointestinal digestion, food processing or fermentation (Gómez-Guillén et al., 2011).

Fish gelatin posseses certain *antioxidant* properties, but the hydrolysis of gelatin chains noticeably increases the antioxidant capacity (Giménez et al., 2009). Fish gelatin peptides show free radical quenching (Mendis et al., 2005) and high metal chelating ability (Alemán et al., 2011). However, the exact mechanism of antioxidant activity has not been fully elucidated. Wiriyaphan et al. (2012) suggested that the antioxidant capacities could rely on the specific amino acid composition, structure and hydrophobicity.

The addition of antioxidants in the formulation of fish gelatin films is a promising technique for developing antioxidant packaging. Consumers are demanding 'clean labelling', so there is an increasing interest in utilizing natural antioxidants, like polyphenolic plant extracts in the formulation of active edible films. Gómez Guillén et al. (2007) and Haddar et al. (2012) proposed the use of extracts of different murta leaves and brown algae extract respectively, to obtain films with antioxidant properties. Some sources of polyphenolic compounds also have antimicrobial activity. Giménez et al. (2013) developed fish gelatin films with green tea aqueous extract, with antioxidant and antimicrobial properties.

Fish gelatin has been shown to be a good source of antihypertensive peptides after enzymatic digestion. However, these peptides have not been extensively studied as other sources. In general, peptides must be of short sequences with low molecular mass because the active site of ACE cannot accommodate large peptides molecules (Gómez-Guillén et al., 2011). Guérard et al. (2010) established that the nature of the matrices, the choice of protease and the extent of hydrolysis determine the ACE inhibitory activity. Furthermore, enzymatically-hydrolysed fish gelatin, which is obtained from skin, showed better antioxidant and antihypertensive effects than peptides derived from fish muscle protein (Kim and Mendis, 2006).

The antimicrobial properties of hydrolysates or peptides from gelatin are scarce. Peptide fractions have been studied in different species and those extracted from tuna and squid skin gelatins reported antimicrobial activities in agar diffusion assay (Gómez-Guillén et al., 2010). The hydrophobic character of amino acids might be responsible for allowing peptides to cross the bacterial membrane, as the positive charge would initiate the peptide interaction with the negatively charged bacterial surface (Wieprecht et al., 1997). Fish gelatin films have been used as antimicrobial packaging material as they can carry and release a variety of active compounds. Essential oils have been widely used as antimicrobial and antioxidant ingredients in fish gelatin films (Kim et al., 2018; Hosseini et al., 2016; Iturriaga et al., 2012; Tongnuanchan et al., 2014).

4. Carob Seed Peel By-products

Carob tree (*Ceratonia siliqua*) grows in Mediterranean countries with Portugal being the first carob-fruit world producer with up to 42.000 tons in 2017, followed by Italy, Morocco, Turkey, Greece, Cyprus, Algeria, Spain, Lebanon and Tunisia (FAOSTAT, 2019).

Recently, the carob tree has been spread to Mediterranean climate areas, such as Mexico, California, Arizona, Chile, Argentina, Australia, South Africa and India. This cultivation is considered as an affordable source for both human and animal nutrition due to its resistance to drought and salinity and its adaptation to poor soils (Battle and Tous, 1997). Economic prosperity has caused a decrease in its production during the last decade (FAOSTAT, 2019). However, carob product research and development has recently received attention due to the nutritional and healthy benefits.

4.1 Processing

The fruit is a brown pod, with a wrinkled surface that becomes leathery when ripe. The carob fruit consists of two main parts: the pod (90 per cent) and the seeds (10 per cent). Upon arrival at a processing facility, the carob pods must be stored at controlled conditions to meet the desired moisture content. Initially, carob pods have between 10–20 per cent of moisture content and the pods need to be at 8 per cent moisture for processing. Firstly, the pods are crushed and milled for both food and feed. Thus, the seeds are separated from the pods (Batlle and Tous, 1997). The carob seed consists of coat or peel (30–33 per cent), endosperm (42–46 per cent) and germ (23–25 per cent) (Gharnit et al., 2006) (Fig. 4).

The first step in locust bean gum extraction is removal of the peel (Fig. 5). Currently, the available technology allows separation of the peel intact by physical treatment with friction. Traditionally, the seeds are dehusked by treatment with dilute sulphuric acid, with a thermo-mechanical treatment, known as acid peeling and thermal peeling by dry roasting (Loullis and Pinakoulaki, 2018; Batlle and Tous, 1997). This by-product has received scant attention. Peel is removed and milled into a fine powder. It is commonly used in the leather industry where it is a tanning agent (Batlle and Tous, 1997). However, carob seed peel, which is a food-grade waste, could represent a good source of natural additive.

Endosperm and germ are separated due to milling because the two fractions have differences in friability (Fig. 5). The germ is much more brittle and reduces in size easily when compared to the endosperm (Batlle and Tous, 1997). The endosperm goes through another milling step to produce a fine powder as locust bean gum (Fig. 5).

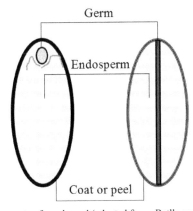

Fig. 4. Components of carob seed (adapted from: Batlle and Tous, 1997).

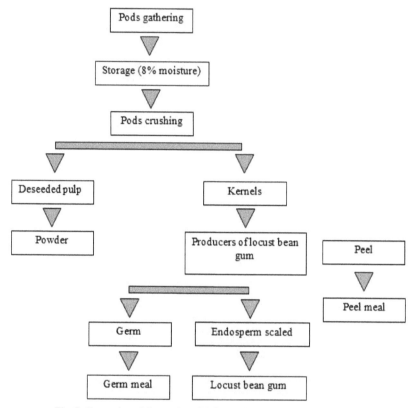

Fig. 5. Processing of the carob pod (adapted from Batlle and Tous, 1997).

4.2 Carob Fractions: Nutritional, Healthy and Technological Properties

Carob pods: Traditionally, the primary uses of carob pods were animal feed and cocoa substitute. Carob can be used as a replacer of cocoa in food products as it has similar sensorial properties (aroma and flavor) as cocoa after roasting due to high sugar content, which makes it a natural sweetener. Furthermore, carob contains various beneficial phytochemicals that are similar to those found in cocoa. In fact, carob shows some advantages over cocoa. One of the main advantages is the absence of caffeine and theobromine. Carob appears to fulfil modern health criteria of consumers of caffeine-free product, which is undesirable in certain products. Another benefit of carob is lower fat content and significantly higher amount of dietary fiber compared to cocoa (Loullis and Pinakoulaki, 2017).

The pod is composed of sugar content (48–56 per cent) (mainly sucrose, glucose and fructose), 2–4 per cent protein, a low-fat content (0.2–0.6 per cent) (Ortega et al., 2009), polyphenols (e.g., tannins, flavonoids, phenolic acids), vitamins (E, D, C, Niacin, B_6 and acid folic) and minerals (K, Ca, Mg, Na, Cu, Fe, Mn and Zn) (Papaefstathiou et al., 2018).

The carob pod has recently received attention due to its health benefits. The pod is rich in bioactive compounds, particularly polyphenols, with potential health benefits in humans against a wide variety of diseases (Goulas et al., 2016). The carob pod has demonstrated anti-hypertensive, triacylglycerol-lowering and anti-inflammatory activities, markers associated with metabolic syndrome, in *in vitro* assays. *In vivo* study, only the highest concentration of carob seed peel showed an anti-hypertensive effect in rats with metabolic syndrome (Martinez-Villaluenga et al., 2018). This effect is attributed to the high content of lignins and polyphenols, especially tannins (Zunft et al., 2003). Carob pods represent a useful therapeutic strategy against non-alcoholic fatty liver disease in animal study (Rico et al., 2019a). Moreover, carob shows anti-carcinogenic properties in cell model studies (Klenow et al., 2009; Klenow et al., 2008; Corsi et al., 2002).

Locust bean gum: The seeds are industrially used for obtaining locust bean gum extraction (E-410) (Roseiro et al., 2013). This gum is contained in the endosperm of the seed. It is a polysaccharide (galactomannan), which consists of 16–20 per cent D-galactose and 80–84 per cent D-mannose (Salinas et al., 2015). Locust bean gum is widely used in the food industry as a thickener, stabiliser, flavoring, gelling and dispersing agent (Roseiro et al., 2013). In this process of gum extraction, different by-products (peel and germ) are generated.

Germ: The germ flour is primarily used as a protein supplement in animal, pet foods and for dietetic supplements for human (Dakia et al., 2007).

Carob germ is rich in protein (32.74 ± 0.97) (Rico et al., 2019b). This protein is highly digestible, as demonstrated by Mamone et al. (2019) *in vitro*-simulated gastrointestinal digestion. Furthermore, carob germ protein can be considered a complete protein because essential amino acids are present in very interesting amounts, according to the FAO standard (Dakia, 2011). The lipid content of carob germ is about 2.92 ± 0.14 per cent (Rico et al., 2019b), being a good source of essential linoleic acid (C18:2, ω-6), which predominates (43.3–50.6 per cent) (Siano et al., 2018), followed by oleic acid (16.2 per cent) and stearic (3.4 per cent) (Dakia et al., 2007).

The production of gluten-free bakery products is the main potential of carob germ (Martin-Diana et al., 2017; Rico et al., 2018). This germ contains a protein which is able to form net structures similar to gluten, although weaker (Miñarro et al., 2012).

Peel: This by-product has been scarcely studied in comparison with other carob fractions. The nutritional composition of carob seed peel is characterized by low fat (0.31 per cent ± 0.01) and high fibre content (61.64 per cent ± 0.32) (Albertos et al., 2015b). The peel has a high polyphenol content, the latter contributing to its antioxidant potential (Albertos et al., 2015b). In comparison with other carob by-products (pods, peel and germ), peel has the highest total phenol content and antioxidant capacity (Rico et al., 2019b). Quercetin and apigenin derivatives are the most abundant polyphenols in this fraction (Rico et al., 2019b).

Carob seed peel has successfully been used as a natural antioxidant in food (Albertos et al., 2015b). Furthermore, gluten-free cracker snacks were developed by Martín-Diana et al. (2017) with the inclusion of carob seed peel. The incorporation of carob seed peel significantly increased the antioxidant capacity of the snacks. Similarly, functional bread was elaborated by replacing wheat with carob seed peel at 8 per cent. Significant correlation between bread composition and bioactive properties was observed. However, the carob seed peel has sensorial limitations due to its high fiber content and pigmentation. The sensory panel was perceived when peel was added as an ingredient in bread (Rico et al., 2018).

The potential *in vitro* bioactivities for alleviating metabolic syndrome was evaluated in carob products. Carob seed peel demonstrated the strongest anti-hypertensive activity and reduced the fat accumulation on mature adipocytes (Rico et al., 2019b).

5. Practical Application: Development of Edible Film Formulated with Fish Gelatin and Carob Seed Peel

Agri-food processing industries generate large quantities of by-products that would cause an environmental problem. Nowadays, efforts focus on reducing waste through the use of more efficient processes and upgrading the waste into added-value by-products. For a long time, the industrial system was based on a traditional linear economy (take-make-dispose). It entails gradually the consumption of finite resources and generation of by-products. Circular economy portends the recuperation of by-products as raw material again. Currently, the locust bean gum (E-410) extraction is the main industrial explotation of carob. Carob seed peel is generated as a by-product and represents 30–33 per cent of the seeds (Gharnit et al., 2006). Carob seed peel has nutritional, antioxidant and bioactive properties. Thus, this by-product would likely be used as an additive for developing functional food. Furthermore, consumers are increasingly demanding convenient and natural food products, free of chemical preservatives. However, carob seed peel directly incorporated in food has organoleptic limitations (Rico et al., 2018). A possible strategy is the inclusion of carob seed peel in edible films to reduce the amounts and consequently minimize its astringency and bitterness.

Objective

Carob seed peel (CSP) has a great potential to be used in fish gelatin films (FGF) as active packaging for food. The gelatin films were prepared with different carob seed peel concentration (0, 2.5, 5 and 10 per cent w/w) by the casting technique (Fig. 6). Antioxidant, color, water vapor permeability, mechanical properties and thermal stability of edible films were evaluated.

Total phenolic content and antioxidant capacity of edible films: Total phenolic content and antioxidant capacity of edible films increases with increasing carob seed peel concentration 0.675 ($\pm$ 0.007) to 1.100 ($\pm$ 0.014) mg GAE/g and from 2799 ($\pm$ 71) to 2934 ($\pm$ 2) mg Trolox/g, respectively. Nevertheless, only the presence of 5 per cent of carob seed peel had a significant ($p < 0.05$) effect in respect to fish gelatin film. Fish gelatin films had 0.515 ($\pm$ 0.021) mg GAE/g and 2720 ($\pm$ 19) mg Trolox/g of films. The antioxidant activity of FGF was also determined by other authors (Gómez-Guillén et al., 2007; Haddar et al., 2012; Bitencourt et al., 2014). FGF had some antioxidant capacity, possibly due to the presence of peptide sequences containing amino acids, such as glycine and proline in the fish gelatin (Mendis et al., 2005). It has good correlations between total phenols and antioxidant activity (Gülçin et al., 2004; Huang et al., 2005). The linear correlation between total phenols (mg GA/g) and DPPH (mgEq of Trolox/g) have been observed in several food matrices. A significant

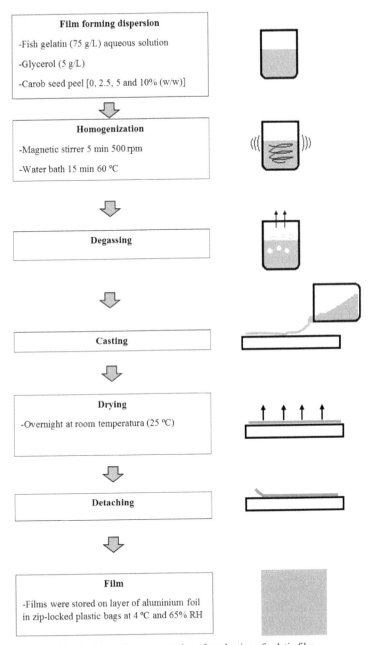

Fig. 6. Schematic representation of production of gelatin film.

correlation between ($R^2 = 0.929$) was observed between polyphenols and antioxidant capacity in films with different carob seed peel concentration.

Color of edible films: The color of fish gelatin films (FGF) with the incorporation of carob seed peel (CSP) changed (Table 1). Lightness was significantly reduced with

Table 1. Colour Parameters of Fish Gelatin Films (FGF) with Different concentrations of Carob Seed Peel (CSP).

	L*	**a***	**b***
FGF	93.92 (± 0.27)$_e$	−0.25 (± 0.04)$_a$	−0.05 (± 0.46)$_a$
2.5% CSP FGF	92.82 (± 1.39)$_e$	0.19 (± 0.02)$_b$	1.90 (± 0.07)$_b$
5% CSP FGF	89.90 (± 2.12)$_d$	0.60 (± 0.09)$_c$	4.28 (± 0.67)$_c$
7.5% CSP FGF	85.83 (± 1.93)$_c$	1.87 (± 0.41)$_d$	8.75 (± 0.89)$_d$
10% CSP FGF	82.84 (± 2.53)$_b$	2.96 (± 0.13)$_e$	10.85 (± 0.09)$_e$

Values (mean ± standard deviation, n = 6) followed by the same lowercase letter in same column are not significantly different (p > 0.05).

increasing CSP addition in the FGF, hereas, a* and b* are increasingly directly related to the concentration of the CSP. Positive a* and b* values of CSP measure redness-yellowness of color respectively. CSP presented a low L* value, which causes darker films, being concentration-dependent. This opacity rise could be very positive for packaging food to prevent the effects of UV light, such as lipid oxidation.

Water vapor barrier properties of edible films: The main limitations of edible proteins films and polysaccharides are the poor water vapor barrier properties (McHugh et al., 2009). Fish gelatin film (FGF) had a water vapor permeability of 0.5666 g-mm/kPA h m². Similar results were obtained by Avena-Bustillos et al. (2006). Nevertheless, fish gelatin presented lower water vapor permeability (WVP) as compared to mammalian (Avena-Bustillos et al., 2006). Fish gelatin contains a minor concentration of proline and hidroxyproline, which provide higher hydrophobicity as compared to mammalian gelatins. Carob seed peel (CSP) did not contribute to improving the water vapor permeability of the films. In fact, the addition of CSP to the FGF significantly increased the WVP of these films as compared to the control film (Table 2). No significant differences were found between 2.5–5 per cent of CSP concentration in FGF. An increase took place from 5 per cent to 7.5 per cent of CSP incorporation in FGF, without a significant rise from this level.

The water barrier properties of films depend on both molecular diffusion coefficient and water solubility in the matrix (McHugh et al., 1993). It can be attributed to the incorporation of CSP rise in the thickness. Hydrophilic films, such as FGF,

Table 2. Effect of Carob Seed Peel (CSP) Concentration on Water Vapor Permeability.

	Film thickness (mm)	**Relative humidity (%RH) at film underside**	**Water vapor permeability (g-mm/kPa-h-m²)**
FGF	0.035 (± 0.003)$_a$	86.08 (± 0.90)$_b$	0.57 (± 0.05)$_a$
2.5% CSP FGF	0.092 (± 0.004)$_b$	81.97 (± 2.18)$_a$	2.01 (± 0.25)$_b$
5% CSP FGF	0.138 (± 0.036)$_c$	82.40 (± 4.76)$_a$	2.54 (± 0.32)$_b$
7.5% CSP FGF	0.159 (± 0.041)$_d$	85.05 (± 3.26)$_b$	2.94 (± 0.43)$_c$
10% CSP FGF	0.168 (± 0.025)$_d$	86.85 (± 1.00)$_b$	3.24 (± 0.92)$_c$

Values (mean ± standard deviation, n = 6) followed by the same lowercase letter in same column are not significantly different (p > 0.05).

often exhibit positive slope relationships between thickness and WVP (McHugh et al., 1993). These films with high permeability show variable percentage of RH at film underside, depending on thickness. Differences in water vapor concentration gradients cause anomalies on WVP if we assume it is 100 per cent RH at the film underside as in polymeric plastic films occur but not for hydrophilic edible films. These anomalies were solved on WVP due to film thickness differences in this work (McHugh et al., 1993). Secondly, CSP induced a plasticization effect, which made the polymer more water permeable. Films with CSP caused less interaction of gelatin with water molecules. Similar results were reported by Núñez-Flores et al. (2012, 2013) with the addition of insoluble compounds, such as lignosulphonate and lignin in fish gelatin. Actually, CSP is a by-product of locust bean gum, which is used as a thickener. WVP generally occurs through the hydrophilic portion of the film and consequently the ratio hydrophilic-hydrophobic is decisive in WVP of the film (Hernandez et al., 1994).

Moisture, water activity and total pure volume of edible films: Similarly, the addition of different CSP concentrations to FGF affects the moisture content of the films (Table 3). Films with 2.5 per cent CSP present the lowest amounts. From this amount, an increase occurs, reducing again the moisture at highest concentration tested (10 per cent). Water activity shows similar behavior. These results can be explained by the capacity of CSP of absorbing water, which acquires the maximum level at 2.5 per cent. Actually, CSP is the by-product of locust bean gum, a thickener in food products (Roseiro et al., 2013). However, CSP is formed mainly by insoluble fiber and consequently it does not present too much water absorption ability. The moisture content and water activity are highly dependent on the CSP concentration. The total pore volume of FGF increases as the concentration of CSP increases (Table 3). A minor total pore volume causes an increase in the tortuosity pathway of solutes through the film matrix (de Moura et al., 2011). Films with higher total pore volume can absorb more water molecules, increasing the moisture content (Yoon, 2014).

Mechanical properties of edible films: The mechanical properties of CSP in FGF were studied by determining the tensile strength and elastic modulus (Fig. 7). Addition of CSP caused a reduction ($P < 0.05$) of tensile strength as compared to FGF (Fig. 7). Our findings were in agreement with those obtained by Gómez-Guillén et al. (2007); Núñez-Flores et al. (2012) and Nuñez Flores et al. (2013). FGF had the maximum tensile strength. CSP caused interference in the aggregations of gelatin helices (Núñez-

Table 3. Effect of CSP Concentration on Moisture, Water Activity and Total Pore Volume of Fish Gelatin Film.

	% Moisture	Water activity	Total pore volume
FGF	10.15 ($\pm$ 0.89)$_b$	0.4497 ($\pm$ 0.0038)$_b$	0.766$_a$
2.5% CSP FGF	6.64 ($\pm$ 0.31)$_a$	0.4277 ($\pm$ 0.0015)$_a$	1.072$_b$
5% CSP FGF	9.56 ($\pm$ 0.57)$_b$	0.4468 ($\pm$ 0.0085)$_b$	2.366$_c$
7.5% CSP FGF	11.35 ($\pm$ 0.45)$_c$	0.4428 ($\pm$ 0.0041)$_b$	2.831$_d$
10% CSP FGF	11.16 ($\pm$ 0.62)$_c$	0.4561 ($\pm$ 0.0090)$_c$	3.063$_e$

Values (mean $\pm$ standard deviation, n = 6) followed by the same lowercase letter in same column are not significantly different (p > 0.05).

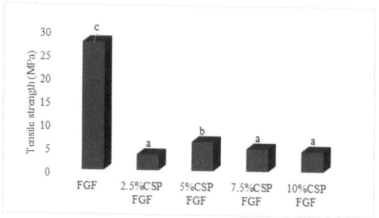

Fig. 7. Effect of CSP concentration on tensile strength (MPa) of fish gelatin films (FGF). Different letters mean significant differences ($p < 0.05$).

Flores et al., 2012). CSP is rich in tannins, mainly 1350 ($\pm$50) mg leucocyanidin which is equivalent to 100 g of dry weight (Albertos et al., 2015b). According to Orliac et al. (2002) tannins give films with lower mechanical properties because they act through weak interactions as compared to other compounds. Tannins have extremely low volatility which prevents the elimination of the non-bound tannin faction. Among CSP concentrations, a significant increase in mechanical properties occurs with 5 per cent CSP in FGF (Fig. 7). This trend also was reported by these authors because the rise of percentage of tannins above a certain limit proves that tannins will not be able to bind to the network.

Elastic modulus is a measure of stiffness of the film. Low values of elastic modulus correspond to an increase in the CSP concentration in FGF (Fig. 8). The reduced value of elastic modulus on CSP-contained films was probably due to the plasticizing effect of CSP when incorporated in FGF. The flexibility in packaging applications is an important factor. In this sense, the use of CSP improves the flexibility.

Thermal properties of edible films: Thermal properties of edible films were analyzed using thermogravimetric analysis (TGA). TGA allowed investigation of structural changes due to temperature increase. The FGF with 2.5 per cent, 5 per cent, 7.5 per cent and 10 pr cent of CSP or without CSP showed mass loss approximately at 100°C, which was mostly caused by film moisture evaporation (Fig. 9). Initially the loss of weight in CSP was observed at 240°C. Meanwhile for gelatin powder this reduction took place later. It is well-documented degradation of gelatin chains that occurs around 270°C (Chiou et al., 2013). All TGA curves of FGF with or without CSP exhibited a large drop in weight when the temperature reached 270°C. In comparison, different CSP concentrations with higher levels of inclusion of CSP (5, 7.5 and 10 per cent) had gelatin decomposition a bit earlier ($\pm$ 270°C). The last composition step was observed at 400°C, with it being more pronounced in CSP-contained films. It can be attributed to thermal decomposition of CSP, composed mainly of dietary fiber (61.64 $\pm$ 0.32%). To sum up, the thermograms showed less weight loss through CSP incorporation. These results were in agreement with Pérez-Espitia et al. (2014), where açai edible

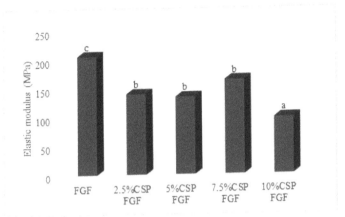

Fig. 8. Effect of CSP concentration on elastic modulus (MPa) of fish gelatin films (FGF). Different letters mean significant differences ($p < 0.05$).

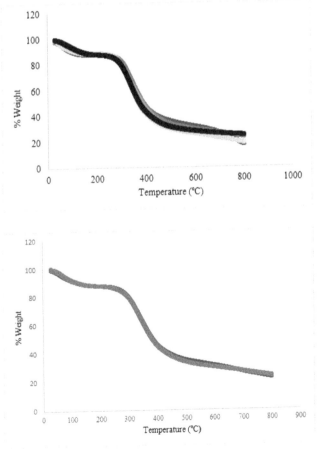

Fig. 9. TGA curves for (a) FGF and (b) FGF with 10 per cent of CSP.

films were studied. Açai was also rich in fiber, mainly 12 per cent, but to a lesser extent as compared to CSP, which contained 61.64 per cent (Albertos et al., 2015). In conclusion, the addition of CSP enhanced the thermal properties of the films.

6. Conclusion and Future Research Trends

European Union promotes circular economy and works towards the goal of all plastic packaging becoming recyclable by 2030. Nowadays, the use of edible films and/or coatings can reduce, contributing to the reduction of plastic packaging.

There is a rising interest in the valorization of agrifood industry. For a long time, these by-products were used for animal feed. Nevertheless, most of them are likely to be used as additives or for developing functional food due to their bioactive compounds. The main limitations in incorporating this into food have sensory aspects. In this sense, the inclusion of these compounds into edible films and/or coatings could be an alternative.

Fish gelatin has the capacity to form edible films with good technological and bioactive properties. This gelatin can be obtained from skin and bones from fish-processing industries. Carob seed peel (CSP) is a by-product from locust bean gum extraction. CSP has not yet received too much attention, though it is a food-grade waste and has good nutritional and bioactive properties.

Further studies and strategies should be implemented to achieve competitive edible packaging, which represent a viable alternative to conventional plastics over a wider range of applications.

References

Albertos, I., D. Rico, A.M. Diez, L. González-Arnáiz, M.J. García-Casas and I. Jaime (2015a). Effect of edible chitosan/clove oil films and high-pressure processing on the microbiological shelf life of trout fillets, *Journal of the Science of Food and Agriculture*, 95: 2858–2865.

Albertos, I., I. Jaime, A. María Diez, L. González-Arnaiz and D. Rico (2015b). Carob seed peel as natural antioxidant in minced and refrigerated (4°C) Atlantic horse mackerel (*Trachurus trachurus*), *LWT – Food Science and Technology*, 64: 650–656.

Albertos, I., R.J. Avena-Bustillos, A.B. Martín-Diana, W. Du, D. Rico and T.H. McHugh (2017). Antimicrobial olive leaf gelatin films for enhancing the quality of cold-smoked salmon, *Food Packaging and Shelf-Life*, 13: 49–55.

Alemán, A., B. Giménez, P. Montero and M.C. Gómez-Guillén (2011). Antioxidant activity of several marine skin gelatins, *LWT—Food Science and Technology*, 44: 407–413.

Association of Plastics Manufacturers (2016). Plastics—The Facts 2016. *An Analysis of European Plastics Production, Demand and Waste Data*. Brussels, Belg: Plastic Eur. <http://www.plasticseurope.org/documents/document/20161014113313-plastics_the_facts_2016_final_version.pdf.>. Downloaded on 20 February 2019.

Avena-Bustillos, R.J., C.W. Olsen, D.A. Olson, B. Chiou, E. Yee, P.J. Bechtel and T.H. McHugh (2006). Water-vapor permeability of mammalian and fish gelatin films, *Journal of Food Science*, 71: E202–E207.

Avena-Bustillos, R.J., B. Chiou, C.W. Olsen, P.J. Bechtel, D.A. Olson and T.H. McHugh (2011). Gelatin-oxygen permeability and mechanical properties of mammalian and fish gelatin films, *Journal of Food Science*, 76: E519–E524.

Arvanitoyannis, I.S. (2002). Formation and properties of collagen and gelatin films and coatings, pp. 275–304. *In*: Gennadios (Ed.). *Protein-based Films and Coatings*, Chap. 11, Boca Raton, Florida, CRC Press.

Baldwin, E.A., M.O. Nisperos, X. Chen and R.D. Hagenmaier (1996). Improving storage life of cut apple and potato with edible coating, *Postharvest Biology and Technology*, 9: 151–163.

Batlle, I. and J. Tous (1997). Carob Tree. *Ceratonia siliqua* L.- *Promoting the Conservation and Use of Underutilized and Neglected Crops*, 17, Institute of Plant Genetics and Crop Plant Research. Gatersleben/International Plant Genetic Resource Institute, Rome.

BenBettaïeb, N., T. Karbowiak, S. Bornaz and F. Debeaufort (2015). Spectroscopic analyses of the influence of electron beam irradiation doses on mechanical, transport properties and microstructure of chitosan-fish gelatin blend films, *Food Hydrocolloids*, 46: 37–51.

Bilbao-Sainz, C., R.J. Avena-Bustillos, D.F. Wood and T.H. McHugh (2010). Composite edible films based on hydroxypropyl methylcellulose reinforced with microcrystalline cellulose nanoparticles, *Journal of Agricultural and Food Chemistry*, 58: 3753–3760.

Bitencourt, C.M., C.S. Fávaro-Trindade, P.J.A. Sobral and R.A. Carvalho (2014). Gelatin-based films additivated with curcuma ethanol extract: Antioxidant activity and physical properties of films, *Food Hydrocolloids*, 40: 145–152.

Burt, S. (2004). Essential oils: Their antibacterial properties and potential applications in foods— A review, *International Journal of Food Microbiology*, 94: 223–253.

Campos, C.A., L.N. Gerschenson and S.K Flores (2011). Development of edible films and coatings with antimicrobial activity, *Food and Bioprocess Technology*, 4: 849–875.

Chiou, B., H. Jafri, R. Avena-Bustillos, K.S. Gregorski, P.J. Bechtel, S.J. Imam and W.J. Orts (2013). Properties of electrospun pollock gelatin/poly(vinyl alcohol) and pollock gelatin/poly(lactic acid) fibers, *International Journal of Biological Macromolecules*, 55: 214–220.

Chiumarelli, M. and M. Hubinger (2012). Stability, solubility, mechanical and barrier properties of cassava starch-carnauba wax edible coatings to preserve fresh-cut apples, *Food Hydrocolloids*, 28: 59–67.

Coma, V., A. Martial-Gros, S. Garreau, A. Copinet, F. Salin and A. Deschamps (2002). Edible antimicrobial films based on chitosan matrix, *Journal of Food Science*, 67: 1162–1169.

Cordeiro de Azeredo, H.M. (2012). Advances in fruit processing technologies. *In:* S. Rodrigues and F.A.N. Fernandes (Eds.), *Edible Coatings*, Boca Raton, CRC Press Inc.

Corsi, L., R. Avallone, F. Cosenza, F. Farina, C. Baraldi and M. Baraldi (2002). Anti-proliferative effects of *Ceratonia siliqua* L. on mouse hepatocellular carcinoma cell line, *Fitoterapia*, 73(7-8): 674–684.

Dangaran, K., P.G. Tomasula and P. Qi (2009). Structure and Function of Protein-Based Edible Films and Coatings, Chap. 2. *In:* M.E. Embuscado and K.C. Huber (Eds.), *Edible Films and Coatings for Food Applications*, Springer, Dordrecht, Heidleberg, London, New York.

Dakia, P.A., B. Wathelet and M. Paquot (2007). Isolation and chemical evaluation of carob (*Ceratonia siliqua* L.) seed germ, *Food Chemistry*, 102: 1368–1374.

Dakia, P.A. (2011). Carob (*Ceratonia siliqua* L.) seeds, endosperm and germ composition, and application to health, *Nuts and Seeds in Health and Disease Prevention*, 293–299.

Da Silva Sabo, S., M. Vitolo, J.M.D. González and R.P.D.S. Oliveira (2014). Overview of *Lactobacillus plantarum* as a promising bacteriocin producer among lactic acid bacteria, *Food Research International*, 64: 527–536.

Debeaufort, F. and A. Voilley (2009). Lipid-based edible films and coatings, Chap. 5, pp. 135–168. *In:* M.E. Embuscado and K.C. Huber (Eds.), *Edible Films and Coatings for Food Applications*, Springer, Dordrecht, Heidleberg, London, New York.

Dehghani, S., S.V. Hosseini and J.M. Regenstein (2018). Edible films and coatings in seafood preservation: A review, *Food Chemistry*, 240: 505–513.

Delves-Broughton, J. (2005). Nisin as a food preservative, *Food in Australia*, 57: 525–527.

De Moura, M.R., R.J. Avena-Bustillos, T.H. McHugh, D.F. Wood, C.G. Otoni and L.H.C. Mattoso (2011). Miniaturization of cellulose fibers and effect of addition on the mechanical and barrier properties of hydroxypropyl methylcellulose films, *Journal of Food Engineering*, 104: 154–160.

Du, W., C.W. Olsen, R.J. Avena-Bustillos, T.H. McHugh, C.E. Levin and M. Friedman (2009). Effects of allspice, cinnamon and clove bud, essential oils in edible apple films on physical properties and antimicrobial activities, *Journal of Food Science*, 74: M372–M378.

European Commission (2018a). A sustainable bioeconomy for Europe: Strengthening the connection between the economy, society and the environment, *Updated Bioeconomy Strategy*. <https://

ec.europa.eu/research/bioeconomy/index.cfm?pg=policy&lib=strategy>. Downloaded on 20 February 2019.

European Commission (2018b). Communication from the Commission to the European Parliament, the Council, the European Economic and Social Committee and the Committe of the Regions, *A European Strategy for Plastics in a Circular Economy*, COM/2018/028 final. Downloaded on 20 February 2019.

European Commission (2018c). Proposal for a Directive of the European Parliament and of the council on the reduction of the impact of certain plastic products on the environment, *COM*, (2018) 340 final. Downloaded on 20 February 2019.

Fabra, M.J., A. Hambleton, P. Talens, F. Debeaufort and A. Chiralt (2011). Effect of ferulic acid and α-tocopherol antioxidants on properties of sodium caseinate edible films, *Food Hydrocolloids*, 25: 1441–1447.

FAOSTAT-Food and Agriculture Organization of United Nations. <http://www.fao.org/faostat/en/#data/QC/visualize>. Downloaded on 25 February 2010.

Folk, J.E. (1983). Mechanism and basis for specificity of transglutaminase-catalyzed ε-(γ-glutamyl) lysine bond formation, *Advances in Enzymology*, 54: 1–57.

Frankel, E.N., S. Huang, E. Prior and R. Aeschbach (1996). Evaluation of antioxidant activity of rosemary extracts, carnosol and carnosic acid in bulk vegetable oils and fish oil and their emulsions, *Journal of the Science of Food and Agriculture*, 72: 201–208.

Ganiari, S., E. Choulitoudi and V. Oreopoulou (2017). Edible and active films and coatings as carriers of natural antioxiants for lipid food, *Trends in Food Science & Technology*, 68: 70–82.

Gharnit, N., N. El Mtili, A. Ennabili and F. Sayah (2006). Pomological characterization of carob tree (*Ceratonia siliqua* L.) from the province of Chefchaouen (NW of Marocco), *Moroccan Journal of Biology*, 2: 1–11.

Giménez, B., A. Alemán, P. Montero and M.C. Gómez-Guillén (2009). Antioxidant and functional properties of gelatin hydrolysates obtained from skin of sole and squid, *Food Chemistry*, 114: 976–983.

Giménez, B., A. López de Lacey, E. Pérez-Santín, M.E. López-Caballero and P. Montero (2013). Release of active compounds from agar and agar-gelatin films with green tea extract, *Food Hydrocolloids*, 30: 264–271.

Gomez-Guillen, M.C., B. Gimenez, M.E. Lopez-Caballero and M.P. Montero (2011). Functional and bioactive proties of collagen and gelatin from alternative sources: A review, *Food Hydrocolloids*, 25: 1813–1827.

Gómez-Guillén, M.C., M. Pérez-Mateos, J. Gómez-Estaca, E. López-Caballero, B. Giménez and P. Montero (2009). Fish gelatin: A renewable material for developing active biodegradable films, *Trends in Food Science and Technology*, 20: 3–16.

Gómez-Guillén, M.C., M. Ihl, V. Bifani, A. Silva and P. Montero (2007). Edible films made from tuna-fish gelatin with antioxidant extracts of two different murta ecotypes leaves (*Ugni molinae* Turcz.), *Food Hydrocolloids*, 21: 1133–1143.

Gómez-Guillén, M.C., M.E. López-Caballero, A. López de Lacey, A. Alemán, B. Giménez and P. Montero (2010). Antioxidant and antimicrobial peptide fractions from squid and tuna skin gelatin, pp. 89–115. *In:* E. Le Bihan and N. Koueta (Eds.), *Sea by-Products as a Real Material: New Ways of Application*, Chap. 7, Kerala, India: Transworld Research Network Singpost.

Goulas, V., E. Stylos, M.V. Chatziathanasiadou, T. Mavromoustakos and A.G. Tzakos (2016). Functional components of carob fruit: Linking the chemical and biological space, *International Journal of Molecular Sciences*, 17: 1875.

Gülçin, I., I. Güngör, S. Beydemir, M. Elmastas and I. Küfrevioglu 2004. Comparison of antioxidant activity of clove (*Eugenia caryophylata Thunb*) buds and lavender (*Lavandula stoechas* L), *Food Chemistry*, 84: 393–400.

Haddar, A., S. Sellimi, R. Ghannouchi, O.M. Alvarez, M. Nasri and A. Bougatef (2012). Functional, antioxidant and film-forming properties of tuna-skin gelatin with a brown algae extract, *International Journal of Biological Macromolecules*, 51: 477–483.

Hernandez, E. (1994). Edible coatings for lipids and resins. pp. 279–304. *In:* J.M. Krochta, E.A. Baldwin and M.O. Nisperos-Carriedo (Eds.), *Edible Coatings and Films to Improve Food Quality*. Technomic Publishing Co, Lancaster.

Holley, R.A. and D. Patel (2005). Improvement in shelf-life and safety of perishable foods by plant essential oils and smoke antimicrobials, *Food Microbiology*, 22: 273–292.

Hosseini, S.F., M. Rezaei, M. Zandi and F.F. Ghavi (2016). Effect of fish gelatin coating enriched with oregano essential oil on the quality of refrigerated rainbow trout fillet, *Journal of Aquatic Food Product Technology*, 25: 835–842.

Hosseini, S.F. and M.C. Gómez-Guillén (2018). A state-of-the-art review on the elaboration of fish gelatin as bioactive packaging: Special emphasis on nanotechnology-based approaches, *Trends in Food Science and Technology*, 79: 125–135.

Huang, D., B. Ou and R.L. Prior (2005). The Chemistry behind antioxidant capacity assays, *Journal of Agriculture Food Chemistry*, 53(6): 1841–56.

Iturriaga, L., I. Olabarrieta and I.M. de Marañón (2012). Antimicrobial assays of natural extracts and their inhibitory effect against *Listeria innocua* and fish spoilage bacteria, after incorporation into biopolymer edible films, *International Journal of Food Microbiology*, 158: 58–64.

Kim, H., S. Beak and K.B. Song (2018). Development of a hagfish skin gelatin film containing cinnamon bark essential oil, *LWT – Food Science and Technology*, 96: 583–588.

Kim, S. and E. Mendis (2006). Bioactive compounds from marine processing byproducts—A review, *Food Research International*, 39: 383–393.

Klenow, S., M. Glei, B. Haber, R. Owen and B.L. Pool-Zobel (2008). Carob fiber compounds modulate parameters of cell growth differently in human HT29 colon adenocarcinoma cells than in LT97 colon adenoma cells, *Food and Chemical Toxicology*, 46: 1389–1397.

Klenow, S., F. Jahns, B.L. Pool-Zobel and M. Glei (2009). Does an extract of carob (*Ceratonia siliqua* L.) have chemoprewentiwe potential related to oxidatiwe stress and drug metabolism in human colon cells? *Journal of Agricultural and Food Chemistry*, 57: 2999–3004.

Kołodziejska, I., K. Kaczorowski, B. Piotrowska and M. Sadowska (2004). Modification of the properties of gelatin from skins of baltic cod (*Gadus morhua*) with transglutaminase, *Food Chemistry*, 86: 203–209.

Krochta, J.M., E.A. Baldwin and M.O. Nísperos-Carriedo (1994). *Edible Coatings and Films to Improve Food Quality*, CRS Press LLC, Boca Ratón, Florida.

Krochta, J.M. and C.D. De Mulder-Johnston (1997). Edible and biodegradable polymer films: Challenges and opportunities, *Food Technologies*, 51: 61–74.

Lee, O. and B. Lee (2010). Antioxidant and antimicrobial activities of individual and combined phenolics in *Olea europaea* leaf extract, *Bioresource Technology*, 101: 3751–3754.

Liu, F., R.J. Avena-Bustillos, C. Bilbao-Sainz, R. Woods, B. Chiou, D. Wood and F. Zhong (2017). Solution blow spinning of food-grade gelatin nanofibers, *Journal of Food Science*, 82: 1402–1411.

Liu, F., F. Türker Saricaoglu, R.J. Avena-Bustillos, D.F. Bridges, G.R. Takeoka, V.C.H. Wu and F. Zhong (2018a). Preparation of fish skin gelatin-based nanofibers incorporating cinnamaldehyde by solution blow spinning, *International Journal of Molecular Sciences*, 19: E618.

Liu, F., F.T. Saricaoglu, R.J. Avena-Bustillos, D.F. Bridges, G.R. Takeoka, V.C.H. Wu and F. Zhong (2018b). Antimicrobial carvacrol in solution blow-spun fish-skin gelatin nanofibers, *Journal of Food Science*, 83: 984–991.

Loullis, A. and E. Pinakoulaki (2018). Carob as cocoa substitute: A review on composition, health benefits and food applications, *European Food Research and Technology*, 244: 959–977.

Lowell, R.T., R.O. Smitherman and E.W. Shell (1978). Factors Determining the Maximum Possible Fish Catch, pp. 225–260. *In*: A.M. Altschul and H.L. Wilcke (Eds.). *New Protein Foods*, vol. 3: *Animal Protein Supplies, Part A*, Academic Press, New York, USA.

Mamone, G., L. Sciammaro, S. De Caro, L. Di Stasio, F. Siano, G. Picariello and M.C. Puppo (2019). Comparative analysis of protein composition and digestibility of *Ceratonia siliqua* L. and *Prosopis* spp. seed germ flour, *Food Research International*, 120: 188–195.

Martin-Belloso, O., M.A. Rojas-Graü and R. Soliva-Fortuny (2009). Delivery of Flavor and Active Ingredients Using Edible Films and Coatings, Chap. 10. pp. 295–309. *In*: M.E. Embuscado and K.C. Huber (Eds.). *Edible Films and Coatings for Food Applications*, Springer, Dordrecht, Heidleberg, London, New York.

Martín-Diana, A.B., N. Izquierdo, I. Albertos, M.S. Sanchez, A. Herrero, M.A. Sanz and D. Rico (2017). Valorization of carob's germ and seed peel as natural antioxidant ingredients in gluten-free crackers, *Journal of Food Processing and Preservation*, 41: E12770.

Martínez-Villaluenga, C., E. Peñas, D. Rico, A.B. Martin-Diana, M.P. Portillo, M.T. Macarulla and J. Miranda (2018). Potential usefulness of a wakame/carob functional snack for the treatment of several aspects of metabolic syndrome: From *in vitro* to *in vivo* studies, *Marine Drugs*, 16: 512.

Martins, J.T., M.A. Cerqueira and A.A. Vicente (2012). Influence of α-tocopherol on physicochemical properties of chitosan-based films, *Food Hydrocolloids*, 27: 220–227.

Mathew, S. and T.E. Abraham (2008). Characterisation of ferulic acid incorporated starch-chitosan blend films, *Food Hydrocolloids*, 22: 826–835.

McHugh T.H., R.J. Avena-Bustillos and J.M. Krochta (1993). Hydrophilic edible films: Modified procedure for water vapor permeability and explanation of thickness effects, *Journal of Food Science*, 58: 899–903.

McHugh, T.H., R.J. Avena-Bustillos and W. Du (2009). Extension of shelf-life and control of human pathogens in produce by antimicrobial edible films and coatings, pp. 225–239, *Microbial Safety of Fresh Produce*.

Mendis, E., N. Rajapakse and S. Kim (2005). Antioxidant properties of a radical-scavenging peptide purified from enzymatically prepared fish skin gelatin hydrolysate, *Journal of Agricultural and Food Chemistry*, 53: 581–587.

Mills, S., L. Serrano, C. Griffin, P.M. O'connor, G. Schaad, C. Bruining et al. (2011). Inhibitory activity of *Lactobacillus plantarum* LMG P-26358 against *Listeria innocua when* used as an adjunct starter in the manufacture of cheese, *Microbial Cell Factories*, 10: S7.

Miñarro, B., E. Albanell, N. Aguilar, B. Guamis and M. Capellas (2012). Effect of legume flours on baking characteristics of gluten-free bread, *Journal of Cereal Science*, 56: 476–481.

Moradi, M., H. Tajik, S.M. Razavi Rohani, A.R. Oromiehie, H. Malekinejad, J. Aliakbarlu and M. Hadian (2012). Characterization of antioxidant chitosan film incorporated with *Zataria multiflora* Boiss. essential oil and grape seed extract, *LWT –Food Science and Technology*, 46: 477–484.

Morrissey, M.T., J. Lin and A. Ismond (2005). Waste management and by-product utilization, pp. 279–323. *In:* J.W. Park. (Ed.). *Surimi and Surimi Seafood*, Florida, CRC Press, Taylor & Francis Group.

Murrieta-Martínez, C.L., H. Soto-Valdez, R. Pacheco-Aguilar, W. Torres-Arreola, F. Rodríguez-Felix and E. Márquez Ríos (2018). Edible protein films: Sources and behavior, *Packaging Technology and Science*, 31: 113–122.

Nieto, M.B. (2009). Structure and function of polysaccharide gum-based edible films and coatings, Chap. 3, pp. 57–112. *In:* M.E. Embuscado and K.C. Huber (Eds.). *Edible Films and Coatings for Food Applications*, Springer, Dordrecht, Heidleberg, London, New York.

Núñez-Flores, R., B. Giménez, F. Fernández-Martín, M.E. López-Caballero, M.P. Montero and M.C. Gómez-Guillén (2012). Role of lignosulphonate in properties of fish gelatin films, *Food Hydrocolloids*, 27: 60–71.

Núñez-Flores, R., B. Giménez, F. Fernández-Martín, M.E. López-Caballero, M.P. Montero and M.C. Gómez-Guillén (2013). Physical and functional characterization of active fish gelatin films incorporated with lignin, *Food Hydrocolloids*, 30: 163–172.

Orliac, O., A. Rouilly, F. Silvestre and L. Rigal (2002). Effects of additives on the mechanical properties, hydrophobicity and water uptake of thermo-molded films produced from sunflower protein isolate, *Polymer*, 43: 5417–5425.

Ortega, N., A. Macià, M. Romero, E. Trullols, J. Morello, N. Anglès and M. Motilva (2009). Rapid determination of phenolic compounds and alkaloids of carob flour by improved liquid chromatography tandem mass spectrometry, *Journal of Agricultural and Food Chemistry*, 57: 7239–7244.

Papaefstathiou, E., A. Agapiou, S. Giannopoulos and R. Kokkinofta (2018). Nutritional characterization of carobs and traditional carob products, *Food Science and Nutrition*, 6: 2151–2161.

Pavlath, A.E. and W. Orts (7878). Edible films and coatings: why, what and how? Chap. 1, pp. 1–23. *In:* M.E. Embuscado and K.C. Huber (Eds.). 2009. *Edible Films and Coatings for Food Applications*, Springer, Dordrecht, Heidleberg, London, New York.

Perez-Gago, M.B., M. Serra and M.A.D. Río (2006). Color change of fresh-cut apples coated with whey protein concentrate-based edible coatings, *Postharvest Biology and Technology*, 39: 84–92.

Popović, S., D. Peričin, Z. Vaštag, V. Lazić and L. Popović (2012). Pumpkin oil cake protein isolate films as potential gas barrier coating, *Journal of Food Engineering*, 110: 374–379.

Rhim, J., H.M. Park and C.S. Ha (2013). Bio-nanocomposites for food packaging applications, *Progress in Polymer Science*, 38: 1629–1652.

Rico, D., A. Alonso de Linaje, A. Herrero, C. Asensio-Vegas, J. Miranda, C. Martínez–Villaluenga, A.B. Martin-Diana (2018). Carob by-products and seaweeds for the development of functional bread, *Journal of Food Processing and Preservation*, 42: E13700.

Rico, D., A.B. Martin-Diana, A. Lasa, L. Aguirre, I. Milton-Laskibar, D.A. de Luis and J. Miranda (2019a). Effect of wakame and carob pod snacks on non-alcoholic fatty liver disease, *Nutrients*, 11: 86.

Rico, D., A.B. Martín-Diana, C. Martínez-Villaluenga, L. Aguirre, J.M. Silván, M. Dueñas and Lasa, A. (2019b). *In vitro* approach for evaluation of carob by-products as source bioactive ingredients with potential to attenuate metabolic syndrome (MetS), *Heliyon*, 5: E01175.

Roseiro, L.B., L.C. Duarte, D.L. Oliveira, R. Roque, M.G. Bernardo-Gil, A.I. Martins et al. (2013). Supercritical, ultrasound and conventional extracts from carob (*Ceratonia siliqua* L.) biomass: Effect on the phenolic profile and antiproliferative activity, *Industrial Crops and Products*, 47: 132–138.

Ruiz-Navajas, Y., M. Viuda-Martos, E. Sendra, J.A. Perez-Alvarez and J. Fernández-López (2013). *In vitro* antibacterial and antioxidant properties of chitosan edible films incorporated with *Thymus moroderi* or *Thymus piperella* essential oils, *Food Control*, 30: 386–392.

Salgado, P.R., C.M. Ortiz, Y.S. Musso, L. Di Giorgio and A.N. Mauri (2015). Edible films and coatings containing bioactives, *COFS Current Opinion in Food Science*, 5: 86–92.

Salinas, M.V., B. Carbas, C. Brites and M.C. Puppo (2015). Influence of different carob fruit flours (*Ceratonia siliqua* L.) on wheat dough performance and bread quality, *Food and Bioprocess Technology*, 8: 1561–1570.

Samelis, J. and J. Sofos (2003). Organic acids, pp. 98–132. *In:* S Roller (Ed.). *Natural Antimicrobials for the Minimal Processing of Foods*, Woodhead Publishing, Cambridge, UK.

Sánchez-González, L., M. Vargas, C. González-Martínez, A. Chiralt and M. Cháfer (2011). Use of essential oils in bioactive edible coatings: A review, *Food Engineering Reviews*, 3: 1–16.

Sánchez-Ortega, I., B.E. García-Almendárez, E.M. Santos-López, A. Amaro-Reyes, J.E. Barboza-Corona and C. Regalado (2014). Antimicrobial edible films and coatings for meat and meat products preservation, *Scientific World Journal*, 2014: 18.

Siano, F., L. Sciammaro, M.G. Volpe, G. Mamone, M.C. Puppo and G. Picariello (2018). Integrated analytical methods to characterize lipids from *Prosopis* spp. and *Ceratonia siliqua* seed germ flour, *Food Analytical Methods*, 11: 3471–3480.

Sheikh, I.A., M. Yasir, I. Khan, S.B. Khan, N. Azum, E.H. Jiffri, M.A. Kamal, G.M. Ashraf, and M.A. Beg (2018). Lactoperoxidase immobilization on silver nanoparticles enhaces its antimicrobial activity, *Journal of Dairy Research*, 85: 460–464.

Silva-Weiss, A., M. Ihl, P.J.A. Sobral, M.C. Gómez-Guillén and V. Bifani (2013). Natural additives in bioactive edible films and coatings: Functionality and applications in foods, *Food Engineering Reviews*, 5: 200–216.

Šuput, D.Z., V.L. Lazić, S.Z. Popović and N.M. Hromiš (2015). Edible films and coatings–Source, properties and application, *Food and Feed Research*, 42: 11–22.

Tongnuanchan, P., S. Benjakul and T. Prodpran (2014). Comparative studies on properties and antioxidative activity of fish skin gelatin films incorporated with essential oils from various sources, *International Aquatic Research*, 6: 1–12.

Wieprecht, T., M. Dathe, R.M. Epand, M. Beyermann, E. Krause, W.L. Maloy and M. Bienert (1997). Influence of the angle subtended by the positively charged helix face on the membrane activity of amphipathic antibacterial peptides, *Biochemistry*, 36: 12869–12880.

Wiriyaphan, C., B. Chitsomboon and J. Yongsawadigul (2012). Antioxidant activity of protein hydrolysates derived from threadfin bream surimi byproducts, *Food Chemistry*, 132: 104–111.

Wong, D.W.S., S.J. Tillin, J.S. Hudson and A.E. Pavlath (1994). Gas exchange in cut apples with bilayer coatings, *Journal of Agricultural and Food Chemistry*, 42: 2278–2285.

Worm, B., H.K. Lotze, I. Jubinville, C. Wilcox and J. Jambeck (2017). Plastic as a persistent marine pollutant, *Annual Review of Environment and Resources,* 42: 1–26.

Yoon, S.D. (2014). Crosslinked potato starch- based blend films using ascorbic acid as a plasticizer, *Journal of Agricultural and Food Chemistry,* 62: 1755–1769.

Zunft, H.J.F., W. Lüder, A. Harde, B. Haber, H.J. Graubaum, C. Koebnick and J. Grünwald (2003). Carob pulp preparation rich in insoluble fibre lowers total and LDL cholesterol in hypercholesterolemic patients, *European Journal of Nutrition,* 42: 235–242.

CHAPTER 7

Biodegradable Polymer Blends for Food Packaging Applications

Vijaya K. Rangari, *Manik C. Biswas, Boniface J. Tiimob*
and *Chibu Umerah*

1. Introduction

Monomer of biopolymers is covalently bonded with each other forming chain-like molecules. The word 'bio' represents the biodegradability of biopolymers. Natural bio-organisms help to promote the degradation or breaking down into small molecules, leaving behind some organic by-products, such as CO_2, H_2O, etc., which are safe for sustainable and eco-friendly environment. Biopolymers are abundant in nature, renewable sources and degradable in nature and these are prime characteristics of biopolymers which are replacing current packaging plastics made from petroleum sources (Sikorska et al., 2017; Siracusa et al., 2008).

Most widely used biopolymers for food packaging applications are natural biopolymers, such as cellulose, chitosan, starch and protein derivatives though they possess poor mechanical and barrier properties. Nowadays, technological advancement triggers researchers to invent synthetic biopolymers with improved mechanical and barrier properties, which not only overcome the drawbacks of natural polymers but also include other properties that help to enhance food safety, quality and shelf-life.

Biopolymers can be categorized into two broad groups: natural and synthetic. Microorganisms based on the origin of raw materials are depicted in Fig. 1. Natural polymers include cellulose, chitosan, starch, carrageenan, agar, alginate, etc., from plant carbohydrates and whey protein, soy protein, gluten, collagen, zein, casein,

Department of Materials Science and Engineering, Tuskegee University, Tuskegee, USA.
* Corresponding author: vrangari@tuskegee.edu

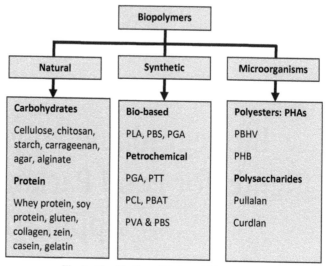

Fig. 1. Classification of biodegradable polymers.

gelatin, etc., from plant or animal protein. Synthetic biopolymers include poly(L-lacticide) (PLA), poly(glycolic acid) (PGA), poly(e-caprolactone) (PCL), poly(vinyl alcohol) (PVA), poly(butylene succinate) (PBS), etc. Finally, biopolymers formed by microbial fermentation include microbial polyesters, such as poly(hydroxyalkanoates) (PHAs) including poly(3-hydroxybutyrate-co-3-hydroxyvalerate) (PBHV), poly(β-hydrocxybutyrate) (PHB), etc., and microbial polysaccharides, such as pullalan and curdlan.

Besides, biodegradable polymers, compostable polymers can be prominent alternatives of current packaging plastics. But the polymers should be disintegrated as a whole within a short period of time, i.e., six months, leaving no eco-toxic residues in the environment (Siracusa et al., 2008). There are so many factors that affect the composting process, such as chemical structure, molecular weight, degree of polymerization, branching, crystallinity, functional groups, hydrophilicity, presence of additives/fillers in the polymers and environmental conditions, such as temperature, humidity, pressure, presence of light, microorganisms, pH and salts which accelerate the composting process (Musiol et al., 2011). There is a possibility of migration of the fillers/additives (plasticizers, antiplasticizers, coupling agents, dyes, modifiers, catalysts, etc.), into the environment, which trigger the necessity to take into consideration the biodegradable additives during polymer blend preparation.

1.1 Natural Polymers

All-natural polymers are biodegradable without leaving any eco-toxic residue in the environment due to their natural origin. There are some enzymes (i.e., depolymerase) which catalyze the degradation process in order to balance the nature. Natural biopolymers are considered as prominent alternatives of plastics for food packaging applications due to their excellent properties, such as biodegradability, renewable origin, abundance, non-toxic, antimicrobial and antioxidant properties and being ecofriendly (Sam et al., 2016).

1.1.1 Cellulose

One of the most abundant biopolymers in nature is cellulose. It consists of hundreds or thousands of β-(1→4)-linked glucose units and is found in the cell walls of all plants, sea animals, bacteria, fungi and some amoeba (Kamel et al., 2008; Ozilgen and Bucak, 2018). The strong inter and intra H-bonding plays a prime role in the high tensile strength, high crystallinity and poor solubility of the polymers. Therefore, it is difficult to use as packaging material in the food industry. Recently, researchers are concentrating on modifying the cellulose structure by replacing the hydroxyl groups with acetate or methyl functional groups. Hence the molecules become ionic and thereby water soluble (Fig. 2).

The most widely used cellulose derivatives are cellulose acetate, methylcellulose (MC), carboxymethylcellulose (CMC), hydroxypropyl cellulose (HPC), hydroxypropyl-methylcellulose (HPMC). Marrez et al. demonstrated the fabrication of sustainable nanocomposites for antimicrobial food-packaging applications using cellulose acetate (CA) and nanosilver. They investigated the antibacterial effect of the composites films against the most common food-borne pathogens and observed high antibacterial activity against *St. aureus*, *B. cereus*, *S. typhi*, *E. coli*, *K. pneumoniae* but low activity against the two strains of *Pseudomonase* spp. They also examined the release of silver nanoparticles (Ag NPs) into the food products from different CA-Ag nanocomposites during the shelf-life of packaging and found that the amount of released Ag NPs is less than the permissible limit which basically increases the possibility of using CA-Ag nanocomposites as a prominent food-packaging material (Marrez et al., 2019). Dairi et al. fabricated ternary hybrid nanocomposite films using plasticized cellulose acetate (CA) with triethyl citrate, incorporating AgNPs/ gelatin-MMt and thymol (Th). The fabricated films showed high tensile strength, UV resistance, oxygen barrier and antimicrobial and antifungal activity, which increase the possibility of the use of ternary blend system as an active food-packaging system. They found that the inclusion of clay into the ternary system enhanced the tensile and oxygen barrier properties, improved UV resistance with the extension of silver release period from nanocomposites, which eventually extended the shelf-life of the food products (Dairi et al., 2019). Researches also studied carboxymethyl cellulose (CMC)-based nanocomposites for active food packaging material reported in

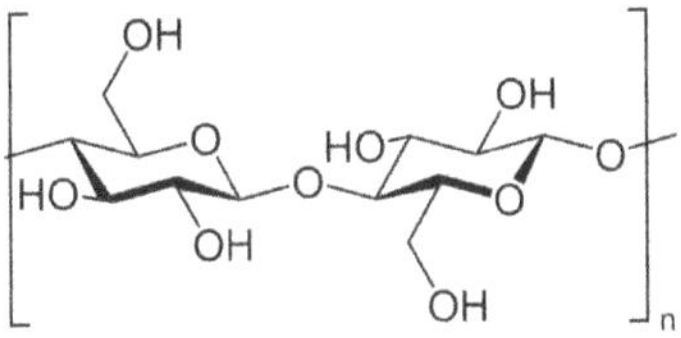

Fig. 2. Structure of cellulose.

the literatures (Chi and Catchmark, 2018; Shahbazi, 2018). Šešlija et al. prepared pectin/carboxymethyl cellulose composite films induced with cross-linking agents (Ca^{+2} ions) via solvent casting method and observed that the films showed enhanced mechanical properties compared to pristine pectin films. The TGA analysis revealed that the presence of CMC in the pectin matrix improved the thermal stability due to high interaction between pectin and CMC polar groups. In addition, the presence of cross-linkers improved water vapor permeability which varied from $1.32 \times 10-7$ to $2.03 \times 10-7 \, g/m \, h \, Pa$ due to increased free volume during fabrication of the films (Šešlija et al., 2018).

1.1.2 Chitosan

Chitosan is another most abundant linear non-toxic biopolymer in nature. It consists of β-(1→4)-linked 2-amino-2-deoxy-D-glucose (D-glucosamine, deacetylated) and 2-acetamido-2-deoxy-D-glucose (N-acetyl D-glucosamine, acetylated) units and is found in the shells of shrimp, lobsters, crabs and oysters (Fig. 3). There are many factors which determine chitosan's properties, such as degree of deacetylation, molecular weight, chemical structure, viscosity, etc. The low degree of acetylation and molecular weight is responsible for its solubility and degradation which is faster than that of its counterparts. The presence of amine group in the structure determines its solubility in many organic solvents. In addition, antimicrobial property of chitosan reported in the literature widens its application in the food industry (Beck et al., 2019; Benhabiles et al., 2012; Chung and Chen, 2008; Gomes et al., 2019; Lin et al., 2019).

Haghighi et al. developed both blend and bi-layer composite films using chitosan and gelatin induced with lauroyl arginate ethyl (LAE). The fabricated blend films exhibited high tensile strength and elastic modulus, but lower water vapor permeability than bilayer films (p<0.05), whereas bilayer films showed high UV resistance with lower transparency values (p<0.05). This is due to the high physical interaction between chitosan and gelatin determined by FTIR. The antimicrobial effect of the films integrated with LAE (0.1 per cent v/v) was investigated against common food-borne pathogens, such as Listeria monocytogenes, *Escherichia coli, Salmonella typhimurium* and *Campylobacter jejuni* and observed excellent inhibition properties (Haghighi et al., 2019).

Indumathi et al. created biodegradable food packaging films from mahua-oil-based polyurethane and chitosan integrated with ZnO nanoparticles. The films induced with ZnO nanoparticles exhibited enhanced tensile strength and stiffness, excellent antibacterial and barrier properties and hydrophobicity of the film. They investigated the shelf-life extension study on wrapped carrot pieces and found that the films extend the shelf-life up to nine days as compared to traditional plastics (Indumathi and Rajarajeswari, 2019).

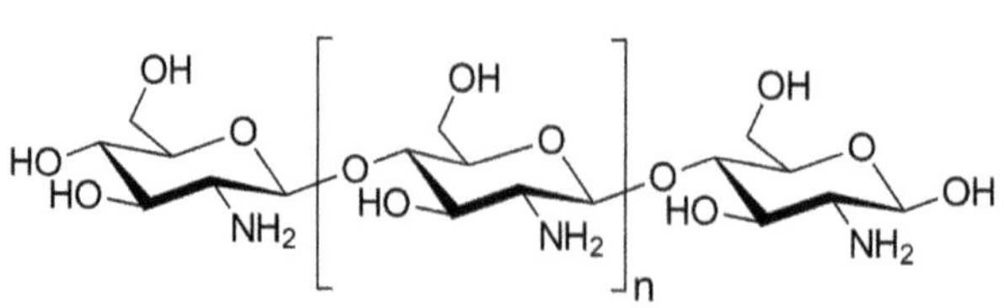

Fig. 3. Structure of chitosan.

1.1.3 Starch

Starch is polymeric carbohydrate consisting of glucose units (amylose and amylopectin) bonded via glycosidic linkage and can be found in different shapes and sizes, such as spheres, polygons, platelets, etc., in rice, corn, potato, wheat, oat, barley and soybean as energy storage materials. The amylose is linear and helical, but amylopectin is branched in structure as shown in Fig. 4. Due to structural heterogeneity, starch with high percentage of amylose forms strong, coherent, colorless and odorless films having more linearity compared to amylopectin. As amylopectin is branched, it forms brittle and anisotropic films with low mechanical strength but can be modified or the poor properties of the films can be improved by using plasticizers.

Starch-based biopolymer films are colorless, odorless, good barrier to oxygen and tasteless, makingthem a prominent candidate for food-packaging applications (Katerinopoulou et al., 2019; Pelissari et al., 2019).

Gopi et al. fabricated potato starch (PS), tapioca starch (TS) and chitosan-based nanocomposites integrated with turmeric nanofiber (TNF) and found that the formation of hydrogen bonds between starch-TNF and chitosan-TNF play a prime role in enhancing the tensile strength and Young's modulus. The composite films showed excellent thermal stability, which can be attributed to the limited mobility of starch polymer chains in the matrix due to chemical interaction of TNF and starch. The fabricated films exhibited excellent antimicrobial activities against *B. cereus*,

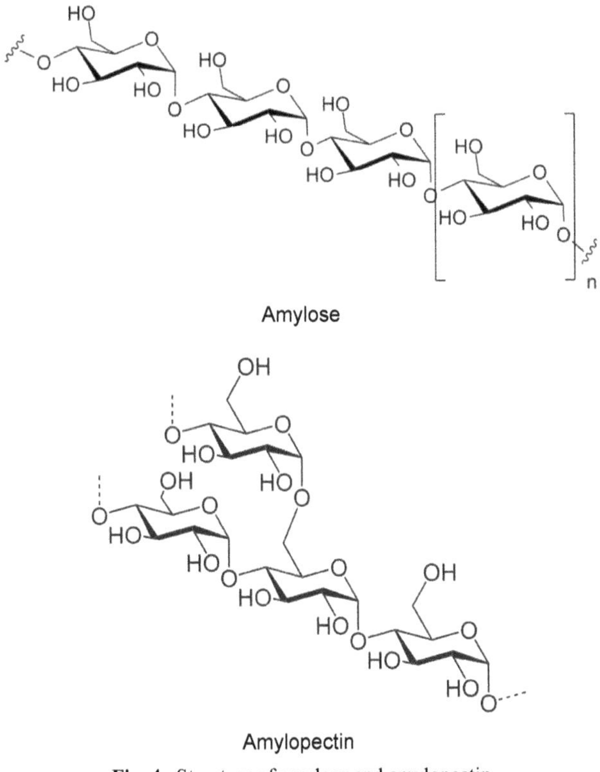

Amylose

Amylopectin

Fig. 4. Structure of amylose and amylopectin.

E. coli, *S. aureus* and *S. typhimurium* due to the inclusion of TNF in the biopolymer matrixes (Gopi et al., 2019).

1.1.4 Carrageenan

Carrageenan has high molecular weight, linear sulfated polysaccharides consisting of α-d-(1→3) and β-(1→4) linked galactopyranosyl and 3,6 anhydrod-galactopyranosyl units via glycosidic linkage and found in red edible seaweeds. It has the ability to form gel and thicken the food products due to their long flexible and helical structures. Based on the position of the ester sulfate group in the monomer, carrageenan isclassified into three different forms, named as k, ι, and λ shown in Fig. 5 (Dul et al., 2015).

   Melanin-capped ZnONP (mZnONP) was prepared with zinc acetate and KOH, using melanin as a capping agent. ZnONP prepared without a capping agent was rod-like, but mZnONP was flake-like. ZnONP and mZnONP were uniformly distributed in the carrageenan polymer matrix to form Carr/ZnONP and Carr/mZnONP nanocomposite films. The Carr/ZnONP nanocomposite films exhibited high UV barrier property with increased mechanical strength, water vapor barrier, and thermal stability properties compared to the neat carrageenan film. However, the Carr/mZnONP composite film showed higher mechanical, water vapor barrier and thermal stability properties than the Carr/ZnONP film. Both nanocomposite films showed strong antibacterial activity against *E. coli* but showed only slight antibacterial activity against *L. monocytogenes*. The Carr/

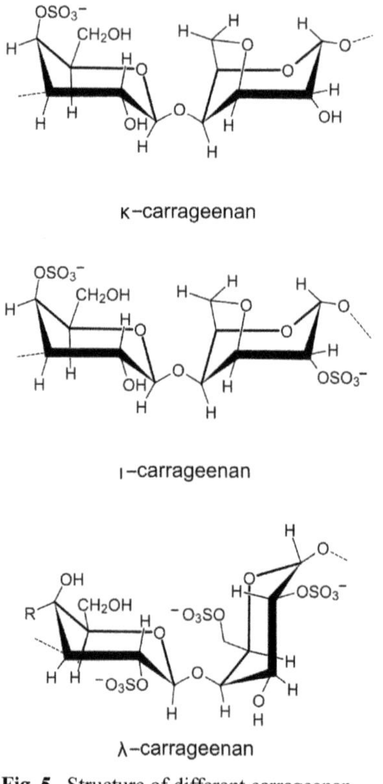

κ–carrageenan

ι–carrageenan

λ–carrageenan

Fig. 5. Structure of different carrageenan.

mZnONP nanocomposite film with increased physical and functional properties can be applied in the food packaging and biomedical fields (Roy and Rhim, 2019).

1.1.5 Agar

Agar consists of linear polysaccharides, agarose and heterogeneous small molecules, agaropectin and is found in the cell walls of certain red algae, such as Gelidium, Pterocladia, Gracilaria, etc. The linear agarose (Fig. 6) is responsible for forming the gel structure of agar with high melting temperature (85°C) as compared to gel formation temperature (32–40°C). It has substantial application in the food industry due to its gelation activity (Hou et al., 2019; Roy et al., 2019; Zhang et al., 2019).

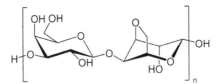

Fig. 6. Structure of agarose.

1.1.6 Alginate

Alginate is copolymer of α-l-guluronic acid and β-d-mannuronic acid units at various ratios, linked through linear β-(1→4) glycosidic linkage and found in some brown algae or bacteria. It is a very promising candidate for biodegradable film formation due to its non-toxicity, biodegradability, compatibility, low cost and simplicity (Vu and Won, 2013). The ratio of two monomeric units in the polymer depends on the source and growing condition, which determines the physical properties, such as solubility. Guluronic acid is more soluble than mannuronic acid. That is why alginates with higher percentage of guluronic acid exhibit more solubility, which helps to widen its application in the food industry (de Oliveira Filho et al., 2019; Fabra et al., 2018; Han et al., 2018).

1.1.7 Protein

Whey protein: Whey is the protein-rich secondary product of cheese and casein production and treated as waste for land filling or feeding materials for pig. It consists of 20 per cent of mild protein and 80 per cent of other four different kinds of proteins, which are β-lactoglobulin (β-LG), α-lactalbumin (α-LA), bovine serum albumin (BSA) and immunoglobulins (Ig). To avoid the disposal cost and environmental pollution, it is necessary to find out alternative uses of whey protein, especially its protein and lactose can be used as food products. Its wide application fields include confectionary, stationary, health food, etc. Whey protein can be used to engineer to form biodegradable films possessingd good barrier properties, odorless, tasteless and flexible edible biofilms. Their barrier property can also be enhanced by using plasticizers which is one of the most attractive properties in food-packaging application. It can be used as a coating on polymeric substrates, having excellent mechanical and physical properties (Galus and Kadzińska, 2019; Hassannia-Kolaee et al., 2016; Oymaci and Altinkaya, 2016).

Soy protein: Soy protein is isolated from soybean via dehulled and defatted techniques. Among the soy proteins, soy protein isolate is a highly refined form with 90 per cent protein content on moisture-free basis whereas the concentrate has 70 per cent flour and 50 per cent protein content. Soy proteins can also be categorized into 2S, 7S, 11S and 15S based on ultracentrifugation sedimentation rate, which is in Svedberg (S) numbers. The highest the number, the larger is the protein. Among them, 7S (conglycinin) and 11S (glycinin) protein fractions are the key elements of soy protein. Glycinin protein fraction contains 20 intramolecular disulfide bonds whereas conglycinin has limited disulfide bonds in its structure. Soy protein-based food packaging films attract much attention nowadays because of it biodegradability, inexpensiveness, non-toxicity and biocompatibility. Unfortunately, the composite films exhibit low mechanical properties, poor resistance to moisture and oxygen and this restricts its application widely in the packaging field. Researchers are trying to develop composite films induced with fillers, plasticizers to improve its poor structural properties. Yu et al. developed soy protein-based composite film for the first time and integrated it with cellulose nanocrystals (CNCs) and pine needle extract (PNE). It showed high tensile strength and high moisture resistance, which is an ideal requirement for active packaging materials and could be attributed to the inclusion of CNC nanocrystals, which minimize the formation of N-H bonds between protein and water molecules. In addition, PNE increases the hydrophobicity of the composites films and eventually decreases the water vapor permeability and shows antioxidant activity (Yu et al., 2018). González et al. also found similar properties when they developed soy protein isolate (SPI)-based active packaging films induced with starch nanocrystals (SNC) (González and Igarzabal, 2015).

Collagen: It is the most abundant fibrous protein found in tendon, ligaments, hides, cartilage, bone, skin and connective tissues. It consists of two different types of covalently bonded intra- and inter-molecular cross-linked chains, such as α and ß chains of amino acids with 100 kDa and 200 kDa molecular weight respectively, forming triple helix. Depending on the sources, collagen is mostly rich in glycine (33 per cent) and proline/hydroproline (20 per cent) among the total amino acids. Collagen-based composite films showed excellent properties, such as high tensile strength, poor water solubility, biodegradability and poor water vapor permeability which helps in getting noticeable attention in the packaging industry (Valencia et al., 2019; Wu et al., 2018). Bhuimbar et al. fabricated collagen-based packaging films extracted from Medusa skin, a black ruff fish using lactic acid with highest yield percentage. They incorporated chitosan and pomegranate peel extract into the matrix to improve the moisture-resistance property, declined water solubility and high antimicrobial activity against common food-borne pathogens. They invested the antibacterial activity against *Bacillus saprophyticus* LNB 333F5, *Bacillus subtilis* NCIM 2635, *Salmonella typhi* NCIM 2501 and *Escherichia coli* NCIM 2832 and found the performance mainly depending on the percentage of pomegranate peel extract in the composite films (Bhuimbar et al., 2019).

Zein: Zein is prolamin protein fraction derived from corn, consisting of amino acids in the structure but rich in hydrophobic leucine, alanine and proline acids, which are responsible for its low solubility in water and high solubility in alcohol. Zein, a prolamin from corn-based films showed similar properties like collagen, such as high tensile strength, poor water solubility, biodegradability, poor water vapor permeability and high thermal stability which help in drawing attention in the packaging industry (Qu et al., 2019; Vahedikia et al., 2019; Xia et al., 2019). Due to the presence of non-polar amino acids, zein-based films exhibit hydrophobic nature, which improves the water-vapor resistance. However, zein-based films show poor mechanical properties, which restrict its broad application. Diverse approaches have been put forward to overcome these drawbacks of corn zein. Addition of plasticizers and hybridization with other polymers can be considered to get the benefits, such as sustainability, barrier to oxygen, moisture, etc., from zein. Vahedikia et al. developed bio-based zein composite films induced with cinnamon essential oil (CEO) and chitosan nano-particles (CNPs) at 2 per cent and 4 per cent (w/w) respectively, either alone or in combination and this improved the mechanical properties and showed excellent antimicrobial activity against *Escherichia coli* and *Staphylococcus aureus*. They observed that the inclusion of CEO alone or with combination of CNPs showed prominent results against the growth of microorganisms, but CNP-induced films showed no significant effect on those food-borne pathogens (Vahedikia et al., 2019). Chen et al. developed zein-chitosan-based films via plasma treatment process. They showed that the plasma treatment makes the zein molecules more orderly and increases the H-bond formation between zein and chitosan, which is the prime reason for its noticeable properties (Chen et al., 2019).

Casein: Casein is one of the milk proteins consisting 80 per cent total milk protein. Three different kinds of casein, such as α, ß, and κ-casein show different molecular weights varying from 19–25 kDa. Due to the presence of non-polar amino acids, casein-based films exhibit hydrophobic nature which improves the water vapor resistance (Herniou et al., 2019; Picchio et al., 2018a). Picchio et al. fabricated poly(n-butyl acrylate) (PBA)-casein-based nanocomposites through emulsion polymerization. The films showed high water vapor permeability with high casein content but high water resistance with low casein content due to favorable grafting with PBA at low percentages. Incorporation of casein in PBA polymer enhances the biodegradability of the synthetic polymers, widening the application area of synthetic polymers in food packaging (Picchio et al., 2018b).

Gelatin: It is the denatured derivatives of collagen possessing protein and polypeptides with different molecular weights that depend on the sources and synthesis process. It consists of three different types of covalently bonded intra- and inter-molecular cross-linked chains, such as α, ß and γ chains of amino acids with 100 kDa, 200 kDa and 300-kDa molecular weight, respectively. Depending on the molecular weights, properties of gelatin vary, including biofilm formation for food packaging application. Kaewprachu et al. developed gelatin-based packaging films induced with nisin and catechin, to investigate *in vitro* antioxidant and antimicrobial effects and extend the shelf-life of the packaged materials. They investigated the shelf-life extension test on pork meat packaged with gelatin-based films induced with nisin and catechin and found that the quality of meat was better maintained over five days more than the control films. This can be attributed to the high oxygen and CO_2 barrier properties of the composite films (Kaewprachu et al., 2018).

1.2 Synthetic Polymers

1.2.1 Poly(lactic acid) (PLA)

PLA is a biodegradable, thermoplastic, aliphatic polyester derived from not only synthetic sources but also renewable resources, such as cornstarch, sugarcane, cassava roots, etc. Researchers have studied PLA extensively in the last few decades for its wide application due to biodegradability in numerous areas, particularly food packaging and medical applications (Mallegni et al., 2018; Shankar et al., 2018). A binary polymer blend comprising agriculture-based PLA and PBAT induced with proteinaceous eggshell nanoparticles (PENP) has been reported in our previous work (Tiimob et al., 2017). In this work, we show that a certain ratio of PBAT and PLA, 70:30 is preferred to overcome the inferior properties of PLA film. This can be attributed to the carboxylic groups of eggshell nanoparticles which enhance the interfacial interaction between PLA and PBAT, eventually improving the structural integrity of the blend system.

1.2.2 Poly(butylene succinate) (PBS)

PLA is a biodegradable thermoplastic aliphatic polyester synthesized via (i) trans-esterification process or (ii) direct esterification process, starting from the diacid. The direct esterification of succinic acid with 1,4-butanediol is the most traditional route to produce PBS through catalytic hydrogenation. PBS exhibits biodegradability due to the presence of ester bonds and promising mechanical properties, which allow its wide application in various areas, such as agricultural mulch films, bags, packaging films and flushable products (Hongsriphan and Pinpueng, 2019; Hongsriphan and Sanga, 2018). To reduce the production cost and improve some poor properties, PBS can be hybridized with other polymers, such as polyethylene succinate (PES) and monomers like adipic acid poly(butylene succinate-co-adipate) (PBSA) (Kataoka et al., 2015; Krishnaswamy and Sun, 2010; Xue and Qiu, 2015). Siracusa et al. reported the behavior of PBS and PBSA food packaging materials with the help of food stimulant. They studied the permeability behavior and found that the crystalline/non-crystalline ratios affect the permeability, which triggers additional studies on PBS for food safety and shelflife extension (Siracusa et al., 2015).

1.2.3 Poly(glycolic acid) (PGA)

PGA or polyglycolide is also a biodegradable thermoplastic aliphatic polyester synthesized by polycondensation or ring-opening polymerization of glycolic acid. PGA can be derived from not only synthetic sources, but also renewable resources and exhibits excellent barrier properties due to high crystallinity to carbon dioxide and oxygen. As a result, PGA can use a shielding layer in multilayer packaging assemblies, such as PET bottle for beer or soft drinks (Coles and Kirwan, 2011).

1.2.4 Poly(trimethylene terephthalate) (PTT)

PTT is a linear thermoplastic polyester synthesized through condensation polymerization or transesterification of 1,3-propanediol and terephthalic acid or dimethyl terephthalate. The monomer 1,3-propanediol can be obtained from plant waste which makes PTT biodegradable and lies between PET and PBT. It exhibits excellent mechanical properties and can improve inferior properties using additives (Bikiaris et al., 2006; Essabti et al., 2018). These properties make PTT a potential alternative to PET or PBT as they are not biodegradable. Kim et al. investigated the PTT films as an alternative of PET and nylon 6 in processing of eco-friendly retort pouch. As PTT films possess high mechanical strength, they are suitable as a face substrate in the process. However, they have high mechanical biodegradability, but show high heat shrinkage and low impact strength, which restrict theirs application widely in the packaging industry. But they suggested that laminated PTT films with aluminum foil can minimize the heat shrinkage activity which overcomes the limitations of processing condition of PTT (Kim et al., 2018a).

1.2.5 Poly(ε-caprolactone) (PCL)

PCL is a biodegradable polyester synthesized through ring-opening polymerization of ε-caprolactone, using appropriate catalysts. It is semicrystalline in nature with low melting temperature, around 60°C, and a glass transition temperature of about −60°C that allows good processability. PCL packaging materials show high flexibility, good resistance to water and organic solvents but are susceptible to microbial attacks which trigger the biodegradability of PCL (Jiang et al., 2018; Lasprilla-Botero et al., 2018; Salević et al., 2019; Tampau et al., 2018).

1.2.6 Poly(butylene adipate-co-terephthalate) (PBAT)

PBAT is a biodegradable copolyester of adipic acid, 1,4-butanediol and terephthalic acid (Fig. 7). The rigid unit consists of 1,4-butanediol and terephthalic acid monomers, is responsible for stiffness whereas the flexible unit consists of 1,4-butanediol and adipic acid monomers and provides the flexibility of the polymer. PBAT is an excellent candidate for food packaging applications due to its flexibility and biodegradability. Nevertheless, it has high water vapor permeability and high cost restrict its application commercially. To reduce the production cost of PBAT-based films and moisture permeability, natural and synthetic additives/fillers can be introduced into the PBAT matrix to fabricate the biocomposites (Biswas et al., 2019; Tavares et al., 2018; Xie et al., 2018).

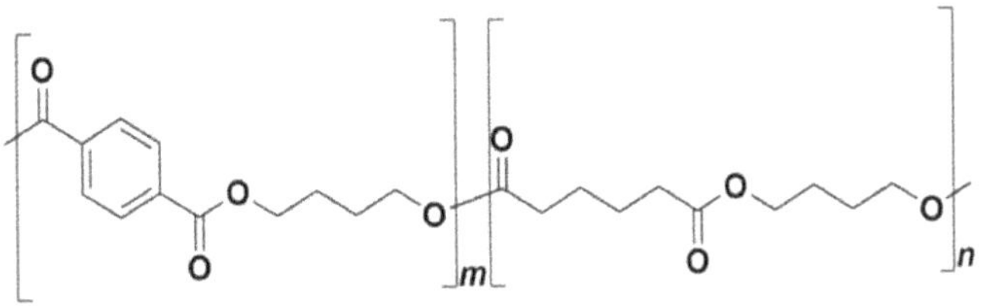

Fig. 7. Structure of PBAT.

1.2.7 Poly(vinyl alcohol) (PVA)

PVA is a nontoxic water-soluble manmade polymer through alkaline hydrolysis of polymerized polyvinyl acetate. It consists mainly of 1,3-diol linkages [-CH2-CH(OH)-CH2-CH(OH)-] with a very low amount of 1,2-diols [-CH2-CH(OH)-CH(OH)-CH2-], depending on the polymerization conditions of the vinyl ester precursor (Ameer, 2016). It exhibits excellent resistance to oxygen, carbon dioxide and aroma, which widen the use of PVA in food packaging applications, specially packaging films. This property, such as tensile strength, barrier properties decrease with the absorption of moisture, which act as plasticizer, restrict the application of PVA films at high humid condition. To enhance the poor mechanical and barrier properties, PVA can be blended with other polymers or filler materials can be added as cross-linkers in the matrix (Kim et al., 2018b; Liu et al., 2018; Tripathi et al., 2018; Zhen Yu et al., 2018).

1.3 Microorganisms

1.3.1 Polyhydroxyalkanoates (PHAs)

PHAs belong to the group of polyesters produced in nature by various microorganisms as a carbon and energy storage source. PHAs exhibit hard crystalline property to elastic nature, depending on the composition of monomers. They are biodegradable and can modify the properties of PHAs by blending with other polymers, fillers or enzymes to widen the use of PHAs in food packaging applications. The chemical structure of most widely used PHAs is shown in Fig. 8.

The microbial synthesis routes of PHAs show some limitations due to the presence of some impurities, like proteins and lipids, which produce asignificant odor and taste problem inthe food product. These limitations prompt the necessity to investigate and develop a standard route and treatment method to avoid the presence of impurities (Koller, 2019; Pérez-Arauz et al., 2019).

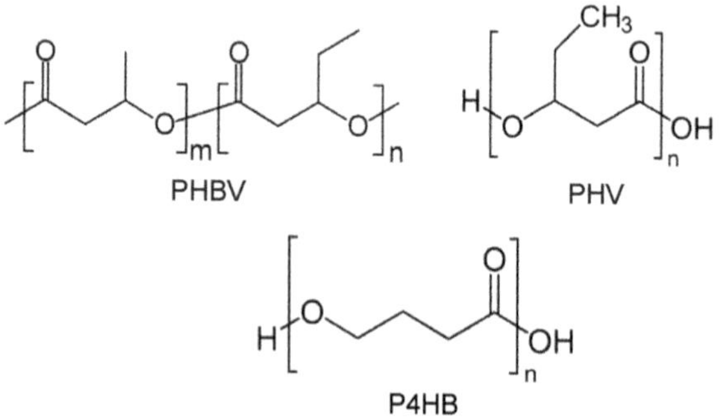

Fig. 8. Structure of different PHAs.

1.3.2 Polysaccharides

Pullulan: Pullulan is a polysaccharide produced from starch by the fungus *Aureobasidium pullulans* and consists of three glucose units known as maltotriose unit and connected through α-1,4 glycosidic linkage whereas connective maltotriose units are connected through α-1,6 glycosidic linkage (Fig. 9). This polymer is mostly tasteless and edible, allowing its application in the food industry for making edible films (Chu et al., 2019; Silva et al., 2018; Wang et al., 2019; Zhao et al., 2019).

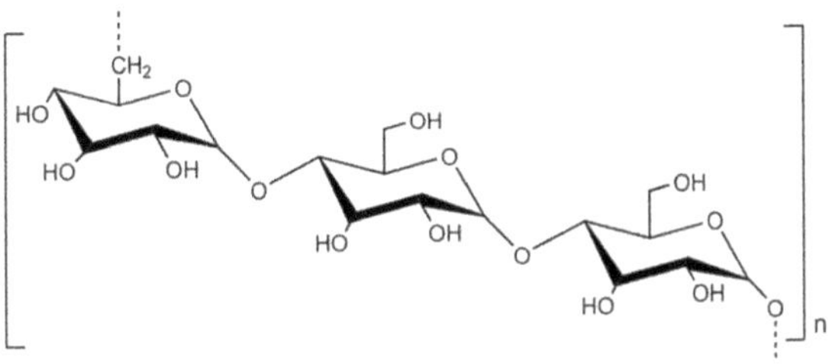

Fig. 9. Structure of pullulan.

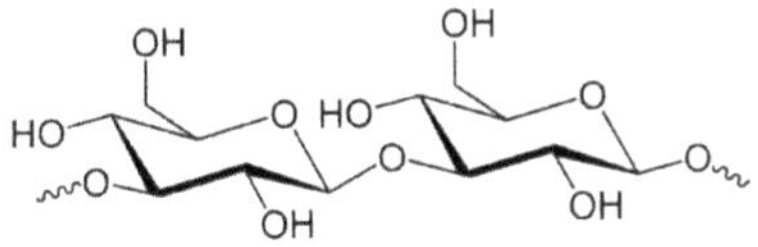

Fig. 10. Structure of curdlan.

Curdlan: Curdlan is water-nsoluble high-molecular weight polymer of glucose consisting of β-(1,3)-linkage in the structure shown in Fig. 10. It is produced as an exopolysaccharide by soil bacteria of the family Rhizobiaceae and shows odorless properties which allow its use as a gelling agent in the food industry (Ahmad et al., 2015; Brodnjak, 2017; Miwa et al., 1994; Sun et al., 2011; Wu et al., 2012).

2. Fabrication Techniques of Polymer Blend Composite Films

Polymer blend composites, especially biodegradable polymer composites, have become the focus over the last few decades in food-packaging application areas due to their ability to degrade in the environment without leaving nontoxic residues. We saw in the previous section that the biopolymers synthesized from bio-renewable resources exhibit flexible, high elongation at break, excellent barrier properties, low melting points but low tensile strength and stiffness. To enhance the poor mechanical and structural properties, the traditional practice is to prepare copolymers through a combination of monomers from synthetic and renewable resources. The most popular and recently developed fabrication techniques for biocomposites are solvent casting, extrusion, blow film extrusion, reactive extrusion, layer-by-layer assembly and 3D printing.

2.1 Solvent Casting

Solvent casting involves the preparation of an homogeneous polymer solution by dissolving the polymer pellets/chips into an organic solvent. The solution is then poured into a mold or substrate and the solvent is evaporated at room temperature or upon heat, leaving polymer films. The newly formed film is then dried under vacuum to remove any residual solvent. This is a very ancient technique but shows some advantages, such as ensuring a uniform film thickness, high optical clarity, high flatness and dimensional stability making it attractive nowadays (Butnaru et al., 2019; 'solvent castting - Google Search,' n.d.). Swaroop et al. fabricated PLA-based biocomposites integrated with magnesium oxide (MgO) nanoparticles via the solvent-casting method. In this method, they used chloroform to dissolve the PLA pellets. The MgO particles dispersed separately through ultra-sonication in an appropriate composition and then added to PLA solution and stirred enough to get an homogeneous solution. Then the mixture was evenly spread on to a glass petri dish with the aid of a manual film applicator to ensure uniform thickness of the film. After the estimated time, the film was peeled off and conditioned further to remove any residual solvent, which may act as a plasticizer. The fabricated film was uniform in thickness ranging from 35 and 45 μm (Swaroop and Shukla, 2018).

2.2 Extrusion

Extrusion is one of most efficient and widely used method to fabricate biobased polymer blends and composites films. Polymer extrusion commonly involves the use of dried polymer pellets or chips. In the fabrication process, the polymer pellets are heated to melt and force the resin through the die by a simultaneous effect of heating and shear from the extrusion screw. The extrudate is cooled and solidified as it is pulled through a water tank/coolant. Mele et al. (2019) showed the fabrication of PLA-based films by the melt extrusion process. They investigated the effect of cardanol oil (CA) on the properties of PLA films. During the process, they mixed the PLA pellets with CA oil at various predetermined weight percentages followed by rotation blades to get the powdered form of the PLA/CA mixer. They used HAAKE Rheomex 19/25 QC single-screw extruder coupled with PolyLab QC drive unit and chill roll system to fabricte the films. The process temperature varied between 140–170°C and the speed was 60 rpm. Depending on the die specifications, the film thickness varied during the extrusion process.

2.3 Blow Film Extrusion

Blow film extrusion is almost similar to the extrusion process, which involves the extrusion of polymer melts through a die while air inflates the melt into composites of thin film. Blow film extrusion has much process stability, control on crystallinity and fabricating of stiff films over extrusion that helps to attract much attention. Many researchers showed the importance of blow film extrusion over extrusion producing PLA-based food packaging films (Garofalo et al., 2018; Karkhanis et al., 2018; Mallegni et al., 2018; Scaffaro et al., 2018; Wang et al., 2018).

2.4 Reactive Extrusion

Reactive extrusion process includes continuous extrusion and reaction simultaneously, usually in a co-rotating twin-screw extruder, with outstanding mixing competency at molecular level. The process is designed in a way that it can adapt to chemical reactions in the viscous media with a good command on the residence time and temperature. The residence time can be varied by changing the input, length of the extruder, screw design and its speed. This method is a prominent candidate as an alternative to traditional extrusion method because of its simplicity, efficiency, less consumption of raw materials, reagents and energy, eco-friendly and more economical. The extruder mainly acts like a horizontal reactor where polymerization of the oligomer or monomers occurs during intermeshing by twin screw extruder (Hopmann et al., 2017). Monika et al. fabricated nanobiocomposites, using PLA/PBS blend at a predetermined ratio 80:20 induced with functionalized chitosan (FCH). During the fabrication process, they used dicumyl peroxide (DCP) as a compatibilizer and found that the presence of FCH and DCP increased the per cent elongation at break and molecular weight due to cross-linking between FCH and polymer. They observed that FCH dispersed uniformly which improves crystallization efficiency and ensures uniform mechanical properties of the composites (Monika et al., 2018).

2.5 3D Printing

The most evolved and attractive technique for biocomposite film formation from polymer solution is three-dimensional (3D) printing, which involves the movement of motorized printer head on a stage, following a pre-programmed design. This technique, also known as rapid prototyping technique, is widely used in food packaging, biomedical and dental applications. In this process, the polymer solution is deposited on to the printing bed in a layer-by-layer fashion to get the final desired part. It is also possible to apply heat in the printing bed to evaporate the solvent fast and get the final part within a couple of minutes. A schematic of 3D printing process is shown in Fig. 11. In the case of a solvent-free approach, the fused deposition modeling (FDM) is now being studied for fabricating composite thin films. In the FDM method, it is necessary to melt the polymer filament at low temperatures to avoid the degradation of polymer and this restricts the use of high-performance polymers due to their high flow

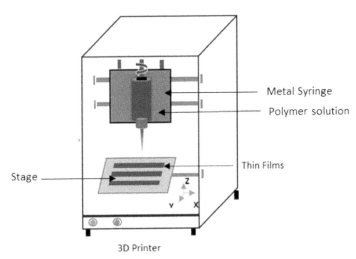

Fig. 11. Schematic of 3D printer.

or melting temperatures. In addition, the polymer must be preprocessed into filaments by extrusion, which limits the isotropic property in the final part. As a result, 3D printing technique attracts attention in composite thin-film fabrication due to low cost, ease of processing, low wastage, isotropic final products, a wide range of material selection and high resolution of the technique reported in the literature (Biswas et al., 2019, 2017; Charlebois and Juhasz, 2018; Sun et al., 2018; Tiimob et al., 2017).

2.6 Layer-by-Layer (LbL) Assembly

Layer-by-Layer Assembly mainly consists of alternative deposition of layers of oppositely charged materials with a washing step in between. The deposition can be done through different ways, such as spray, immersion, spin, electromagnetism or fluidics ('Layer-by-Layer Assembly - Google Search,' n.d.). LbL method offers various advantages over the traditional deposition methods, such as being inexpensive, simple, of uniform thickness, has great control on thickness, bilayers and fabrication technique, which gain much attention. Researchers nowadays use this method over the other fabrication techniques to develop biocomposite films with high moisture and gas barrier (oxygen and carbon dioxide) (Heo et al., 2019; Li et al., 2019; Subbiahdoss et al., 2019, 2019; Sun et al., 2011; Takma and Korel, 2019; Ummi and Siddiquee, 2019).

3. Biodegradable Polymer Blends

The use of large amounts of recalcitrant polymers in the packaging industry is a cause for much concern in waste management. Though plastics constitute only 11 per cent of total municipal waste in the United States and Canada, it takes several decades for a majority of them to degrade in landfills. In addition, their high carbon footprint significantly aggravates the issues confronting the environment (Bohlmann, 2004). In order to tackle these issues, researchers have sought alternatives in bio-based and biodegradable plastic designs for similar applications in the market (Jiang et al., 2006; Pakravan et al., 2011). These efforts, which are geared toward the development of renewable substitutes to petroleum-sourced plastics, require the adoption of more economical and industrially-scalable manufacturing processes, which can improve the weak structural properties of biodegradable polymers (Jiang et al., 2006; Garlotta, 2001).

Poly(lactic acid) (PLA) has shown good biodegradability and continues to be relevant in the field of soft biomaterials. It is among the top biopolymers with high strength and modulus obtainable from renewable sources and has good suitability for processing with other biopolymers in a close range of melting temperature. However, the brittleness of PLA is a drawback in its scope for application and large-scale processing techniques, such as blown-film extrusion, blow molding and foaming, for which melt stability is critically required (Li and Micheal, 2011; Garlotta, 2001; Liu et al., 2010; Nampoothiri et al., 2010; Theryo et al., 2010; Al-Itry et al., 2014; Zhang et al., 2008). Hence, in most polymer designs, PLA is blended with a softer polymer (elastomers and plasticizers) to stiffen or strengthen it while toughening the PLA to offset its inherent extreme brittleness and low elongation (Tabasi et al., 2015; Jiang et al., 2006). PLA is synthesized from starch by lactic acid bacteria conversion of starch monomers to lactic acid, which is then polymerized to PLA. Also, poly(butylene adipate-co-terephthalate) (PBAT), another biodegradable synthetic aliphatic-aromatic copolyester polymerized from monomers of adipic acid, 1, 4-butanediol, and dimethyl terephthalic acid, is an elastomer which has close melting temperature than that of PLA, and can potentially reduce the brittleness of PLA with proper design. The polymerization of PBAT involves the formation of aliphatic and aromatic polyesters from adipic acid and dimethyl terephthalate respectively, using 1, 4-butanediol. These are then polymerized into PBAT, using tetrabutylorthotitanate (TBOT) catalyst. While the aromatic monomers in PBAT contribute to good thermal stability and mechanical properties, the aliphatic region provides biodegradability, flexibility and toughness due to the carbon-carbon single-bond backbone (Ki and Park, 2001). However, high cost, low strength and stiffness limit its scope in application. Blending PBAT with natural occurring polymers, such as soy meal (Zhou et al., 2013), starch (Brandelero et al., 2010) and corn gluten meal (Reddy et al., 2014) has shown balanced properties which can widen its scope in usage.

In a study of toughened blends of PLA and polycaprolactone (PCL), development of laminar morphology with dispersed PCL phases was observed. This led to an improved elongation at break by almost 150 per cent due to 40 per cent PCL and over 400 per cent due to 60 per cent PCL content compared to pristine PLA (Tabasi et al., 2015). In addition, a different study on PLA/PCL blends also revealed good impact toughness and improved elongation, due to the effect of the PCL (Matta et al., 2014). Again, biodegradable blends of poly(3-hydroxybutyrate) (PHB) and PLA bio-plasticized with trybutyrin showed distinct glass transition temperatures, improved elongation and biphasic melts morphology, suggesting immiscibility between the PHB and PLA (D'Amico et al., 2016). Also, soy protein isolate (SPI)/PBAT blend, in which the effect of SPI on thermal and mechanical properties of PBAT was investigated (Gou et al., 2015), showed that the hydrophilic SPI adversely affected most mechanical properties of the system, yet improved the storage modulus and glass transition temperature, due to the reinforcing effect of SPI. Furthermore, the effect of poly(3-hydroxybutyrate-co-3-hydroxyvalerate) (PHBV) on the structural morphology and softening of PLA revealed that the PLA/PHBV blend remarkably improved in dimension stability while softening the PLA to form pliable structures (Li et al., 2015). These emphasize the importance of microstructural tuning in polymer blends to tailor the properties of individual polymers in a blend to desirable properties not attainable in single polymers. Hence, different ratios of compostable PBAT/PLA (agro-based) extruded blend sheets are studied to assess the effect of blending on the morphology, thermal and mechanical properties of extremely brittle PLA.

The following sections present some methods and results regarding the use and characterization of polymer-blend composites and polymer filler application.

3.1 Extrusion of the Polymer Blend Composites

About 150 g of each of the blend pellets were dried for 12 hours at 140°F in a hopper (DRI-AIR Industries Inc., model RH5). This was then fed into a 19-mm (diameter) tabletop single-screw extruder (Wayne SN: 8001) which is driven by 2 hp motor. Thermostat controls five heating zones and screw rotational speed facilitated the melting, mixing and the formation of continuous viscous melt for the extrudate. Three heating zones are located in the barrel while two are in the die zone. The optimum working temperatures were maintained at 320°F, 320°F, 320°F, 315°F and 312°F, for the barrel and die zones respectively. About 40-mm wide and 0.1–0.3 mm thick blend specimens were obtained at a screw speed of 20 rpm and a feed rate of 4.4 g/ min. and collected at the die orifice. The continuous hot molten sheets were passed through water stationed at the orifice of the die for quenching. These blends were stored in a high vacuum desiccator (JOEL, EMDSC-U10A) and only removed during characterization.

3.2 Characterization of the Polymer Blend Composites

3.2.1 Raman Analysis

Molecular vibrational spectroscopic analysis of the pristine and blended polymer systems was carried out with DXR Raman microscopy (thermo scientific). This test was done using 532 nm laser (5.0 mW power), filter and grating. OMNIC Software was used for data acquisition and analysis. The essence of the vibrational analysis is to assist in the identification of functional groups identical to each pure polymer in the blend, any form of phase change and type of interactions occurring between the blend systems.

3.2.2 X-Ray Diffraction

X-ray diffraction (XRD) analysis was performed on all specimens, using a Rigaku diffractometer (DMAX 2100) equipped with Cu Kα radiation, operated at a step size of 0.02°, scan rate of 1°/min., 3° to 80° Braggs angle of diffraction and 40 kV to 30 mA.

3.2.3 Differential Scanning Calorimetry

A TA Q 2000 differential scanning calorimeter (DSC) was used to study thermal profiles of the specimens. Samples of 10.0 ± 0.1 mg were used in the test. Each sample was sealed in an aluminum pan and the test run against an empty reference pan. Each specimen was cooled at 20.0°C/min. from room temperature to –40°C and subsequently heated from –40°C to 200°C at 5°C/min. and held at constant temperature for 2.0 min. to erase previous thermal history. It was again cooled from 200°C to –40°C at 20.0°C/min. before finally scanning at 5.0°C/min. from –40°C to 200°C to determine the various heat transitions in each specimen.

3.2.4 Thermogravimetric Analysis

Thermogravimetric analysis (TGA) was carried out with TA Q 500 equipment. Samples of 14 ± 0.2 mg were placed in platinum pans. An empty platinum pan was used as a reference. Each sample was heated from 30°C to 600°C in a 50 mL/min. flow of N_2. A heating rate of 5°C/min. was used and the continuous records of sample temperature and mass were taken.

3.2.5 Tensile Testing

Measurement of tensile mechanical properties was done by using Zwick Roell Z2.O mechanical testing system in accordance with ASTM D 882, using a crosshead speed of 500 mm/min. and 2.5 kN load cells and wedge grips. Specimens were cut from the extruded sheets of polymer systems, 19 mm x 0.3 mm x 120 mm. The test was conducted at 20 mm gauge length with TestXpert data acquisition and analysis software. At least 15 specimens were tested, averaged and reported as mechanical properties.

3.2.6 Scanning Electron Microscopy

Microstructure and blend morphologies were probed, using Joel JSM-5800 Scanning electron microscopy (SEM). Film samples were cut and placed on a carbon tape of 4-inch wide sample holder of the SEM. This was then sputter-coated with gold-palladium for 5 min. in Hummer 6.2 sputtering system purged with N_2 gas and operated at 20 millitorr 5 volts and 15 milliamps. Fractured surfaces of the tensile specimens were examined, using Hitachi S-3400N SEM.

3.2.7 Results

Raman spectroscopy helps in the investigation of the structure and interactions of molecules at the functional group level (Furukawa et al., 2006). Here, the structural elucidation of the individual and blend polymers was carried out with Raman-microanalysis to identify functional groups in the polymer systems.

Figure 12 shows that the micro-Raman spectra of PBAT (ecoflex)/PLA blends and the pristine polymers yielded bands specific to each polymer in the region of 3250 to 750 cm^{-1} vibration frequencies. The assignment of these bands on the Raman spectra is specified in Table 1. The PBAT co-polyester contains a benzene ring in its structure. Evidence of this is seen in the spectra with the appearance of vibrational bands at 1618 and 3085 cm^{-1} for aromatic –C=O and –C-H stretching respectively. The rest of the vibrational frequencies for PBAT are presented in Table 1. This conforms to the finding of different investigations (Furukawa et al., 2006; Kister et al., 1998). In 100 per cent amorphous PLA, the –C=O stretching mode is observed at 1768 cm^{-1}. Kister et al. (1998) reported a similar observation on PLA. The rest of the frequencies due to Raman excitations of the PLA molecules in the pure and blended PBAT/PLA polymer are specified in Table 1. This matches the results of past studies which revealed that the frequencies at 1050 are due to –C-CH$_3$ stretching, while 1395 and 2950 cm^{-1} are due to –CH$_3$ symmetric deformation. In addition, those at 1461 and 3004 cm^{-1} are attributed to –CH$_3$ asymmetric deformation and 1290 cm^{-1} is a result of -CH bending in PLA (Furukawa et al., 2006; Kister et al., 1998). The presence of specific vibrations of each polymer in the 50/50 and 70/30 blends in the representative spectra at their expected frequencies suggests that the blends are immiscible and that no strong interactions or new bonds were formed.

The morphology of the polymer systems analyzed by X-ray diffraction is presented in Fig. 13. This shows the diffraction patterns of the pristine polymers (Fig. 13a and b) and those of PBAT/PLA 90/10 and 70/30 (Fig. 13c and d respectively). The broad halo of the pure PLA in Fig. 13b extends from $2\theta° = 3$–40; this pattern is typical of amorphous PLA, as observed elsewhere (Kister et al., 1998). For pure PBAT, a semi-crystalline diffraction pattern was observed with five prominent crystalline peaks distributed on the amorphous curve at $2\theta° = 15.8, 17.5, 20.0, 23.1, 24.3$ and 29.3.

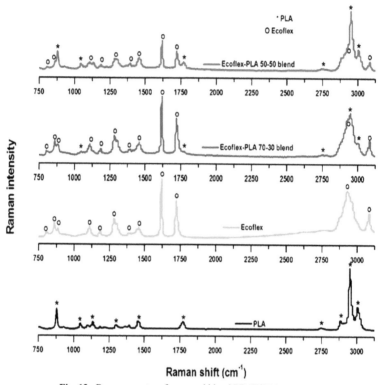

Fig. 12. Raman spectra of pure and blend PBAT/PLA systems.

Table 1. Raman Shift and Assignments of the Bands to PLA and PBAT.

Shift (cm⁻¹)	Functional group
3085	–C-H aromatic stretching (PBAT)
3003	–CH$_3$ asymmetric stretching (PLA)
2950	–CH$_3$ asymmetric stretching (PLA)
1768	–C=O stretching (PLA)
1721	–C=O stretching (PBAT)
1618	-C=C aromatic (PBAT)
1461	–CH$_3$ asymmetric deformation (PLA)
1392	–CH$_3$ asymmetric deformation (PLA)
1298	–C-H bending (PLA)
1186	–C-O-C stretching (PBAT)
1132	–CH$_3$ rocking (PLA)
1046	–C-CH$_3$ stretching (PLA)
927	–C-C back bone stretching with CH$_3$ rocking (PLA)
879	–C-COO stretching (PLA)

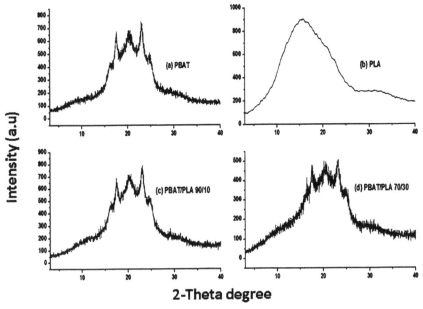

Fig. 13. X-ray diffraction patterns of neat and blend PBAT/PLA polymer systems.

The diffraction pattern of the blends show a semi-crystalline morphology, quite similar to that of pure PBAT; however, it is evident from Fig. (13c and d) that the base of the amorphous region on the blends are wider than those of the PBAT, due to the extension resulting from the contribution of amorphous PLA around the region of $2\theta° = 3–12$. Similar diffraction pattern was observed in a different study for PBAT/ PLA 60/40 and 40/60 blends (Arruda et al., 2015). The semi-crystalline structure of the PBAT in the blend is very critical to morphology-related structural properties of the blend, especially stiffness and flexibility. Crystallinity has been found to alter microstructure to improve mechanical and thermal properties of various polymer blends (Bai et al., 2012; Li et al., 2015; Zhang et al., 2012; Samuel et al., 2013). Hence, in some situations, semi-crystalline structured blends can be advantageous in properties and application as compared to individual polymers in isolation.

The morphologies of neat polymers, 70/30 and 50/50 blends, are shown in Fig. 14. The neat PBAT (Fig. 14a) and PLA (Fig. 14b) systems show smooth single-phase matrices, due to the existence of chemically similar molecules in each pure polymer. However, the morphologies of PBAT/PLA 70/30 and 50/50 blends revealed a heterogeneous phase, with ellipsoid domains of PLA dispersed in a continuous phase of PBAT (Fig. 14c and d).

The domain sizes range from 10–200 µm. The observation, of course, domains in the phase-separated 'sea island' structured system indicating that the PBAT and PLA are immiscible. This is similar to findings in previous reports on immiscible polymer blends. For example, morphological analysis on PPC/PMMA blend revealed a 'sea island' structure due to immiscibility between the two polymers as a result of difference in chemical structure (Li and Shimazu, 2009). Other immiscible blends with phase-segregated morphologies include PLLA/ABS (Dong et al., 2015), PCL/PLA (Chen

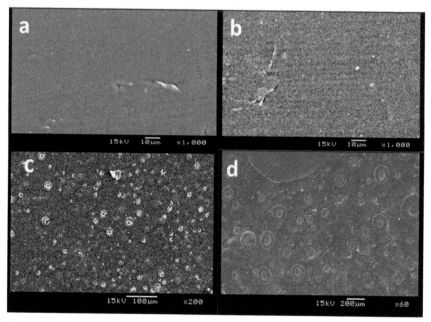

Fig. 14. SEM morphology analysis (a) PBAT (b) PLA (c) PBAT/PLA 70/30 and (d) PBAT/PLA 50/50.

et al., 2014; Goffin et al., 2012), PLA/PBSA (Ojijo et al., 2013), PLA/NR (Chen et al., 2014), PLA/PHBV/PBS (Zhang et al., 2012) and PLA/SOY/ILLA (Chang et al., 2009). The mechanical properties of immiscible blends are highly dependent on their microstructures. Blends consisting of droplet domains of one component dispersed in the matrix of another microstructure affect the impact strength and toughness based on the aspect ratio of the droplets and the matrix thickness in between the droplets (Chang et al., 2009).

DSC heating curves of PBAT, PLA and PBAT/PLA blends after crystallization from melts at 5°C/min. are shown in Fig. 15. The neat PLA displayed a glass transition temperature at 60.20°C, cold crystallization at 104.83°C and a melting point at 148.75°C with a shoulder at 156.48°C. Also, the PBAT revealed a melting point at 122.74°C. On comparing the curve in Fig. 15, it becomes evident that the cold crystallization temperature of PLA slightly shifted by about 2°C higher in the blends with much broader peaks. This indicates that the PLA remained amorphous and was not able to crystallize during 20°C/min. cooling rate, as observed elsewhere (Jiang et al., 2006). In addition, two distinct melting temperatures were observed in all the blends, attributed to the individual polymers. This affirms the immiscibility of the blends revealed in the microstructural analysis (XRD, Raman microscopy and SEM). Different studies reported similar findings in DSC dynamic scans for PBAT/PLA blends (Arruda et al., 2015; Jiang et al., 2006). A high amount of PBAT led to significant changes in the PLA melting and shoulder peaks, suggesting the presence of a new crystalline structure, induced by PBAT. This double melting peak is attributed to less perfect crystals, which had enough time to melt and reorganize into crystals with higher structural perfection and requiring higher fusion temperature (Jiang et al., 2006).

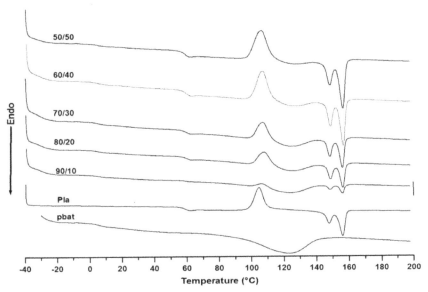

Fig. 15. DSC curves for neat and blend PBAT/PLA polymers.

Thermal stability of PLA, PBAT and their blend systems determined by thermogravimetric analysis is shown in Fig. 16 and Table 2. The amount of remaining material in the various neat and different blends of PBAT and PLA polymers are used to determine the thermal stability or resistance to temperature-induced degradation in the materials. In Fig. 16, the PBAT curve showed more thermal stability than PLA. Blending of the two led to a significant improvement in the thermal stability of the blend to levels comparable to those of pristine PBAT at 50 per cent decomposition (T_{d50}). The onset of degradation temperature (T_{donset}) also improved considerably, compared to the neat PLA system (about 321 in 90/10 blend as compared to 295 in neat PLA). However, the increasing presence of PLA made the materials more susceptible to combustion, leading to much reduced thermal properties, especially in the 50/50 blend. The residual yields of the various blends as revealed in Table 2 indicate reduction in char yield of the blends to almost equal that of PLA, thus emphasizing the fact that PLA is more combustible and less thermally stable.

The high thermal stability in the PBAT significantly influenced that of PLA in the blends, except in residual. The higher thermal stability of PBAT and those of 90/10, 80/20, 70/30 and 60/40 are attributed to the thermal resistance of the aromatic ring in the PBAT structure (Ki and Park, 2001) as well as its semi-crystalline nature which requires more heat energy to change state before degrading. The appearance of a minor degradation in 70/30, 60/40 and 50/50 is due to the decomposition of PLA (308–312°C) in the blends while major degradation (T_{d50}) is due to the co-polyester. The minor degradation is not quite evident in the 90/10 and 80/20 blends, probably due to the low content of PLA. Essentially, blending significantly improved the thermal degradation of PLA without compromise on the onset of degradation temperatures, T_{d50} and T_{dmax} of PBAT in blends with 10–40 per cent of PLA. But at 50 per cent PLA, there was a general reduction in thermal stability of the blend significantly, except for the T_{dmax}.

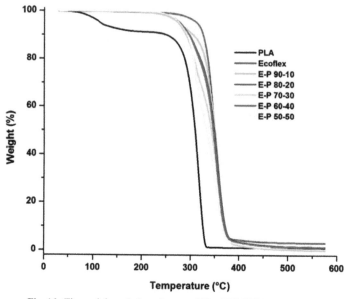

Fig. 16. Thermal degradation of neat and blend PBAT/PLA polymers.

Table 2. Thermal Profiles of Pure and Blended PBAT/PLA Polymers.

Specimen	T_{donset}	T_{d50}	T_{dmax}	% Residue
PBAT-PLA 0/100	294.80 [aeg]	316.44 [a]	322.02 [a]	1.04 [adef]
PBAT-PLA 100/0	326.30 [bcd]	349.00 [bcdef]	351.90 [bdg]	3.29 [bc]
PBAT-PLA 90/10	321.03 [bcf]	350.10 [bcdf]	355.71 [cdef]	0.62 [cg]
PBAT-PLA 80/20	320.81 [abcdf]	349.64 [bcdf]	354.13 [bcdefg]	1.37 [ade]
PBAT-PLA 70/30	288.90 [aeg]	342.10 [bef]	355.59 [cdef]	1.17 [adef]
PBAT-PLA 60/40	311.43 [cdf]	346.91 [bcdef]	356.34 [cdef]	1.50 [adef]
PBAT-PLA 50/50	291.70 [aeg]	332.90 [g]	352.62 [bdg]	0.46 [cg]

[a] Mean values with different letters in the same column represent significant differences (p < 0.05) among the samples according to a one-way analysis of variance (ANOVA) and Tukey's multiple comparison tests.

This is due to the increased content of highly combustible PLA in the polymer. The improvement in thermal stability can be attributed to the aromatic portion of PBAT, which absorbs most of the thermal energy and the early development of char from highly combustible PLA component to impede the infiltration of volatiles, resulting in delays in thermal degradation of the blend.

A study found that PPLA, the stereoisomer of PLA was about 50°C less thermally stable than polyester (Lebarbe et al., 2014), while PLA/PHB blend exhibited improved thermal stability over their neat systems (Jandas et al., 2014). This supports the results on thermal stability with the PBAT/PLA blends.

Tensile analysis of the pristine and binary blend polymer systems is presented in Fig. 17. The figure shows the stress-strain curve representing the mechanical behavior of the polymeric systems under tensile load of 2.5 KN, while Fig. 17b,c,d summarize the percentage elongation, elastic modulus and tensile strength respectively. The pure PLA is very stiff and brittle under tensile load, while the tensile strength of PLA is above 40 MPa (Fig. 17d) and the elongation at break is just about 4 per cent. However, pure PBAT showed excellent elongation ($\sim$ 2072 per cent) and low tensile strength of about 14 MPa. Compared to the neat PLA, the ductility of the PLA improved significantly in the blends with higher content of PBAT elastomer. But, significant reduction in modulus and strength was observed in those blends because of high content of soft PBAT (Fig. 17c,d).

The 90/10, 80/20 and 70/30 blends showed distinct yielding, followed by considerable cold drawing during the tensile test, indicating that the brittle fracture of PLA was transformed into ductile fracture due to blending with PBAT.

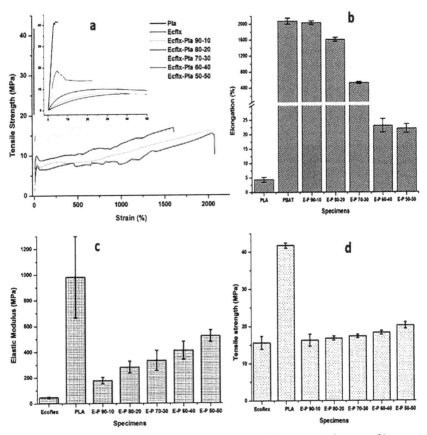

Fig. 17. Tensile analysis of neat and blend PBAT/PLA systems (a) stress vs strain curves, (b) per cent elongation, (c) elastic modulus and (d) tensile strength.

The reduction in tensile strength and modulus in blend is due to the high amount of soft PBAT component, which has high plasticizing effects. Similar trends in property changes were reported in poly(lactic acid)/poly(ether-b-amide) PLA/PEBA blends (Zhang et al., 2014), PLA/PBAT blends (Arruda et al., 2015; Jiang et al., 2006), PLLA/PCL blends (Bai et al., 2012), PLA/castor oil (Robertson et al., 2011), PLA/PCL (Tabasi et al., 2015) and PLA/PHB (Jandas et al., 2014) due to the soft and toughening effects induced by the second polymer serving as an elastomer in each case of PLA in these blends.

The toughening effect of the PBAT/PLA binary polymer blends, investigated through SEM analysis of the fractured surfaces after tensile tests, is shown in the micrographs in Fig. 18. The fractured surface of the PBAT reveals a microstructure which suggests a ductile pull to failure, leaving out fibrils of the PBAT polymer (Fig. 18a) but those of PLA show a nearly smooth and fractured surface virtually without interruptions, indicating a typically brittle behavior. However, the fractured surface of the binary blend with 70/30 (Fig. 18 c,d) altered the smooth fractured surface in the pure PLA to one with multiphase and rough surface morphology. The fractured surface (Fig. 18 c,d) showed a pull out of one phase from the other at the interface of the blend systems. This explains the slight increase in tensile strength of the PBAT and an enormous improvement in the toughness of PLA. The improvement in tensile strength remained moderate because of the poor compatibility between the two phases. The efficiency of an elastomer in toughening a polymer in a blend is highly dependent on the interfacial adhesion and cavitation. It is known that the interfacial interaction between the dispersed domains and the matrix play a very delicate role in

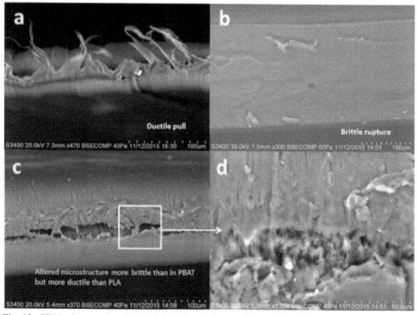

Fig. 18. SEM micrographs of fractured surfaces after tensile analysis (a) PBAT (b) PLA (c) PBAT/PLA 70/30 and (d) high magnification of PBAT/PLA 70/30 blend.

toughening the blend. High toughening effect is achieved when the dispersed phase favors strong interfacial interactions with the matrix to inhibit the crack growth (Zhang et al., 2014). The phase segregation in immiscible blend helps in toughening of the materials through the crack path interruption to delay failure. This effect was observed in the blends which revealed ellipsoid phase segregations of PLA in PBAT matrix, as shown in Fig. 18, leading to increased toughness in their structures.

Similar surface morphologies were reported in various toughened blends of PLA/ PEBA blends (Zhang et al., 2014), PLA/PBAT blends (Arruda et al., 2015; Jiang et al., 2006), PLLA/PCL blends (Bai et al., 2011), PLA/castor oil (Robertson et al., 2011), PLA/PCL (Tabasi et al., 2015) and PLA/PHB (Jandas et al., 2014).

4. Carbon Nanoparticles Derived from Waste Coconut Shell (CCSP) for Polymer Filler Application

The coconut shell powder (CCSP) is derived from the dry coconut shell, which is a waste product from coconut fruit. Billions of coconuts are grown in tropical areas, such as Sri Lanka, India, Indonesia, Nigeria, Ghana, Puerto Rico and many other countries in the world. Coconut shells are broken down into powder and used in industries specifically for its high strength and modulus properties. When mixed with PLA and other biocomposites, the elastic modulus is enhanced but the tensile strength and elongation decrease (Chun et al., 2012). CSP also contains high carbon content which acts as precursor for creating other carbon-based crystalline materials. Useful materials created from CCSP can potentially demonstrate electrical properties either exclusively or when blend with polymers to create an electrically conductive polymer with improved thermal stability and stiffness. Hence, this study explored the high temperature and pressure, and microwave-assisted techniques of synthesizing carbon material from coconut shell and embedding it into a biodegradable polymer blend to study its effects on the structural properties of the blend.

4.1 Synthesis and Characterization of Coconut Shell-derived Resource Carbon Material

In this work, we have shown the conversion of renewable waste amorphous material (coconut shell) into crystalline carbon materials with tailored properties suitable for application in the design of low-cost filter, energy and structural materials (Liu et al., 2014; Cheng et al., 2010; Khalil et al., 2010). Approximately 10 grams of >150 μm coconut shell powder (CSP) was heated in a custom-designed high pressure reactor (Parr 4838) at 5°C/min. to 800°C. The reaction was held isothermally at 800°C for two hours under autogenic isobaric pressure of 124 bar (1800 Psi) emanating from the combustion of the CSP in one set of experiment, while in another the pressure was released. A yield of approximately 3 grams was obtained in both the cases. These two carbonized materials were divided into three (about 1 gram each) into a 20 mL aluminum oxide crucible and microwaved (Microwave Research and Applications Inc.) at 50 μP for 15, 30 and 60 minutes at 1100°C. The specimens were characterized by using multipoint BET surface area measurement and Raman microscopy.

Fig. 19. 0.3 per cent CSP 75/25 PBAT/PLA-blended film.

The polymer ratios of PBAT and PLA 75/25 (24/8 g/g) were respectively weighted into a 260 mL plastic beaker and about 100 mL of chloroform was added and magnetically stirred for at least eight hours. Once the solution was created it was loaded into a micro syringe and printed from the Hyrel System 30M 3D printer. Five strips each were printed in each cycle (Fig. 19). The dimensions of the strips were about 200 mm x 25 mm x 100 microns. This procedure was repeated for printing of the neat polymer films and the blend composites. In the blend composite, 0.3 per cent of CCSP was added to the polymer mixture and dissolved together before printing the films. They were left to dry at room temperature for 10 min. after printing was complete.

They were then used to test its tensile strength by using the Zwick/Roell Z2.5 universal mechanical testing-machine. The microstructure of the samples was also characterized with Raman spectroscopy, using the Thermo Scientific DXR Raman Microscope and X-ray diffraction techniques as previously mentioned.

4.2 Results

The surface areas measured by 11-point BET analysis of the coconut shell derived carbon materials are presented in Table 3. These revealed increase in the surface

Table 3. BET Surface Areas of Coconut Shell Carbon Materials.

Specimen	Microwave time (min.)	BET surface area (m²/g)	
		0 pressure	124 bars
CCSP	-	0.4	95.1
CCSP-15	15	417.7	242.4
CCSP-30	30	832.3	286.5
CCSP-60	60	77.2	165.3

areas in both cases due to microwave heating but this is significant at 15 and 30 minutes heating of the 0 pressure specimen. Also the autogenic pressure showed a significant influence on the carbonized material which was not microwave heated; in that, the surface area was ~ 238 fold more. These results show great promise for further experimentation and are higher than those reported for birchwood-derived porous carbon (Tsyganova et al., 2014).

Raman spectroscopy is a very powerful technique for characterizing carbon-based materials, such as graphene's derivatives, like graphene-oxide. Figure 20 shows the Raman spectra of the raw cellulosic coconut shell powder as received and those of its derived materials processed through autogenic high pressure reaction and microwave irradiation.

The spectra show the typical disordered (D) and graphite (G) bands found in most carbon materials due to Raman vibrational excitations. The D band occurs at 1350 cm^{-1} while the G is at 1591 cm^{-1}, suggesting the formation of very small crystallites of graphene-oxide from the coconut shell powder. This is very significant because graphene oxide can be used for making graphitic polymer composites, which retain some of the electrical properties of graphene while maintaining the physical properties of the polymer.

The tensile test results for the various samples are shown in Table 4. The averages and standard deviations are given for the elastic modulus, tensile strength and strain at force load (ε Fmax per cent). PBAT, as a neat polymer, displays the highest elongation of about 1445.1 per cent as compared to lowest elongation percentage PLA which is 6.8 per cent. However, PLA as a neat polymer scored the highest in the elastic modulus along with the tensile strength as expected. PBAT possessed the lowest in

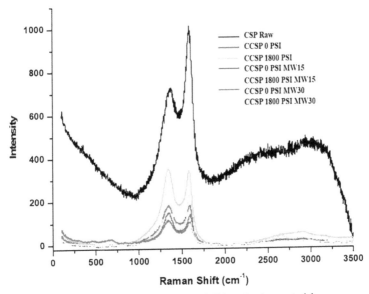

Fig. 20. Raman spectra of coconut shell-based carbon materials.

Table 4. Tensile Analysis of Neat Blend and 0.3% CCSP with Blended Polymers.

Specimen	Elastic modulus (MPa)	Tensile strength (MPa)	Ɛ Fmax %
PBAT	48.6 ± 12.5	11.5 ± 1.2	1445.1 ± 416.2
PLA	405.7 ± 176.3	22.3 ± 4.0	6.8 ± 1.5
75/25 PBAT/PLA	68.9 ± 6.3	7.7 ± 0.5	85.4 ± 12.1
0.3% CCSP 0Psi 50MP	88.2 ± 8.0	8.7 ± 0.6	48.9 ± 13.1
0.3% CCSP 1800 Psi 0 MP @ 800C	70 ± 4.9	6.0 ± 0.6	53.8 ± 7.8
0.3% CCSP 1800 Psi @ 800C 50 MP	82.2 ± 8.2	6.7 ± 1.0	46.8 ± 8.0
0.3% CCSP 0Psi 0 MP @ 800C	90.3 ± 9.2	7.5 ± 0.7	48.6 ± 5.8

elastic modulus, yet was second highest in tensile strength. The 75/25 blend of the two polymers performed poorly in mechanical properties. The 75/25 PBAT/PLA scored 85.4 per cent for the elongation percentage, which is the second highest of the samples. The PBAT's flexibility influenced the blend whereas PLA's brittleness influenced it as well. The addition of 0.3 per cent CCSP led to a low performance in the elongation compared to the blended polymers, yet it was higher than the PLA in this regard. As expected, the CCSP influenced the polymers which were used to decrease the elongation percentages and the tensile strengths. The CCSP, however, has influenced the polymers to achieve a higher elastic modulus as compared to the 75/25 PBAT/PLA blend. The brittleness of the CCSP and PLA has overcome the flexibility of PBAT. Our expectations for CCSP to improve mechanical properties were fulfilled in the elastic modulus, due to the highly stiff CCSP particles, which influenced the stiffness of the 25/75 blend. However, flexibility and strength were compromised due to larger aspect ratios of the CCSP particles, which potentially created highly stressed weak points on the films.

The Raman graph in Fig. 21 identifies the different functional groups in the neat and blend polymers Raman spectroscopy analyzes and displays frequency vibrations and intensities unique to a material. The functional groups are present in the polymers to differentiate one material from the other and serve as a means to identifying the material. PBAT possess the benzene ring, unlike PLA. PLA, like PBAT, possess the ester group. There are heavy vibrations taking place around 900 cm^{-1} of the PLA due to the presence of the ester group attached to a methyl $-(CH_3)$ group. Activity takes place around 1400 cm^{-1} due to the presence of a $-CH_3$ and its asymmetric deformation. Around 1700 cm^{-1} there is a peak due to a ketone. For PBAT there are peaks located around 600 cm^{-1}, 800 cm^{-1}, 1100–1500 cm^{-1}, 1600–1750 cm^{-1}. These are due to the ketones, ether group and a methyl group (add the vibration frequency due to the peak of the aromatic group in PBAT, it is a key indicator of PBAT).

X-ray diffraction analysis of the polymer systems is shown in Fig. 22. This reveals the crystallinity and other microstructural characteristics of the materials. Pure PLA shows a predominantly amorphous peak, indicating that the molecules in the material are highly disordered in the arrangement, while PBAT displays a semi-crystalline, suggesting that its molecules are partially aligned in the polymer structure. The 0.3 per cent of CCSP shows a peak around 7–8 in the 2-theta degree axis which shows the presence of carbon material in the polymer-blend composite.

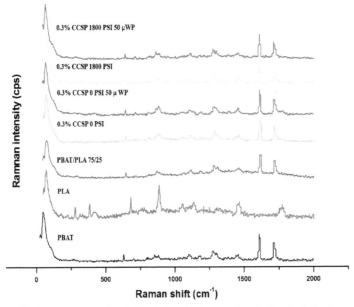

Fig. 21. Raman spectra of neat, blended, and 0.3 per cent CSP with blended polymers.

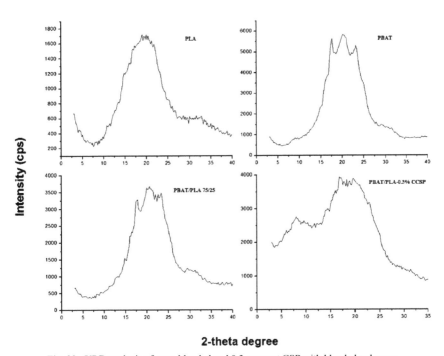

Fig. 22. XRD analysis of neat, blended and 0.3 per cent CSP with blended polymers.

5. Conclusion

In this chapter, various biodegradable polymers, used for food packaging applications and their possible fabrication methods, were presented. In applications the extruded compostable poly(butylene adipate-co-terephthalate) (PBAT)/agro-based polylactic acid (PLA) blend sheets were studied to determine the effect of blending on the microstructure, thermal and tensile properties of PLA. These blends (90/10, 80/20, 70/30, 60/40 and 50/50) were characterized by using DSC, TGA, Raman spectroscopy, XRD, SEM and tensile testing. The DSC and SEM results revealed that the two polymers are immiscible due to the presence of distinct melting points and phase-segregated morphologies in the blend structure. X-ray diffraction revealed that the PLA is amorphous while PBAT is semi-crystaline, resulting in a semi-crystalline immiscible-blend Raman spectroscopy showing frequency vibrations and intensities unique to the individual polymers. The tensile test showed that PLA led to improvement in tensile strength and modulus, while PBAT led to a significant enhancement in train-to-failure of the pristine blends systems. Also, SEM microanalysis of the fractured surfaces showed heterogeneous mixtures of the two matrices which interrupt crack paths to divert propagation and delay failure, leading to improvement in toughness. The results suggest that the 70/30 blend possessed the desirable balance strength and flexibility for flexible designs and applications. In another study, the carbon from the coconut shell powder reinforcement of 0.3 wt per cent led to increase in the tensile strength (12 per cent) and elastic modulus (27 per cent) of the 75/25 immiscible polymer blend. Further study will be conducted to determine the electrical properties of the carbon-embedded polymer blend for sensing applications. The longevity of the polymer before it biodegrades will also take place in the future.

Acknowledgements

The authors sincerely acknowledge financial support from NSF-CREST # 1735971, NSF-RISE #1459007 and AL/NSF EPSCoR # 1655280 grants.

References

Ahmad, M., N.P. Nirmal and J. Chuprom (2015). Blend film based on fish gelatine/curdlan for packaging applications: spectral, microstructural and thermal characteristics, *RSC Advances*, 5: 99044–99057.

Al-Itry, R., K. Lamnawar and A. Maazouz (2014). Reactive extrusion of PLA, PBAT with a multi-functional epoxide: Physico-chemical and rheological properties, *European Polymer Journal*, 58: 90–102.

Arruda, L.C., M. Magaton, R.E.S. Bretas and M.M. Ueki (2015). Influence of chain extender on mechanical, thermal and morphological properties of blown films of PLA/PBAT blends, *Polymer Testing*, 43: 27–37.

Bai, H., H. Xiu, J. Gao, H. Deng, Q. Zhang, M. Yang and Q. Fu (2012). Tailoring impact toughness of poly(L-lactide)/poly(ε-caprolactone)(PLLA/PCL) blends by controlling crystallization of PLLA matrix, *ACS Applied Materials & interfaces*, 4: 897–905.

Beck, B.H., M. Yildirim-Aksoy, C.A. Shoemaker, S.A. Fuller and E. Peatman (2019). Antimicrobial activity of the biopolymer chitosan against *Streptococcus iniae*, *Journal of Fish Diseases*, 42: 371–377.

Benhabiles, M.S., R. Salah, H. Lounici, N. Drouiche, M.F.A. Goosen and N. Mameri (2012). Antibacterial activity of chitin, chitosan and its oligomers prepared from shrimp shell waste, *Food Hydrocolloids*, 29: 48–56.

Bhuimbar, M.V., P.K. Bhagwat and P.B. Dandge (2019). Extraction and characterization of acid soluble collagen from fish waste: Development of collagen-chitosan blend as food packaging film, *Journal of Environmental Chemical Engineering*, 102–983.

Bikiaris, D., V. Karavelidis and G. Karayannidis (2006). A new approach to prepare poly(ethylene terephthalate)/silica nanocomposites with increased molecular weight and fully adjustable branching or cross-linking by SSP, *Macromolecular Rapid Communications*, 27: 1199–1205.

Biswas, M.C., S. Jeelani and V. Rangari (2017). Influence of bio-based silica/carbon hybrid nanoparticles on thermal and mechanical properties of biodegradable polymer films, *Composites Communications*, 4: 43–53.

Biswas, M.C., B.J. Tiimob, W. Abdela, S. Jeelani and V.K. Rangari (2019). Nano silica-carbon-silver ternary hybrid induced antimicrobial composite films for food packaging application, *Food Packaging and Shelf-Life*, 19: 104–113.

Bohlmann, G.M. (2004). Biodegradable packaging life-cycle assessment, *Environmental Progress*, 23: 342–346.

Brandelero, R.P.H., F. Yamashita and M.V.E. Grossmann (2010). The effect of surfactant Tween 80 on the hydrophilicity, water vapor permeation and the mechanical properties of cassava starch and poly(butylene adipate-co-terephthalate) (PBAT) blend films, *Carbohydrate Polymers*, 82: 1102–1109.

Brodnjak, U.V. (2017). Experimental investigation of novel curdlan/chitosan coatings on packaging paper, *Progress in Organic Coatings*, 112: 86–92.

Butnaru, E., E. Stoleru, M.A. Brebu, R.N. Darie-Nita, A. Bargan and C. Vasile (2019). Chitosan-based bio-nanocomposite films prepared by emulsion technique for food preservation, *Materials*, 12: 373.

Chang, K., M.L. Robertson and M.A. Hillmyer (2009). Phase inversion in polylactide/soybean oil blends compatibilized by poly(isoprene-b-lactide) block copolymers, *ACS Applied Materials & Interfaces*, 1: 2390–2399.

Charlebois, S. and M. Juhasz (2018). Food futures and 3D printing: Strategic market foresight and the case of structure 3D, *International Journal on Food System Dynamics*, 9: 138–148.

Chen, G., S. Dong, S. Zhao, S. Li and Y. Chen (2019). Improving functional properties of zein film via compositing with chitosan and cold plasma treatment, *Industrial Crops and Products*, 129: 318–326.

Chen, J., L. Lu, D. Wu, L. Yuan, M. Zhang, J. Hua and J. Xu (2014a). Green poly(ε-caprolactone) composites reinforced with electrospun polylactide/poly(ε-caprolactone) blend fiber mats, *ACS Sustainable Chemistry & Engineering*, 2: 2102–2110.

Chen, Y., D. Yuan and C. Xu (2014b). Dynamically vulcanized bio-based polylactide/natural rubber blend material with continuous cross-linked rubber phase, *ACS Applied Materials & Interfaces*, 6: 3811–3816.

Chi, K. and J.M. Catchmark (2018). Sustainable development of polysaccharide polyelectrolyte complexes as eco-friendly barrier materials for packaging applications, pp. 109–123, *Green Polymer Chemistry: New Products, Processes and Applications*, American Chemical Society.

Chu, Y., T. Xu, C. Gao, X. Liu, N. Zhang, X. Feng, X. Liu, X. Shen and X. Tang (2019). Evaluations of physicochemical and biological properties of pullulan-based films incorporated with cinnamon essential oil and Tween 80, *International Journal of Biological Macromolecules*, 122: 388–394.

Chung, Y.-C. and C.-Y. Chen (2008). Antibacterial characteristics and activity of acid-soluble chitosan, *Bioresource Technology*, 99: 2806–2814.

Coles, R. and M.J. Kirwan (2011). *Food and Beverage Packaging Technology*, John Wiley & Sons.

Dairi, N., H. Ferfera-Harrar, M. Ramos and M.C. Garrigós (2019). Cellulose acetate/AgNPs-organoclay and/or thymol nano-biocomposite films with combined antimicrobial/antioxidant properties for active food packaging use, *International Journal of Biological Macromolecules*, 121: 508–523.

D'amico, D.A., M.I. Montes, L.B. Manfredi and V.P. Cyras (2016). Fully bio-based and biodegradable polylactic acid/poly(3-hydroxybutirate) blends: Use of a common plasticizer as performance improvement strategy, *Polymer Testing*, 49: 22–28.

de Oliveira Filho, J.G., J.M. Rodrigues, A.C.F. Valadares, A.B. de Almeida, T.M. de Lima, K.P. Takeuchi, C.C.F. Alves, H.A. de Sousa Falcão, E.R. da Silva and F.H. Dyszy (2019). Active food packaging: Alginate films with cottonseed protein hydrolysates, *Food Hydrocolloids*, 92: 267–275.

Dong, W., M. He, H. Wang, F. Ren, J. Zhang, X. Zhao and Y. Li (2015). PLLA/ABS blends compatibilized by reactive comb polymers: Double Tg depression and significantly improved toughness, *ACS Sustainable Chemistry & Engineering*, 3: 2542–2550.

Dul, M., K.J. Paluch, H. Kelly, A.M. Healy, A. Sasse and L. Tajber (2015). Self-assembled carrageenan/ protamine polyelectrolyte nanoplexes—Investigation of critical parameters governing their formation and characteristics, *Carbohydrate Polymers*, 123: 339–349.

Essabti, F., A. Guinault, S. Roland, G. Régnier, S. Ettaqi and M. Gervais (2018). Preparation and characterization of poly(ethylene terephthalate) films coated by chitosan and vermiculite nanoclay, *Carbohydrate Polymers*, 201: 392–401.

Fabra, M.J., I. Falcó, W. Randazzo, G. Sánchez and A. López-Rubio (2018). Antiviral and antioxidant properties of active alginate edible films containing phenolic extracts, *Food Hydrocolloids*, 81: 96–103.

Furukawa, T., H. Sato, R. Murakami, J. Zhang, I. Noda, S. Ochiai and Y. Ozaki (2006). Raman microspectroscopy study of structure, dispersibility and crystallinity of poly(hydroxybutyrate)/ poly(l-lactic acid) blends, *Polymers*, 47: 3132–3140.

Galus, S. and J. Kadzińska (2019). Gas barrier and wetting properties of whey protein isolate-based emulsion films, *Polymer Engineering & Science*, 59: E375–E383.

Garlotta, D. (2001). A literature review of poly(lactic acid), *Journal of Polymers and the Environment*, 9: 63–84.

Garofalo, E., P. Scarfato, L. Di Maio and L. Incarnato (2018). Tuning of co-extrusion processing conditions and film layout to optimize the performances of PA/PE multilayer nanocomposite films for food packaging, *Polymer Composites*, 39: 3157–3167.

Goffin, A.-L., Y. Habibi, J.M. Raquez and P. Dubois (2012). Polyester-grafted cellulose nanowhiskers: a new approach for tuning the microstructure of immiscible polyester blends, *ACS Applied Materials & Interfaces*, 4: 3364–3371.

Gomes, L., H. Souza, J. Campiña, C. Andrade, A. Silva, M. Gonçalves and V. Paschoalin (2019). Edible chitosan films and their nanosized counterparts exhibit antimicrobial activity and enhanced mechanical and barrier properties, *Molecules*, 24: 127.

González, A. and C.I.A. Igarzabal (2015). Nanocrystal-reinforced soy protein films and their application as active packaging, *Food Hydrocolloids*, 43: 777–784.

Gopi, S., A. Amalraj, S. Jude, S. Thomas and Q. Guo (2019). Bionanocomposite films based on potato, tapioca starch and chitosan reinforced with cellulose nanofiber isolated from turmeric spent, *Journal of the Taiwan Institute of Chemical Engineers*.

Gou, G., C. Zhang, Z. Du, W. Zou, H. Tian, A. Xiang and H. Li (2015). Structure and property of biodegradable soy protein isolate/PBAT blends, *Industrial Crops and Products*, 74: 731–736.

Gou, Z., B. Liu, Q. Zhang, W. Deng, Y. Wang and Y. Yang (2014). Recent advances in heterogeneous selective oxidation catalysis for sustainable chemistry, *Chemical Society Reviews*, 43: 3480–3524.

Haghighi, H., R. De Leo, E. Bedin, F. Pfeifer, H.W. Siesler and A. Pulvirenti (2019). Comparative analysis of blend and bilayer films based on chitosan and gelatin enriched with LAE (lauroyl arginate ethyl) with antimicrobial activity for food packaging applications, *Food Packaging and Shelf-Life*, 19: 31–39.

Han, Y., M. Yu and L. Wang (2018). Physical and antimicrobial properties of sodium alginate/ carboxymethyl cellulose films incorporated with cinnamon essential oil, *Food Packaging and Shelf-Life*, 15: 35–42.

Hassannia-Kolaee, M., F. Khodaiyan, R. Pourahmad and I. Shahabi-Ghahfarrokhi (2016). Development of ecofriendly bionanocomposite: Whey protein isolate/pullulan films with nano-SiO$_2$, *International Journal of Biological Macromolecules*, 86: 139–144.

Heo, J., M. Choi and J. Hong (2019). Facile surface modification of polyethylene film via spray-assisted layer-by-layer self-assembly of graphene oxide for oxygen barrier properties, *Scientific Reports*, Nature Publisher Group, 9: 1–7.

Herniou, C., J.R. Mendieta and T.J. Gutiérrez (2019). Characterization of biodegradable/non-compostable films made from cellulose acetate/corn starch blends processed under reactive extrusion conditions, *Food Hydrocolloids*, 89: 67–79.

Hongsriphan, N. and S. Sanga (2018). Antibacterial food packaging sheets prepared by coating chitosan on corona-treated extruded poly(lactic acid)/poly(butylene succinate) blends, *Journal of Plastic Film & Sheeting*, 34: 160–178.

Hongsriphan, N. and A. Pinpueng (2019). Properties of agricultural films prepared from biodegradable poly(butylene succinate) adding natural sorbent and fertilizer, *Journal of Polymers and the Environment*, 1–10.

Hopmann, C., M. Adamy and A. Cohnen (2017). Introduction to Reactive Extrusion, *Reactive Extrusion: Principles and Applications*, 1–10.

Hou, X., Z. Xue, J. Liu, M. Yan, Y. Xia and Z. Ma (2019). Characterization and property investigation of novel eco-friendly agar/carrageenan/TiO_2 nanocomposite films, *Journal of Applied Polymer Science*, 136: 47–113.

Indumathi, M.P. and G.R. Rajarajeswari (2019). Mahua oil-based polyurethane/chitosan/nano ZnO composite films for biodegradable food packaging applications, *International Journal of Biological Macromolecules*, 124: 163–174.

Inventory of Effective Food Contact Substance (FCS) Notifications (n.d.). https://www.accessdata.fda.gov/scripts/fdcc/?set=FCN.

Jiang, D., X. Hu, Z. Lin, A. De'sousa, J. Xiao, W. Jin and Q. Huang (2018). Mechanical properties and crystallization behaviors of oriented electrospun nanofibers of zein/poly(ε-caprolactone) composites, *Polymer Composites*, 39: 2151–2159.

Jiang, L., M.P. Wolcott and J. Zhang (2006). Study of biodegradable polylactide/poly(butylene adipate-co-terephthalate) blends, *Biomacromolecules*, 7: 199–207.

Kaewprachu, P., C.B. Amara, N. Oulahal, A. Gharsallaoui, C. Joly, W. Tongdeesoontorn, S. Rawdkuen and P. Degraeve (2018). Gelatin films with nisin and catechin for minced pork preservation, *Food Packaging and Shelf-Life*, 18: 173–183.

Kamel, S., N. Ali, K. Jahangir, S.M. Shah and A.A. El-Gendy (2008). Pharmaceutical significance of cellulose: A review, *Express Polym. Lett.*, 2: 758–778.

Karkhanis, S.S., N.M. Stark, R.C. Sabo and L.M. Matuana (2018). Water vapor and oxygen barrier properties of extrusion-blown poly(lactic acid)/cellulose nanocrystals nanocomposite films, *Composites Part A: Applied Science and Manufacturing*, 114: 204–211.

Kataoka, T., T. Abe and T. Ikehara (2015). Crystalline layered morphology in the phase-separated blend of poly(butylene succinate) and poly(ethylene succinate), *Polymer Journal*, 47: 645.

Katerinopoulou, K., A. Giannakas, N.M. Barkoula and A. Ladavos (2019). Preparation, characterization, and biodegradability assessment of maize starch-(PVOH)/Clay nanocomposite films, *Starch-Stärke*, 71: 1800076.

Ki, H.C. and O.O. Park (2001). Synthesis, characterization and biodegradability of the biodegradable aliphatic–aromatic random copolyesters, *Polymers*, 42: 1849–1861.

Kim, J.M., I. Lee, J.Y. Park, K.T. Hwang, H. Bae and H.J. Park (2018a). Applicability of biaxially oriented poly(trimethylene terephthalate) films using bio-based 1, 3-propanediol in retort pouches, *Journal of Applied Polymer Science*, 135: 46251.

Kim, J.M., M.H. Lee, J.A. Ko, D.H. Kang, H. Bae and H.J. Park (2018b). Influence of food with high moisture content on oxygen barrier property of polyvinyl alcohol (PVA)/vermiculite nanocomposite coated multilayer packaging film, *Journal of Food Science*, 83: 349–357.

Kister, G., G. Cassanas and M. Vert (1998). Effects of morphology, conformation and configuration on the IR and Raman spectra of various poly(lactic acid), *Polymers*, 39: 267–273.

Koller, M. (2019). Switching from petro-plastics to microbial polyhydroxyalkanoates (PHA): The biotechnological escape route of choice out of the plastic predicament? *The Euro Biotech Journal*, 3: 32–44.

Krishnaswamy, R.K. and X. Sun (2010 September). PHA compositions comprising PBS and PBSA and methods for their production (WO/2010/151798).

Lasprilla-Botero, J., S. Torres-Giner, M. Pardo-Figuerez, M. Álvarez-Láinez and J.M. Lagaron (2018). Superhydrophobic bilayer coating based on annealed electrospun ultrathin poly(ε-caprolactone) fibers and electrosprayed nanostructured silica microparticles for easy emptying packaging applications, *Coatings*, 8: 173.

Lebarbé, T., E. Grau, B. Gadenne, C. Alfos and H. Cramail (2014). Synthesis of fatty acid-based polyesters and their blends with poly(L-lactide) as a way to tailor PLLA toughness, *ACS Sustainable Chemistry & Engineering*, 3: 283–292.

Li, H. and M.A. Huneault (2011). Effect of chain extension on the properties of PLA/TPS blends, *Journal of Applied Polymer Science*, 122: 134–141.

Li, H., Y. He, J. Yang, X. Wang, T. Lan and L. Peng (2019). Fabrication of food-safe superhydrophobic cellulose paper with improved moisture and air barrier properties, *Carbohydrate Polymers*.

Li, L., W. Huang, B. Wang, W. Wei, Q. Gu and P. Chen (2015). Properties and structure of polylactide/poly(3-hydroxybutyrate-co-3-hydroxyvalerate)(PLA/PHBV) blend fibers, *Polymers*, 68: 183–194.

Li, Y. and H. Shimizu (2009). Compatibilization by homopolymer: Significant improvements in the modulus and tensile strength of PPC/PMMA blends by the addition of a small amount of PVAc, *ACS Applied Materials & Interfaces*, 1: 1650–1655.

Lin, L., Y. Gu and H. Cui (2019). Moringa oil/chitosan nanoparticles embedded gelatin nanofibers for food packaging against *Listeria monocytogenes* and *Staphylococcus aureus* on cheese, *Food Packaging and Shelf-Life*, 19: 86–93.

Liu, H., F. Chen, B. Liu, G. Estep and J. Zhang (2010). Super toughened poly(lactic acid) ternary blends by simultaneous dynamic vulcanization and interfacial compatibilization, *Macromolecules*, 43: 6058–6066.

Liu, Y., S. Wang and W. Lan (2018). Fabrication of antibacterial chitosan-PVA blended film using electrospray technique for food packaging applications, *International Journal of Biological Macromolecules*, 107: 848–854.

Mallegni, N., T. Phuong, M.B. Coltelli, P. Cinelli and A. Lazzeri (2018). Poly(lactic acid) (PLA)-based tear resistant and biodegradable flexible films by blown film extrusion, *Materials*, 11: 148.

Marrez, D.A., A.E. Abdelhamid and O.M. Darwesh (2019). Eco-friendly cellulose acetate green synthesized silver nano-composite as antibacterial packaging system for food safety, *Food Packaging and Shelf-Life*, 20: 100302.

Matta, A.K., R.U. Rao, K.N.S. Suman and V. Rambabu (2014). Preparation and characterization of biodegradable PLA/PCL polymeric blends, *Procedia Materials Science*, 6: 1266–1270.

Mele, G., E. Bloise, F. Cosentino, D. Lomonaco, F. Avelino, T. Marcianò, C. Massaro, S.E. Mazzetto, L. Tammaro and A.G. Scalone (2019). Influence of cardanol oil on the properties of poly(lactic acid) films produced by melt extrusion, *ACS Omega*, 4: 718–726.

Miwa, M., Y. Nakao and K. Nara (1994). Food applications of curdlan, pp. 119–124, *Food Hydrocolloids*, Springer.

Mohanty, A.K., M. Misra and G.I. Hinrichsen (2000). Biofibres, biodegradable polymers and biocomposites: An overview, *Macromolecular Materials and Engineering*, 276: 1–24.

Monika, A.K. Pal, S.M. Bhasney, P. Bhagabati and V. Katiyar (2018). Effect of dicumyl peroxide on a poly(lactic acid)(PLA)/poly(butylene succinate) (PBS)/functionalized chitosan-based nanobiocomposite for packaging: A reactive extrusion study, *ACS Omega*, 3: 13298–13312.

Musiol, M., J. Rydz, W. Sikorska, P. Rychter and M. Kowalczuk (2011). A preliminary study of the degradation of selected commercial packaging materials in compost and aqueous environments, *Polish Journal of Chemical Technology*, 13: 55–57.

Nampoothiri, K.M., N.R. Nair and R.P. John (2010). An overview of the recent developments in polylactide (PLA) research, *Bioresource Technology*, 101: 8493–8501.

Niaounakis, M. (2014). *Biopolymers: Processing and Products*, William Andrew.

Ojijo, V., S. Sinha Ray and R. Sadiku (2013). Toughening of biodegradable polylactide/poly(butylene succinate-co-adipate) blends via in situ reactive compatibilization, *ACS Applied Materials & Interfaces*, 5: 4266–4276.

Oymaci, P. and S.A. Altinkaya (2016). Improvement of barrier and mechanical properties of whey protein isolate based food packaging films by incorporation of zein nanoparticles as a novel bionanocomposite, *Food Hydrocolloids*, 54: 1–9.

Ozilgen, S. and S. Bucak (2018). Functional Biopolymers in Food Manufacturing, pp. 157–189, *Biopolymers for Food Design*, Elsevier.

Pakravan, M., M.C. Heuzey and A. Ajji (2011). A fundamental study of chitosan/PEO electrospinning, *Polymers*, 52: 4813–4824.

Pelissari, F.M., D.C. Ferreira, L.B. Louzada, F. dos Santos, A.C. Corrêa, F.K.V. Moreira and L.H. Mattoso (2019). Starch-based edible films and coatings: An eco-friendly alternative for food packaging, pp. 359–420, *Starches for Food Application*, Elsevier.

Pérez-Arauz, A.O., A.E. Aguilar-Rabiela, A. Vargas-Torres, A.I. Rodríguez-Hernández, N. Chavarría-Hernández, B. Vergara-Porras and M.R. López-Cuellar (2019). Production and characterization of biodegradable films of a novel polyhydroxyalkanoate (PHA) synthesized from peanut oil, *Food Packaging and Shelf-Life*, 20: 100297.

Picchio, M.L., Y.G. Linck, G.A. Monti, L.M. Gugliotta, R.J. Minari and C.I.A. Igarzabal (2018a). Casein films cross-linked by tannic acid for food packaging applications, *Food Hydrocolloids*, 84: 424–434.

Picchio, M.L., L.I. Ronco, M.C. Passeggi, R.J. Minari and L.M. Gugliotta (2018b). Poly(n-butyl acrylate)–casein nanocomposites as promising candidates for packaging films, *Journal of Polymers and the Environment*, 26: 2579–2587.

Qu, L., G. Chen, S. Dong, Y. Huo, Z. Yin, S. Li and Y. Chen (2019). Improved mechanical and antimicrobial properties of zein/chitosan films by adding highly dispersed nano-TiO$_2$, *Industrial Crops and Products*, 130: 450–458.

Reddy, M.M., M. Misra and A.K. Mohanty (2014). Biodegradable blends from corn gluten meal and poly(butylene adipate-co-terephthalate) (PBAT): Studies on the influence of plasticization and destructurization on rheology, tensile properties and interfacial interactions, *Journal of Polymers and the Environment*, 22: 167–175.

Robertson, M.L., J.M. Paxton and M.A. Hillmyer (2011). Tough blends of polylactide and castor oil, *ACS Applied Materials & Interfaces*, 3: 3402–3410.

Roy, S. and J.W. Rhim (2019). Carrageenan-based antimicrobial bionanocomposite films incorporated with ZnO nanoparticles stabilized by melanin, *Food Hydrocolloids*, 90: 500–507.

Roy, S., J.W. Rhim and L. Jaiswal (2019). Bioactive agar-based functional composite film incorporated with copper sulfide nanoparticles, *Food Hydrocolloids*, 93: 156–166.

Salević, A., C. Prieto, L. Cabedo, V. Nedović and J.M. Lagaron (2019). Physicochemical, antioxidant and antimicrobial properties of electrospun poly(ε-caprolactone) films containing a solid dispersion of sage (*Salvia officinalis* L.) extract, *Nanomaterials*, 9: 270.

Sam, S.T., M.A. Nuradibah, K.M. Chin and N. Hani (2016). Current application and challenges on packaging industry based on natural polymer blending, pp. 163–184. *In:* O. Olatunji (Ed.). *Natural Polymers: Industry Techniques and Applications*, Springer International Publishing, Cham.

Samuel, C., J. Cayuela, I. Barakat, A. J. Müller, J.M. Raquez and P. Dubois (2013). Stereocomplexation of polylactide enhanced by poly(methyl methacrylate): Improved processability and thermo-mechanical properties of stereocomplexable polylactide-based materials, *ACS Applied Materials & Interfaces*, 5: 11797–11807.

Scaffaro, R., F. Sutera and L. Botta (2018). Biopolymeric bilayer films produced by co-extrusion film blowing, *Polymer Testing*, 65: 35–43.

Šešlija, S., A. Nešić, M.L. Škorić, M.K. Krušić, G. Santagata and M. Malinconico (2018). Pectin/Carboxymethylcellulose Films as a Potential Food Packaging Material, pp. 1600163 *Macromolecular Symposia*, Wiley Online Library.

Shahbazi, Y. (2018). Application of carboxymethyl cellulose and chitosan coatings containing *Mentha spicata* essential oil in fresh strawberries, *International Journal of Biological Macromolecules*, 112: 264–272.

Shankar, S., L.F. Wang and J.-W. Rhim (2018). Incorporation of zinc oxide nanoparticles improved the mechanical, water vapor barrier, UV-light barrier and antibacterial properties of PLA-based nanocomposite films, *Materials Science and Engineering*, C 93: 289–298.

Sikorska, W., M. Musiol, B. Zawidlak-Węgrzyńska and J. Rydz (2017). Compostable polymeric ecomaterials: Environment-friendly waste management alternative to landfills, *Handbook of Ecomaterials*, 1–31.

Silva, N.H., C. Vilela, A. Almeida, I.M. Marrucho and C.S. Freire (2018). Pullulan-based nanocomposite films for functional food packaging: Exploiting lysozyme nanofibers as antibacterial and antioxidant reinforcing additives, *Food Hydrocolloids,* 77: 921–930.

Siracusa, V., P. Rocculi, S. Romani and M. Dalla Rosa (2008). Biodegradable polymers for food packaging: A review, *Trends in Food Science & Technology,* 19: 634–643.

Siracusa, V., N. Lotti, A. Munari and M. Dalla Rosa (2015). Poly(butylene succinate) and poly(butylene succinate-co-adipate) for food packaging applications: Gas barrier properties after stressed treatments, *Polymer Degradation and Stability,* 119: 35–45.

Subbiahdoss, G., G. Zeng, H. Aslan, J. Ege Friis, J. Iruthayaraj, A.N. Zelikin and R.L. Meyer (2019). Antifouling properties of layer by layer DNA coatings, *Biofouling,* 1–14.

Sun, J., W. Zhou, D. Huang and L. Yan (2018). 3D Food Printing: Perspectives, pp. 725–755, *Polymers for Food Applications,* Springer.

Sun, Y., Y. Liu, Y. Li, M. Lv, P. Li, H. Xu and L. Wang (2011). Preparation and characterization of novel curdlan/chitosan blending membranes for antibacterial applications, *Carbohydrate Polymers,* 84: 952–959.

Swaroop, C. and M. Shukla (2018). Nano-magnesium oxide reinforced polylactic acid biofilms for food packaging applications, *International Journal of Biological Macromolecules,* 113: 729–736.

Tabasi, R.Y., Z. Najarzadeh and A. Ajji (2015). Development of high performance sealable films based on biodegradable/compostable blends, *Industrial Crops and Products,* 72: 206–213.

Takma, D.K. and F. Korel (2019). Active packaging films as a carrier of black cumin essential oil: Development and effect on quality and shelf-life of chicken breast meat, *Food Packaging and Shelf-Life,* 19: 210–217.

Tampau, A., C. González-Martínez and A. Chiralt (2018). Release kinetics and antimicrobial properties of carvacrol encapsulated in electrospun poly-(ε-caprolactone) nanofibres: Application in starch multilayer films, *Food Hydrocolloids,* 79: 158–169.

Tavares, L.B., N.M. Ito, M.C. Salvadori, D.J. dos Santos and D.S. Rosa (2018). PBAT/kraft lignin blend in flexible laminated food packaging: Peeling resistance and thermal degradability, *Polymer Testing,* 67: 169–176.

Theryo, G., F. Jing, L.M. Pitet and M.A. Hillmyer (2010). Tough polylactide graft copolymers, *Macromolecules,* 43: 7394–7397.

Tiimob, B.J., G. Mwinyelle, W. Abdela, T. Samuel, S. Jeelani and V.K. Rangari (2017). Nanoengineered eggshell–silver tailored copolyester polymer blend film with antimicrobial properties, *Journal of Agricultural and Food Chemistry,* 65: 1967–1976.

Tripathi, R.M., R.N. Pudake, B.R. Shrivastav and A. Shrivastav (2018). Antibacterial activity of poly(vinyl alcohol)-biogenic silver nanocomposite film for food packaging material, *Advances in Natural Sciences: Nanoscience and Nanotechnology,* 9: 025020.

Ummi, A.S. and S. Siddiquee (2019). Nanotechnology applications in food: opportunities and challenges in food industry, pp. 295–308, *Nanotechnology: Applications in Energy, Drug and Food,* Springer.

Vahedikia, N., F. Garavand, B. Tajeddin, I. Cacciotti, S.M. Jafari, T. Omidi and Z. Zahedi (2019). Biodegradable zein film composites reinforced with chitosan nanoparticles and cinnamon essential oil: Physical, mechanical, structural and antimicrobial attributes, *Colloids and Surfaces B: Biointerfaces,* 177: 25–32.

Valencia, G.A., C.G. Luciano, R.V. Lourenço, A.M.Q.B. Bittante and P.J. do Amaral Sobral (2019). Morphological and physical properties of nano-biocomposite films based on collagen loaded with laponite®, *Food Packaging and Shelf-Life,* 19: 24–30.

Vu, C.H.T. and K. Won (2013). Novel water-resistant UV-activated oxygen indicator for intelligent food packaging, Food Chemistry, 140: 52–56.

Wang, B., C. Yang, J. Wang, S. Xia and Y. Wu (2019). Effects of combined pullulan polysaccharide, glycerol, and trehalose on the mechanical properties and the solubility of casted gelatin-soluble edible membranes, *Journal of Food Processing and Preservation,* 43: e13858.

Wang, W., H. Zhang, R. Jia, Y. Dai, H. Dong, H. Hou and Q. Guo (2018). High performance extrusion blown starch/polyvinyl alcohol/clay nanocomposite films, *Food Hydrocolloids,* 79: 534–543.

Wu, C., S. Peng, C. Wen, X. Wang, L. Fan, R. Deng and J. Pang (2012). Structural characterization and properties of konjac glucomannan/curdlan blend films, *Carbohydrate Polymers*, 89: 497–503.

Wu, X., A. Liu, W. Wang and R. Ye (2018). Improved mechanical properties and thermal-stability of collagen fiber based film by cross-linking with casein, keratin or SPI: Effect of crosslinking process and concentrations of proteins, *International Journal of Biological Macromolecules*, 109: 1319–1328.

Xia, C., W. Wang, L. Wang, H. Liu and J. Xiao (2019). Multilayer zein/gelatin films with tunable water barrier property and prolonged antioxidant activity, *Food Packaging and Shelf Life*, 19: 76–85.

Xie, J., Z. Wang, Q. Zhao, Y. Yang, J. Xu, G.I. Waterhouse, K. Zhang, S. Li, P. Jin and G. Jin (2018). Scale-up fabrication of biodegradable poly(butylene adipate-co-terephthalate)/organophilic–clay nanocomposite films for potential packaging applications, *ACS Omega*, 3: 1187–1196.

Xue, P. and Z. Qiu (2015). Synthesis, thermal properties and crystallization kinetics of novel biodegradable poly(ethylene succinate-co-diethylene glycol succinate) copolyesters, *Thermochimica Acta*, 606: 45–52.

Yu, Z., B. Li, J. Chu and P. Zhang (2018a). Silica *in situ* enhanced PVA/chitosan biodegradable films for food packages, Carbohydrate Polymers, 184: 214–220.

Yu, Z., L. Sun, W. Wang, W. Zeng, A. Mustapha and M. Lin (2018b). Soy protein-based films incorporated with cellulose nanocrystals and pine needle extract for active packaging, *Industrial Crops and Products*, 112: 412–419.

Zhang, K., A.K. Mohanty and M. Misra (2012). Fully biodegradable and biorenewable ternary blends from polylactide, poly(3-hydroxybutyrate-co-hydroxyvalerate) and poly(butylene succinate) with balanced properties, *ACS Applied Materials & Interfaces*, 4: 3091–3101.

Zhang, N., Q. Wang, J. Ren and L. Wang (2009). Preparation and properties of biodegradable poly(lactic acid)/poly(butylene adipate-co-terephthalate) blend with glycidyl methacrylate as reactive processing agent, *Journal of Materials Science*, 44: 250–256.

Zhang, R., W. Wang, H. Zhang, Y. Dai, H. Dong and H. Hou (2019). Effects of hydrophobic agents on the physicochemical properties of edible agar/maltodextrin films, *Food Hydrocolloids*, 88: 283–290.

Zhao, Z., X. Xiong, H. Zhou and Q. Xiao (2019). Effect of lactoferrin on physicochemical properties and microstructure of pullulan-based edible films, *Journal of the Science of Food and Agriculture*, 99: 4150-4157.

Zhou, X., A. Mohanty and M. Misra (2013). A new biodegradable injection molded bioplastic from modified soy meal and poly(butylene adipate-co-terephthalate): Effect of plasticizer and denaturant, *Journal of Polymers and the Environment*, 21: 615–622.

CHAPTER 8

New Advances in Active Packaging Incorporated with Seaweeds for Food Preservation

João Reboleira, Andreia Miranda, Susana Bernardino
and *Maria Manuel Gil**

1. Introduction

Throughout the last century, the booming world population and rapid economic development have led the food sector towards a consistent trend of retail distribution through supermarkets (Schumann and Schmid, 2018). By the year 2000, around 80 per cent of the retail food sales in Europe and the United States were provided with large, convenient store chains, with the rest of the world following this ever-growing trend (Popkin et al., 2012). The nature of this distribution system by itself, with high demand for bulk production and large distance transportation of food supplies, favors the adoption of highly processed and very shelf-stable products. It is a format of distribution that is, therefore, heavily reliant on packaged foods and in constant demand for new packaging solutions. In addition, food waste is still a growing problem in most food supply-chains. In 2012, the European Commission estimated a yearly average of 89 million tonnes of waste generated, varying according to the country and sector. Certain perishable products, such as raw meat, can see a waste ratio of up to 40 per cent across producers, retailers and consumers (Realini and Marcos, 2014).

In order to minimize this impact, there is a lot of pressure for innovation in the fields of food packaging and storage, leading to technologies, such as modified

MARE – Marine and Environmental Sciences Centre, ESTM, Polytechnic Institute of Leiria, Campus 4 | Edifício CETEMARES, Avenida do Porto de Pesca, Peniche 2520 – 630, Portugal.
* Correponding author: maria.m.gil@ipleiria.pt

atmosphere packaging (MAP), vacuum packaging and incorporation of traditional preservatives, ensuring relatively high standards of food safety and shelf-life (Schumann and Schmid, 2018).

This development is, however, limited to the scope of the 20th century. Over the last two to three decades, there has been a shift in consumer awareness regarding food quality, the use of food additives and the sustainability of product and package (Ghaani et al., 2016). Yet despite this, the concept of food packaging has remained mostly unchanged, regardless of the many attempts at innovation. The purpose of packaging continues to be containment, protection, convenience and communication about its contents. While by no means immune to the passage of time, as new technologies and materials make their way into shelves through their presence in food packaging, it remains open to innovation as its multifaceted function encourages new developments. Tackling the current limitations of food packaging is also a direct means of economizing business procedures, reducing food waste and solving food quality and safety issues, addressing both the needs of producer and consumer alike. It should be no wonder then that research in the fields of active, intelligent and smart packaging has been increasing.

2. New Packaging Trends: Active, Intelligent and Smart Packaging

The formats of active, intelligent and smart packaging have gone beyond the traditional functions of food packaging, with the goal of having the product and its immediate environment react to provide increased information, safety, shelf-life and overall quality to the consumer. All of these enhancements have been implemented as a response to market trends and consumer demands, and as a whole, they summarize the bulk of scientific and technical innovation.

Over the course of the last decade, research on active packaging has steadily risen and has kept ahead in a number of publications when compared against intelligent and smart packaging, as can be seen in Fig. 1. Ghaani et al. have commented on these

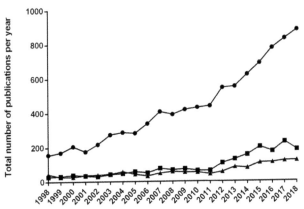

Fig. 1. Academic research on active (●), intelligent (▲), and smart (■) packaging, from 1998 to 2018 as per the total number of publications with the mentioned keywords in their titles. *Source*: apps. webofknowledge.com.

metrics in 2016, stating that the disparity was mostly due to the nature of research involved in each field, with active packaging needing constant innovation in new packaging materials, active compounds and incorporation methodologies. Intelligent packaging on the other hand, demands more sophisticated and robust systems, whose reliability, cost efficiency and efficacy are more difficult to evaluate.

The volume of research regarding active packaging is by no means a discouraging factor, as its steady growth amidst academia means that many innovation opportunities remain open and unexplored, as well as the economic systems that surround their implementation, the understanding of which is ever more important in modern, sustainability-focused research.

2.1 Active Packaging

Active packaging technology is based on the concept of deliberate altering of interactions between the product and the packaging environment in order to maintain or improve food quality and stability. European legislation has also defined the terms 'active' and 'intelligent packaging' in both Regulation 1935/2004/EC and in Regulation 450/2009/EC. In these, a broad definition of objects used in active packaging is given as 'materials and articles that are intended to extend the shelf-life or to maintain or improve the condition of packaged food' (Restuccia et al., 2010). This is achieved through the use of additives, enhancers, absorbers or combinations of these, either by the presence within the packaging environment, or as part of the packaging structure itself. Examples of active agents used in this technology include compounds of both synthetic and natural origin, such as organic acids, enzymes, bacteriocins, fungicides, natural extracts, ionic compounds and ethanol (Realini and Marcos, 2014; Sanches-Silva et al., 2014; Lee et al., 2015).

The substances responsible for the active function of the package can be kept in separate containers within it, such as the inside of a small paper sachet. Alternatively, they can be part of the packaging material itself, mixed or complexed with the polymer encasement, or as part of a film solution. In addition, these materials can be packed in layers which expose different materials to product and outside environment, depending on the design (Restuccia et al., 2010; Fang et al., 2017).

Active packaging technologies are classified according to their underlying mechanisms and their prevalence is dependent on the industrial sector where they are applied. Moisture absorbers, antimicrobial releasers, carbon dioxide emitters and releasers, oxygen scavengers and antioxidant releasers are amongst the most common mechanisms.

2.1.1 Moisture Absorbers

A very significant cause for the increased perishability of high water-activity (a_w) food product is the accumulation of water in the inner package, promoting the growth of molds and bacteria. Control of the moisture content is of utmost importance to increase the stability of these food products (Restuccia et al., 2010; Gomaa et al., 2018b). Moisture-absorbing materials absorb the in-package moisture and free water, leading to a lowering of relative humidity to a point at which condensation no longer takes place and can be used to increase the shelf-life of high a_w food products.

These types of packaging are most often used in dry and dehydrated foods, as well as general meat, poultry and fish. For this purpose, packages often include silica gel, propylene glycol, polyvinyl alcohol or diatomaceous earth. These absorbent compounds are often used as inner layers between two micro-porous or non-woven polymers. This format is highly flexible regarding package shape and size, and is thus compatible with a wide range of food products (Restuccia et al., 2010; Realini and Marcos, 2014).

2.1.2 Antimicrobial Releasers

The deployment of antimicrobial agents inside food packaging is used to delay spoilage and suppress the activity of certain microorganisms. Gradual diffusion of agents is often the goal in antimicrobial active packaging and is achieved either with the presence of an emitter, or through direct contact with immobilized substances on a contact surface (Lee et al., 2015).

Antimicrobial agents often include chlorine dioxide, organic and inorganic acids, bacteriocins and chelating agents. The choice of the agent is more dependent on the product, taking into account that some antimicrobials are only effective within limited pH and a_w limits. The main spoilage organisms must also be accounted for (Lee et al., 2015; Liu et al., 2017). As an example, the high antimicrobial activity of chlorine dioxide at neutral pH makes it an ideal agent against common strains of *Listeria monocytogenes*, thus making it an ideal agent for preservation of raw poultry and seafood (Dehghani et al., 2018). Furthermore, recent developments in seafood preservation have made use of several organic acids to enhance food safety with great success. Interesting synergistic effects between citric acid and edible coatings, such as chitosan, have yielded very high antimicrobial and antioxidant activities, albeit at a high production cost considering the nature of the packaging materials (Qiu et al., 2014). Organic acids have also been frequently used in the beverage industry, as part of inner linings of juice packages, with both antimicrobial and antioxidant capabilities (Lee et al., 2015).

Recently, bacteriocins are gathering attention as natural sources of focused antimicrobial activity. These are microorganism-produced peptides that can inhibit closely related microorganisms. They are often regarded as non-toxic, and can be selectively synthesized to be heat-resistant, though their production is still very limited, and their application scope quite small, with only nisin having worldwide approval as a safe additive (McMillin, 2017).

2.1.3 Carbon Dioxide Releasers/Emitters

Also possessing antimicrobial capabilities, but often regarded as a separate class of preservative, carbon dioxide is a colorless, combustible gas that can reduce bacterial and fungal growth, while also having the side effect of lowering the pH of certain food matrixes (Lee et al., 2015).

High CO_2 concentrations can inhibit the growth of most Gram-negative aerobic bacteria, not just through reduced relative oxygen levels, but also due to the inherent antimicrobial activity of this molecule. Carbon dioxide releasers are often present in the form of sachets or absorbent pads and they allow for smaller packaging headspaces compared to conventional MAP. Ascorbic acid and sodium bicarbonate mixtures, as well as ferrous carbonate, are often used in these systems (Fang et al., 2017; McMillin, 2017). Synergistic effects of CO_2 emitters, along with oxygen scavengers and antimicrobial agent releasers, have been reported by different studies and show a promising outcome for the use of all of these systems in food applications (Chen and Brody, 2013).

2.1.4 Oxygen Scavengers and Antioxidant Packaging.

The effect of oxygen in the preservation and storage of foodstuffs is a well-known obstacle and one of the main limitations of both the short shelf-life of perishable products and long-lasting, highly stable ones. It is responsible for the oxidation of food ingredients and products and allows the growth of aerobic, degradative and pathogenic bacteria and molds (Bolumar, Andersen and Orlien, 2011; Lee et al., 2015). This effect can be lessened with the use of oxygen absorbing systems that can both absorb residual oxygen after packaging and minimize the damage of any addition of O_2 molecules that permeate the physical barriers. When the use of these scavengers is combined with MAP or vacuum packaging, an almost complete removal of oxygen can be achieved, though there are many situations in which cumulative processing is either unfeasible or undesirable (Bolumar et al., 2011; Realini and Marcos, 2014; Lee et al., 2015; Veiga-Santos et al., 2018).

The packaging of processed meat represents an important example where the concern with oxidation is second to none and oxygen levels are reduced to as low as possible during storage. Photo-oxidation is possible with even trace amounts of this molecule and so complete removal is achieved with the use of several oxygen-scavenging technologies, including iron powder, ascorbic acid, photosensitive dye oxidation, antioxidant enzymes (glucose oxidase and alcohol oxidase), oleic and linoleic acids and immobilized yeast (Realini and Marcos, 2014).

Sharing the same goal as oxygen-scavenging package additives, antioxidant-releasing packages also inhibit or delay oxidative damage on food. These can either be reducing agents, or free radical scavengers of both synthetic and natural origin (Veiga-Santos et al., 2018). The use of these compounds as regular food additives has a long history in the food industry, as oxidative damage is one of the main causes for loss of quality. However, direct application of antioxidants can have its drawbacks, as it often casts a negative effect on color and flavor. Public rejection of food additives also makes it increasingly difficult to include antioxidant agents as part of a product's formulation. Active packaging with antioxidant properties has proven to be an effective strategy in preventing oxidative degradation in select food products, such as fresh poultry, red meats and sausages (Fang et al., 2017; Schumann and Schmid, 2018; Holman et al., 2018a).

Several other advantages can be associated with the use of antioxidant active packaging. As pointed out by Bolumar et al. (2011), the use of antioxidants incorporated in packaging polymers can lower the amount of active substance needed to reach a specific shelf-life increase in processed chicken meat. This technology also allows for localized activity, a controlled release of the active agent and the simplification of processing by removing mixing and incorporation procedures (Realini and Marcos, 2014). Common agents include butylated hydroxytoluene (BHT), butylated hydroxyanisole (BHA) and nisin. Pure standards of natural antioxidants have increasingly been used as alternatives to these agents and include α-tocopherol, caffeic acid, catechin, quercetin and carvacrol. These compounds are sometimes used after complexingn with other molecules. This is especially the case with α-tocopherol, which is frequently used after complexing with β-cyclodextrin, quercetin or with γ-cyclodextrin (Sanches-Silva et al., 2014).

2.2 Intelligent Packaging

It is impossible to discuss innovation in food packaging technologies without mention of the emerging trend of improving the communication with consumers through the use of intelligent packaging. Defined as any item or material which monitors the condition of packaged food or its surrounding medium, this type of package effectively enhances consumer safety, not by promoting a product's shelf life, but by communicating its limit and/or expiration. This information is not limited to the consumer, as it can also help retailers and manufacturers performing quality control (Restuccia et al., 2010; Lee et al., 2015; Schumann and Schmid, 2018).

Intelligent packaging has long seen representation in supermarket store shelves, as even the ubiquitous UPC barcode can be classified as a smart device—an integral part of intelligent packaging. Over the last two decades, the use of sensors and indicators has become very common in packaging of highly perishable foods, such as meat, poultry and fish. These include freshness and pathogen indicators, with numerous patents submitted over the last few years for the meat industry alone (Holman et al., 2018a).

2.3 Smart Packaging

Intelligent packaging and active packaging can work synergistically to yield what is defined as 'smart' packaging, i.e., a total packaging concept that combines the benefits arising from active and intelligent technology. As more and more active and intelligent products reach the market, there is a natural tendency of manufacturers and retailers to combine these technologies into unique final products. Innovation within the field of smart packaging can always be distinguished between new active technologies and innovative smart devices, and as such, many authors will report their work in either one field or the other (Vanderroost et al., 2014; Ghaani et al., 2016; Holman et al., 2018b).

2.4 Sustainability of Packaging Materials

Recent life-cycle assessment methodologies performed by Siracusa et al. in 2015 have identified problems related to the production of polystyrene foam packaging and stated the difficulties in minimizing the environmental impact of traditional packaging industries without major overhauls. This conclusion was mirrored by authors Siracusa et al. in 2015 and Schumann and Schmid in 2018, who stated that not only is the disposal of commercial plastics very difficult, but also is their sustainable production, which stems from limited natural resources. Increasing public awareness of the issue regarding food package waste has also led to greater investment in the development of biodegradable packaging materials.

The use of edible films, sheets and coatings is gaining rapid traction in Western markets as a means to address environmental and consumer issues. Currently, the variety of packaging material, which meets these requirements, is vast, with lipids (fatty acids, waxes, acylgycerol), polysaccharides (alginates, arabic gum, xanthan gum, chitosan), proteins (whey protein, gelatin) and composites widely available and compatible with a wide range of industries and products (Dehghani et al., 2018).

3. The Role of Natural Bioactives in New Active Packaging Solutions

3.1 Growing Trend of Natural Products as a Source of Bioactives for Active Packaging

As stated previously, the growing amounts of waste generated from food packaging and widespread awareness over the long-term environmental issues of plastic proliferation have tipped the consumer preference towards edible, organic and natural origin package materials (Chen and Brody, 2013; Realini and Marcos, 2014; Fang et al., 2017). Edible films and coatings are a good alternative for the partial or total substitution of plastic packaging due to their properties—they are biodegradable, non-toxic, environmental-friendly and, on many occasions, made with by-products of the food industry. The use of edible films and coatings can be application of active food packaging as the edibility and biodegradability of the films are extra functions not present in conventional packaging systems and it can be applied for the continuous delivery of active compounds and additives to the food matrix.

Natural, well-studied bioactives are major contenders for use in active systems, as they commonly have scientific consensus regarding their toxicological and antimicrobial potential. Often, these compounds are applied directly as their minimally or non-processed source. Examples of whole plant sources include cinnamon (*Cinnamomum zeylanicum* L.), oregano (*Origanum vulgare* L.), clove (*Syzygium aromaticum* L.), rosemary (*Rosmarinus officinalis* L.), ginger (*Zingiber officinale* Rosc.) and lemongrass (*Cymbopogon citratus (DC.)* Stapf.). Alternatively, minimally processed extracts of barley husks (*Hordeum vulgare* L.), green tea (*Camellia sinensis* L.), mint (*Mentha spicata* L.) and pomegranate peel (*Punica granatum* L.) have seen successful use as natural sources of antioxidant and antimicrobial bioactives with effective shelf-life increasing when used in active systems (Kadam et al., 2015; Sanches-Silva et al., 2015).

There are, however, some drawbacks in the use of these sources when looking for powerful antioxidant food-grade agents. Most of the extracts used are concentrated hydrophobic phases resulting from solid-liquid extractions with organic solvents, using techniques, which are not environmentally sound or completely risk-free in regard to human health risks. Many also struggle with being financially sound, given the very low concentrations of these compounds and the relatively high price of the raw materials (Sanches-Silva et al., 2014; Tavassoli-Kafrani et al., 2016). It is here that the search for alternative sources of bioactives can yield significant benefits to the packaging industry, if it can provide cheaper sources either of valuable biomass, greener extraction procedures, or a combination of both.

Throughout time, marine seaweeds have developed complex mechanisms to promote adaptation to external factors (e.g., UV radiation, salinity and temperature stress) as well as to defend themselves from biological pressures, such as competitors, grazers or parasites. To do so, these organisms divert resources into producing unique bioactive compounds that, when adequately processed, have various applications for humankind. Adding the fact that approximately half of the global biodiversity exists in marine environments, the sea and its inhabitants provide a large source of novel and potentially revolutionary bioactive compounds (Abdul Khalil et al., 2017; Roohinejad et al., 2017; Saravana et al., 2018). As such, using commonly available seaweeds as the source for both the main polymers and the supplementing antioxidant/ antimicrobial agents in bioactive films can potentially reduce production costs and create a safer, more sustainable product.

3.2 Seaweeds as a Source of Natural Bioactive Compounds

In recent years, marine resources have been widely studied due to their composition in bioactive substances with high potential of application in several areas. In particular, seaweeds, being rich in polysaccharides, phenolic compounds and other antioxidant compounds, are an important and commercially valuable renewable resource for the development of active packaging (Abdul Khalil et al., 2017).

Seaweeds are aquatic photosynthetic organisms and belong to the Eukaryota domain and two distinct kingdoms, namely the Plantae and the Chromista. In the Plantae kingdom are included green algae (Chlorophyta) and red algae (Rhodophyta) and in the Chromista, brown algae (Phaeophyta) (Pereira, 2018). The classification of seaweeds into three distinct classes, namely: green algae, red algae and brown algae depends on the combination of different photosynthetic present in their cells. The color, in case of green seaweeds, is due to the presence of chlorophyll a and b in the same proportion as higher plants. Phycoerythrin and phycocyanin mask the pigments, such as chlorophyll a and β-carotene, and are responsible for the color of red seaweeds. The dominance of the xanthophyll pigments and fucoxanthin is responsible for the color of brown seaweeds. This compound masks the other pigments, such as chlorophyll a and c, β-carotenes and other xanthophylls. The chemical composition of seaweeds between species is highly variable, with large differences in polysaccharides, protein, vitamins, minerals and bioactive compounds with antibacterial, antiviral and antifungal properties, as well as many others (Gupta and Abu-Ghannam, 2011).

3.3 Seaweeds in Detail: Main Structural Compounds

Seaweeds contain large amounts of polysaccharides, which range between 4–76 per cent of dry weight. Alginates, carrageenan and agar are the ones with highest economic and commercial significance, since these polysaccharides exhibit high molecular weights, high viscosity and excellent gelling, stabilizing and emulsifying properties. Besides this, seaweed polysaccharides are abundant and renewable resources and have properties, such as biocompatibility, biodegradability, high water retention capacity and excellent film characteristics that contribute to their increasing use in the development of biopolymers for production food packaging. These polysaccharides, due to the previously described properties, have been widely used in the food industry, as described in Table 1.

Table 1. Functional Claims of Seaweed Polysaccharides Used in the Food Industry.

Product categories	Functional activities of major seaweed polysaccharides	
Baked goods Beverages Confectionery Dairy Desserts Dressings and dips Fried foods Frozen foods Meat analogs Meat products Pasta Restructured products Sauces and gravies Snack foods Soups	Alginate	Forms gel Emulsifier Fat replacer Water binding agents Controls syneresis Provides smooth texture Creates creamy mouth feel Enhances fibre content Antioxidant activity Antimicrobial activity Increase yield Reduce production costs
	Carrageenan	Forms heat stable gel Controls syneresis Emulsifier Adds to mouth feel Adds viscosity Antioxidant activity Antimicrobial activity Anti-browning activity Moisture retention Enhances texture Forms creamy and smooth gel Enhances fiber content Increase yield Reduce production costs
	Agar	Forms gel Syneresis control Emulsifier Adds texture Reduces sugar bloom Enhances fiber content Increase yield Reduce production costs

3.3.1 Alginate

Alginates, also known as alginic acids or algins, are commonly found in brown algae as the most abundant structural polysaccharide (Davis et al., 2003). They are unbranched polymers consisting of $(1\rightarrow4)$-linked β-D-mannuronic acid (M) and α-L-guluronic acid (G) rich regions with alternate G- and M-rich regions (Fig. 2). The linear alginate polymer is composed of homopolymorphic M or G regions, interspersed with alternate M and G regions, where a regular repeating pattern is absent (Draget et al.,

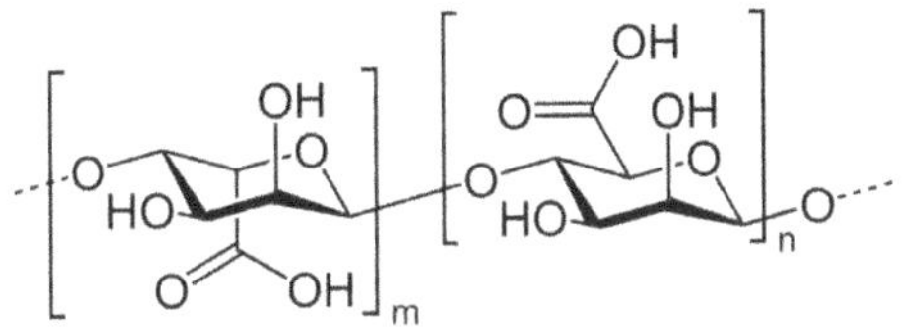

Fig. 2. Structure of alginate monomers, β-D-mannuronic acid and α-L-guluronic acid.

2005). The free hydroxyl and carboxyl groups of alginate, which are distributed along the polymer chain backbone, make the chemical modification, which can alter the characteristics of native alginate (Yang et al., 2011).

3.3.2 Carrageenans

Carrageenan-extracted certain species of red algae possess also good film-forming properties. They have alternate monomers of β-D-galactose and α-D-galactose (O'Sullivan et al., 2010). Commercially, carrageenans are classified as lamda (λ), kappa (κ) and iota (ι) carrageenans (Fig. 3), depending on the gel-forming ability (Kariduraganavar et al., 2014).

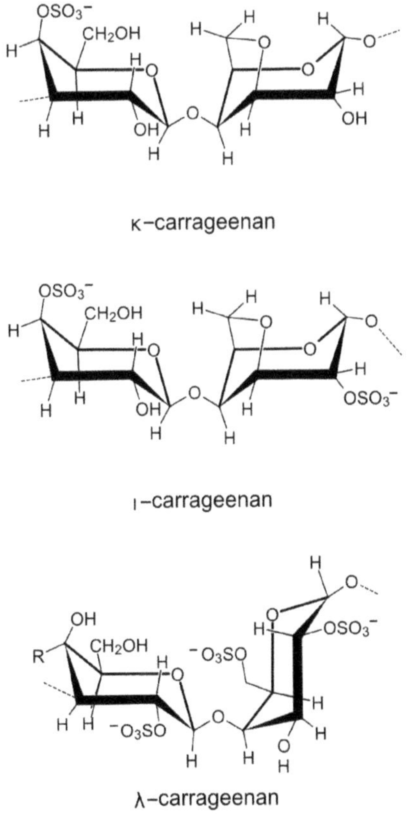

κ–carrageenan

ι–carrageenan

λ–carrageenan

Fig. 3. Basic structures of kappa, iota and lambda carrageenan.

3.3.3 Agar

Similar to carrageenan, agar can be extracted from certain families of Rhodophyta. Agar is a linear polymer with altering 3-linked β-D-galactopyranosyl and 4-linked 3,6-anhydro-α-L-galactopyranosyl units (Glicksman, 1987). The structure of monomers of agar liner polymer is given in Fig. 4.

Agar can be divided into two types—agarose and agaropectin—where agarose is the fraction having the greatest gelling capacity. It is an altering copolymer of 3-linked β-D-galactopyranose and 4-linked 3,6-anhydro-α-L-galactopyranose units (Glicksman, 1987). Agaropectin is structurally similar to agar, except that some units have 4, 6-O-(1-carboxyethylidene)-D-galactopyranose or by sulfated or methylated sugar residues (O'Sullivan et al., 2010).

The properties and applications in food packaging of the polysaccharides obtained from algae are summarized in Table 2 (Abdul Khalil et al., 2017).

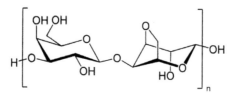

Fig. 4. Structure of agar constituints β-D-galactose and α-L-galactose.

Table 2. Applications of Seaweed Polysaccharides in Food Packaging.

Types of composites	Applications	References
Alginate/Nanocrystalline cellulose	Polymeric packaging films for food	Huq et al., 2012
Alginate/Starch	Packaging of Precooked Ground Beef Patties	Wu et al., 2001
Carrageenan/Locust bean gum/ Organically modified nanoclay	Shelf-life extension of food	Martins et al., 2013
Carrageenan/Grapefruit seed extract	Active food packaging	Kanmani and Rhim, 2014
Agar/Nanoclay	Biodegradable food packaging	Rhim, 2011
Alginate/Polycaprolactone/Oregano or savory or cinnamon essential oils	Edible film for food	Salmieri and Lacroix, 2006
Alginate/Oregano or cinnamon or savory essential oils	Preservative coating on bologna and ham slices	Oussalah et al., 2016
Alginate/Apple puree/Lemongrass or oregano or vanillin essential oils	Prolong shelf-life of fresh cut 'Fuji' apples	Rojas-Graü et al., 2007
Alginate/Lemongrass essential oil	Quality retention of fresh cut pineapples	Azarakhsh et al., 2014
κ-carrageenan/Montmorillonite/ *Zataria multiflora* Boiss essential oil	Antimicrobial packaging for food	Shojaee-Aliabadi et al., 2014
Agar/Nanocrystalline cellulose/ Savory essential oil	Active packaging for improving the safety and shelf-life of foodstuff	Atef et al., 2015

3.4 Seaweeds in Detail: Bioactive Compounds

Another important topic is the possibility to incorporate natural bioactive compounds, adding functionality to food packaging. Seaweeds are reported to be a valuable source of bioactive compounds, such as antioxidants and antimicrobials (Table 3) (Murray et al., 2013). Nowadays, it is possible to obtain seaweeds from aquaculture throughout the year, making them a sustainable and cost-effective source of natural bioactive compounds.

Table 3. Sources of Bioactive Marine Compounds.

Active Compound	Seaweed	References
Fucoidan	*Fucus vesiculosus*	Rocha de Souza et al., 2007
	Saccharina japonica	Saravana et al., 2016
	Ascophyllum nodosum	Yuan and Macquarrie, 2015
Sulfated polysaccharides	*Dictyota cervicornis*	Costa et al., 2010
	Laminaria japonica	Zhang et al., 2010
Phlorotannins	*Eisenia bicyclis*	Shibata et al., 2008
	Ecklonia cava	Ahn et al., 2007
	Saccharina japonica (Laminaria japonica)	Saravana et al., 2018
	Carpophyllum plumosum, Carpophyllum flexuosum, Ecklonia radiata	Zhang et al., 2018
Fucoxantina	*Saccharina japonica (Laminaria japonica)*	Saravana et al., 2018
Polyphenols	*Fucus vesiculosus*	Wang et al., 2009
	Kappaphycus alvarezii	Kumar et al., 2008
	Gelidella acerosa	Devi et al., 2008
	Palmaria palmate	Yuan et al., 2005
	Sargassum siliquastrum	Lim et al., 2002

3.4.1 Sulfated Polysaccharides

Seaweeds are an important source of sulphated polysaccharides, whose chemical structure varies according to species. The major polysaccharides present in brown (Phaeophyceae), red (Rhodophyceae) and green (Chlorophyceae) algae are, respectively, fucoidan and laminarans, carrageenan and ulvan (Fig. 5) (Kim, 2011).

Sulphated polysaccharides derived from seaweeds present high antioxidant potential and are widely used in the food industry. The antioxidant activity of these compounds depends on their structural characteristics, namely the degree of sulfation, molecular weight, type of major sugar and glycosidic branching (Zhang et al., 2003; Qi et al., 2005).

3.4.2 Phlorotannins

Phlorotannins are unique phenolic compounds that belong to a large class of marine secondary metabolites exclusively produced by brown algae. They are often

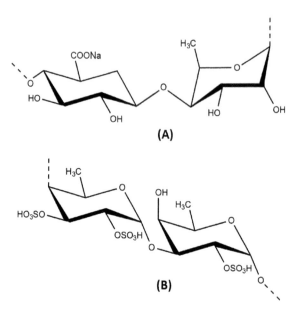

Fig. 5. Antioxidant sulphated polysaccharides derived from marine algae: (A) ulvan and (B) fucoidan.

considered to act as a chemical defense against herbivores and possess primary functions, such as contributing to cell wall structure and reproduction. Several studies have reported the high antioxidant potential of these compounds on free radicals (Kang et al., 2003; Heo et al., 2005; Shibata et al., 2008). They have demonstrated radical sequestration activities against superoxide and DPPH radicals compared to those of ascorbic acid and α-tocopherol. These potent antioxidant compounds are oligomers or polymers of phloroglucinol (1,3,5-trihydroxybenzene), connected by aryl–aryl bonds (fucols), ether bonds (phlorethols, hydroxyphlorethols, fuhalols), or both (fucophlorethols), or with a dibenzodioxin linkage (eckols and carmalols) (Fig. 6) (Shibata et al., 2008). These phlorotannins demonstrate radical sequestration

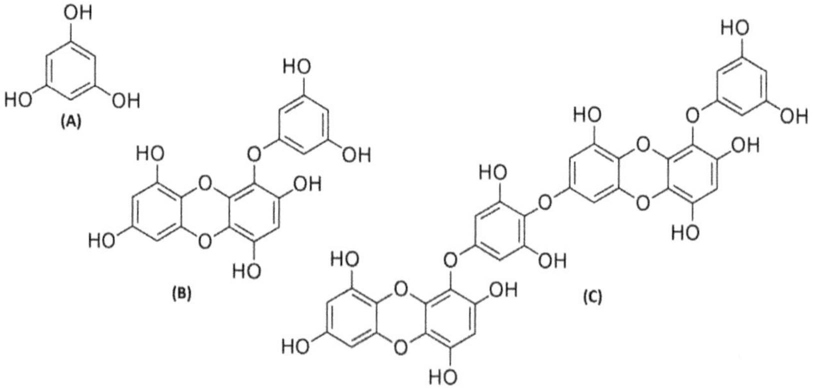

Fig. 6. Phlorotannins derived from marine algae: (A) phloroglucinol, (B) eckol, and (C) dieckol with antioxidant activity.

activities against superoxide and DPPH radicals comparable to those of ascorbic acid and α-tocopherol. Thus, phlorotannins can be used as potential antioxidants in the food industry.

3.4.3 Carotenoids

Carotenoids, namely fucoxanthin and astaxanthin (Fig. 7) are pigmented compounds that are synthesized by seaweeds and have strong antioxidant activity due to quenching of singlet oxygen and scavenging of free radicals. Heo et al. (2008) studied the cytoprotective effect of fucoxanthin isolated from brown algae, *Sargassum siliquastrum* against H_2O_2-induced cell damage. The results showed that fucoxanthin effectively inhibited intracelular ROS formation, DNA damage and apoptosis induced by H_2O_2. On the other hand, astaxanthin is effective as α-tocopherol in inhibiting free radical-initiated lipid peroxidation in rat liver microsomes and is 100 times higher than α-tocopherol in protecting rat mitochondria against Fe^{2+}catalyzed lipid peroxidation *in vivo* and *in vitro*. Thus, these results indicate that fucoxanthin and astaxanthin can be used as a source of natural antioxidant and ingredients in functional food related to the prevention and control oxidative stress (Kim, 2011).

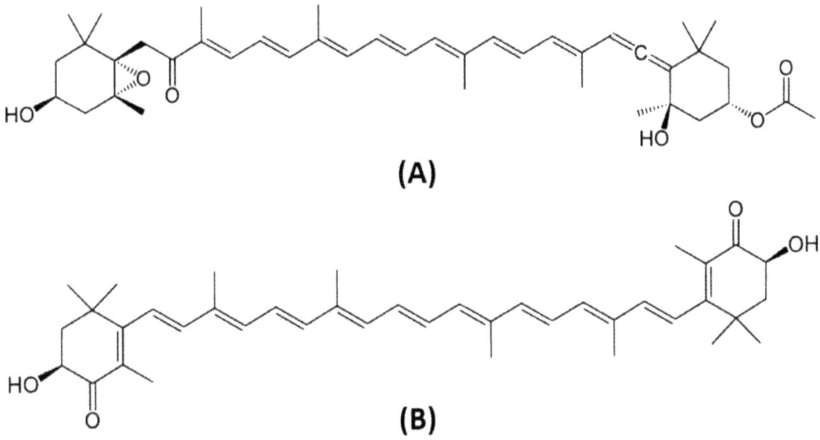

(A)

(B)

Fig. 7. Antioxidant carotenoids derived from marine algae: (A) fucoxanthin, and (B) astaxanthin.

4. The Extraction of Seaweed Compounds for Packaging Applications

The challenge of exploiting seaweed as a resource lies in the extraction of the nutrients and bioactive components, because some substances are not easily extractable, display low yields, thermolability, bioactivity loss and other problems. Accordingly, any promising research in this field will have its innovative core in the nature and articulation of extraction methodologies. The selection of appropriate methodologies must follow economic, technological complexity, environmental sustainability and process harmlessness for human and animal health criteria. The state-of-the-art extraction technologies encompass enzyme-assisted processes, ultrasounds, high pressure, new solvents, aqueous solutions with specific pH, temperature and ionic strength, centrifugation and membrane processes, ranging from molecular weight cut-off dialysis to ultrafiltration and nanofiltration.

4.1 Conventional Extraction Methods

The most extended way of extracting seaweeds' bioactive compounds is by using conventional extraction, which consists of a solid-liquid extraction, using huge quantities of solvents and a long time for the extraction. This technique could be carried out by maceration, applying heat and/or agitation, or with a Soxhlet extractor. Nevertheless, this technique is not an environmentally friendly technique due to the huge amount of solvent and energy consumption.

The extraction efficiency of any conventional method depends mainly on the choice of solvents and one of the factors of more relevance in the choice of solvent is the polarity of the target. Not only is this factor important but there are other factors that should also be considered in solvent selection for bioactive compound extraction as is the case of the molecular affinity between solvent and solute, mass transfer, co-solvent use, environmental safety, human toxicity and financial viability (Kadam et al., 2015).

Soxhlet extraction is carried out in a Soxhlet extractor. During Soxhlet extraction, the suitable solvent evaporates, condenses and then passes through the homogenized and ground solid sample located in the extraction thimble. After that, the solvent returns to the flask together with the extract. The principal fragilities attributed to Soxhlet extraction are the long duration of process (10–24 hours), environmental pollution due to the high consumption of organic solvent (300 mL/sample) and the fact that the compounds that are extracted must be stable at the boiling point of the extraction solvent. Due to the large volume of extract that is equal to the initial solvent volume, it must be concentrated mostly by evaporation (Azmir et al., 2013).

Maceration became a popular and inexpensive way to get essential oils and bioactive compounds. For small-scale extraction, maceration can be a method widely used; however, it usually consists in several steps and the time spent in all this process may be considered as a disadvantage in the choice of this method. The first step is grinding the sample into small particle to increase the surface area for proper mixing with solvent. Then, the appropriate solvent is added in a closed vessel and thirdly, the liquid is strained off but the marc, which is the solid residue of this extraction process, is pressed to recover large amount of occluded solutions. The strained and the press out liquid that is obtained is mixed and separated from impurities by filtration. Occasional shaking in maceration facilitates extraction in two ways: (a) increases diffusion, (b) removes concentrated solution from the sample surface for bringing new solvent to the menstruum for more extraction yield (Azmir et al., 2013).

Hydrodistillation is a traditional method for extraction of bioactive compounds and essentials oils. It can be divided in three types: water distillation, water and steam distillation and direct steam distillation. In hydrodistillation, the samples are packed in a still compartment, water is added in sufficient amount and then brought to boil. In an alternative method, direct steam can be injected into the sample and hot water and steam will act as the main influential factors to free bioactive compounds. Then indirect cooling by water condenses the vapor mixture of water and oil and the condensed mixture flows from the condenser to a separator, where oil and bioactive compounds separate automatically from the water. Hydrodistillation involves three main physicochemical processes: hydrodiffusion, hydrolysis and decomposition by heat. At a high extraction temperature, some volatile components may be lost. This drawback limits its use for thermo labile compound extraction (Azmir et al., 2013).

4.2 Green Extraction Methods

Green extraction is based on principles that contemplate (i) the use of renewable plant resources (e.g., seaweeds), (ii) the use of alternative solvents replacing petrochemicals ones, (iii) the reduction of both energy consumption, (iv) business unit operations and (v) the transformation of waste residue into co-products and by-products via processes that (vi) do not degrade nor contaminate the raw material. These specifications are in accordance with the concept of circular economy, which requests eco-friendly innovations and sustainable feedstock as macroalgae to close the loop of the products' lifecycle. In this context, supercritical fluid extraction (SFE), microwave-assisted extraction (MAE), ultrasound-assisted extraction (UAE) and pressurized liquid extraction (PLE) are good examples of ecological technologies that have been used in seaweed compounds of interest research (Table 4) (Falkenberg et al., 2019).

4.2.1 Supercritical Fluid Extraction (SFE)

SFE is based on the principle of extraction with fluids in their supercritical conditions, i.e., temperature and pressure are raised above their critical point with characteristics of both liquids and gases. The fluid density is similar to the values found for liquids, while its viscosity is close to values of gas. In SFE method, the carbon dioxide (CO_2) is the most used solvent due to its nontoxicity, safety and low cost. Conditions during the extraction, especially pressure and temperature, are responsible for selectivity and solubility of the various compounds in the supercritical fluid. As CO_2 has low critical temperature and pressure, bioactive compounds stay preserved and no degradative changes can occur (Michalak et al., 2016). The main advantages of supercritical fluid extraction are as follows:

1) supercritical fluid has a higher diffusion coefficient, lower viscosity and surface tension than a liquid solvent causing more penetration to sample matrix and higher mass transfer so the extraction time is significantly reduced;

2) complete extraction is possible due to the repeated reflux of supercritical fluid to the sample;

3) solvation power of supercritical fluid is adjustable by varying the temperature and/or pressure leading to the higher selectivity of supercritical fluid than liquid;

4) in supercritical fluid extraction, the supercritical fluid can be depressurized and the separation of solute from solvent becomes easier, resulting in significant time savings;

5) in supercritical extraction, the ideal fluids have their critical temperature points near room temperature, which is suitably for thermolabile compounds (vitamin, polyphenols, etc.);

6) in SFE, small amounts of the sample can be extracted, which will save time and further costs of experiment;

Table 4. Green Exctraction Techniques for Seaweeds with Associated Bioactivities.

Technology	Seaweed species	Highlights	Experimental conditions	References
Supercritical Fluid Extraction (SFE)	Cladophora glomerata, green	Indole acetic acid (IAA), indole butyric acid (IBA), phenyleacetic acid (PAA), naphtyleacetic acid (NAA), trans-zeatin (TZ), kinetin (KA), isopentenyladenine (IA), 6-benzylaminopurine (6-BA), and abscisic acid (ABA)	Solvent, CO_2; temp., 40°C; pres., 500 bar	Górka and Wieczorek, 2017
	Gracilaria mammillaris, red	Carotenoids, phenolic compounds and antioxidant activity (TBARS)	Solvent, CO_2; temp., 40 to 60°C; pres.,100 to 300 bar; flow rate, 0.4 kg h^{-1}; co-solvent 2 to 8 wt% EtOH	Ospina et al., 2017
	Saccharina japonica, brown	Total carotenoids and fucoxanthin-rich extracts with antioxidant activities	Solvent, CO_2 plus sunflower oil 2 wt%; temp., 50.62°C; pres., 300 bar	Saravana et al., 2018
	Sargassum muticum, brown	Extracts with antioxidant activities	Solvent, CO_2; temp., 60°C; pres., 152 bar;co-solvent, EtOH	Anaëlle et al., 2013
	Ulva prolifera, green	Polysaccharides extracted exhibited antioxidant and pancreatic lipase activities	Water plus HCl 0.1 M at 150°C, 500 W, 15 min	Lee et al., 2015; Li et al., 2019
Microwave-assisted extraction (MAE)	Solieria chordalis, red	Carrageenan-rich extract presented antiviral activity against Herpes simplex virus type 1 (HSV-1)	0.5% KOH at 105°C for 25 min	Boulho et al., 2017
	Ecklonia radiata, brown	Antioxidant extract produced by microwave-assisted enzymatic extraction. Carbohydrases (Viscozyme, Celluclast and Ultraflo) and proteases (Alcalase, Neutrase and Flavourzyme) were used	Buffers specific for each enzyme tested, 50 °C for 3 h	Charoensiddhi et al., 2015
	Fucus vesiculosus, brown	Extract with sulfated polysaccharides (fucoidan), known to present anti-leishmania, antiviral,and antibacterial effects besides the antioxidant activities	Pressure, 120 psi; biomass/solvent volume, 1/25 for 1 min	Rodriguez-Jasso et al., 2011
	Hormosira banksii, brown	Polyphenol-rich extract with antioxidant activity	Solvent, EtOH 70% (v/v); 30°C, 50 Hz, and 150 W for 60 min	Dang et al., 2017

Extraction method	Seaweed species	Description / activity	Conditions	Reference
Ultrasound-assisted extraction (UAE)	*Ascophyllum nodosum*, brown	Aqueous extract with polyphenols and carbohydrate and antioxidant activity	Solvent, water, 35°C, 20 kHz and <1000 W for 4 min	Moreira et al., 2017
	Gelidium pusillum, red	Phycobili protein-rich extracts wherein part of the extraction process is ultrasonication	Solvent, buffer; ultrasonication amplitude (60 to 120 μm), 30–40°C for 1 to 10 min	Mittal et al., 2017
	Sargassum muticum, brown	Fucoidan-rich and phlorotannin-containing extracts with antioxidant activities	Solvent: water, 25°C, 40 kHz, and 150 W for 5 to 30 min	Flórez-Fernández et al., 2017
	Laurencia obtusa, red	Extracts with polyphenols and antioxidant activities	Solvent: EtOH 95%, 30–50°C, 40 kHz, and 250 W for 30 to 60 min	Topuz et al., 2016
	Ascophyllum nodosum and *Laminaria hyperborean*, brown	Aqueous extract with antioxidant activities and anti-bacterial activities against *Staphylococcus aureus*, *Listeria monocytogenes*, *Escherichia coli*, and *Salmonella typhimurium*	Solvent: water or water plus HCl 0.1 M, 20 kHz, and 750 W for 15 min	Kadam et al., 2015
Pressurized liquid Extraction (PLE)	*Saccharina japonica*, brown	Fucoidan-rich extracts with antioxidant activities	Solvents: water, EtOH (25–70%), NaOH (0.1%), or formic acid (0.1%); temp., 80–200°C; pres., 5–100 bar	Saravana et al., 2016
	Saccharina longicruris and *Ascophyllum nodosum*, brown; *Ulva lactuca*, green	Aqueous extract with antioxidant activities and anti-bacteria activities against *Escherichia coli*, *Micrococcus luteus* and *Brochothrix thermosphacta*	Solvent, EtOH; temp., 50°C; pres., 69 bar for 5 min	Boisvert et al., 2015
	Fucus serratus and *Laminaria digitata*, brown; *Gracilaria gracilis*, red and *Codium fragile*, green	Aqueous and ethanolic extracts with antioxidant activities	Solvent, water and EtOH 80%, temp., 100 or 120°C; pres., 90 or 103 bar for 25 min	Heffernan et al., 2014
	Ascophyllum nodosum, *Pelvetia canaliculata*, and *Fucus spiralis*, brown; *Ulva intestinalis*, green	Aqueous and ethanolic extracts with antioxidant activities	Solvent, water and EtOH 80%; temp., 100 or 120°C; pres., 90 or 103 bar for 25 min	Tierney et al., 2013
	Sargassum muticum, brown	Extracts with antioxidant activities	Solvent, EtOH (25 and 75%); temp., 120°C; pres., 103 bar for 20 min	Anaëlle et al., 2013
	Himanthalia elongata, brown	Aqueous and ethanolic extracts with antiviral activity against Herpes simplex virus type 1 (HSV-1)	Solvent, water or EtOH; temp., 100°C; pres., 103 bar for 20 min	Santoyo et al., 2011

7) SFE assumes small amounts of solvent and is considered as environmentally friendly;
8) supercritical fluid can be reused, thus minimizing waste generation.

Typical instrumentation intended for supercritical fluid extraction consists of the following basic parts: CO_2 tank, mobile phase tank, heat exchanger, flow meter, a pump to pressurize the gas, cosolvent vessel, cosolvent pump and mixer (Žlabur et al., 2018).

4.2.2 Microwave-Assisted Extraction (MAE)

The main mechanism in microwave-assisted extraction is the use of microwave energy. Microwaves are electromagnetic fields in the frequency range from 300 MHz to 300 GHz. The plant material is bombarded with electromagnetic radiation in the microwave range (300 MHz to 300 GHz), which causes rapid rotation of the polarized molecules (e.g., water), leading to the accelerated movement of molecules (vibration), creating friction and heat generation. In general, the mechanism of microwave heating can be classified into two major categories: dipolar rotation and ionic conduction. Dielectric constant is the most important characteristic of solvents and plant materials in the application of microwaves. Solvents with lower dielectric constant absorb less microwave energy. Dielectric properties of plant materials depend on chemical composition, temperature and frequency. In ionic conduction, all charged particles of plant matrices under microwave frequency will be accelerated in one direction and then in the opposite, resulting in collision with nearby particles. The particles are set into more agitated motion and heat is generated. Microwave assisted extraction shows a number of advantages compared to conventional extraction methods. The extraction of bioactive compounds assisted by microwaves presupposes the possibility of using different types of solvents, polar and non-polar, the extraction time is significantly reduced and the recovery of the analyte is significantly higher than the conventional methods of isolation (Žlabur et al., 2018).

Non-solvent microwave extraction (SFME) is a combination of microwave heating and dry distillation, performed at atmospheric pressure without the addition of solvent or water. The SFME method provides easy application and principle of operation, which involves placing vegetable material in a microwave reactor without addition of any solvent. Under microwave energy, the water molecules in the sample cells begin to rotate, heat is generated, which distends the sample cells and leads to the rupture of the sample cell structure (Žlabur et al., 2018).

4.2.3 Ultrasound-assisted Extraction (UAE)

UAE uses ultrasound waves with a frequency above 20 kHz to 100 kHz. These waves cause the creation of bubbles and zones of high and low pressure. When bubbles collapse in the strong ultrasound field, cavitation occurs. The implosive collapse, cavitation, near liquid-solid interfaces causes breakdown of particles, which means that mass transfer is increased and bioactive compounds are released from biological matrix (Žlabur et al., 2018).

Ultrasound equipment can be ultrasonic bath (indirect sonification) or ultrasonic probe (direct sonification). The differences between these two are operating conditions and the way the ultrasound waves affect the sample. Ultrasonic bath operates at frequency of 40–50 kHz and at power of 50–500 W, but ultrasonic probe can operate only by the frequency of 20 kHz. The samples are immersed in the ultrasonic bath, whereas, the ultrasonic probe is inserted into the sample. Costs of the equipment are lower than the other alternative extraction techniques and a wide variety of solvents can be used (Tiwari, 2015). UAE operates at low temperatures which enable preservation of thermolabile compounds and prevents complete damage of the structure. Low amounts of solvent are used and the working time of extraction is reduced, which makes UAE a fast, inexpensive method as compared to traditional methods (Chemat et al., 2010).

4.2.4 Pressurized Liquid Extraction (PLE)

Pressurized liquid extraction (PLE) has several known synonyms: pressurized fluid extraction (PFE), accelerated fluid extraction (ASE), enhanced solvent extraction (ESE) and high-pressure solvent extraction (HSPE). The main principle of PLE is application of high pressure, which uses liquids at temperatures higher than their normal boiling point. In PLE, high pressure facilitates the extraction of bioactive compounds.

The basic function of PLE is to improve and accelerate the classical solvent extraction, using pressures in range from 3–20 MPa and organic liquid solvents at high temperature (50–200°C), which ensures rapid extraction of bioactive compounds. In pressurized liquid extraction, the high pressure allows faster penetration of solvent (liquid) into the solid matrix and the increase of temperature in the system lowers the polarity of the solvent and decreases its dielectric constant, so that the solvent can be used for the match the polarity of a solvent and the bioactive compounds intended for extraction (Otero et al., 2018). The final result of high-temperature effect is the promotion of higher analyte solubility, increasing the solubility and mass transfer rate while simultaneously decreasing the viscosity and surface tension of solvents and thereby improving the extraction rate. The following are the advantages of pressurized liquid extraction: faster extraction process, reduced extraction time, use of less solvents, higher yields of extracted compounds are achieved compared to the conventional solvent extraction techniques. The instrumentation of the PLE application is simple but still little commercial equipment is available (Žlabur et al., 2018).

5. Application of Seaweed Bioactives in Active Packaging

Seaweeds' bioactive compounds have a long history of application in the food industry alone. The market for functional foods is still in increasing demand and the huge variety of seaweed species means that the potential for interesting compounds is virtually unending (Roohinejad et al., 2017). Insight into how these compounds are extracted and used in miscellaneous food applications can be a useful first step when attempting to find ways to tap the potential of seaweeds in active packaging applications.

The modern food industry is facing challenges and requires specific approaches to overcome them. As referred before, one of these challenges is related to the packaging of food products with a short shelf-life period (such as raw fish and meat). Although the use of conventional packaging materials such as plastics and their derivatives is effective for food transportation and distribution, it compromises product shelf-life and creates serious environmental problems. Therefore, the use of active bio-based films as packaging material is still one of the most promising ways for maintaining food quality and safety. In addition, the demand for active packaging systems, comprised of a low-environmental-impact polymer matrix and natural antimicrobial/antioxidant agents, is growing significantly. The incorporation of marine natural antioxidants into films and edible coatings modifying their structure, improving their functionality and applicability in foods is gaining attention. In these films, the incorporated seaweeds bioactive compounds are released in a slow controlled process to maintain an adequate concentration of these compounds for a certain period. Furthermore, the use seaweeds bioactive extracts for the production of the nanofibers to be used in the ultrathin thermoplastic films, taking into account their antioxidative properties that may contribute to increase the shelf-life of the packed products is also gaining attention.

Listed in Table 5 are a variety of recent publications regarding seaweed bioactives used in active packaging, as well as some other studies, not related to packaging technologies themselves, but whose insight and conclusions can be applied to the principles of active packaging research.

6. Legal Aspects of Active Packaging in the EU

For several years, European legislation on food contact materials for packaging has provided guidelines to ensure the protection of consumer's health and to ensure that no food contact material, due to undesirable chemical reactions, would modify the composition or organoleptic properties of the food product (Restuccia et al., 2010). However, the legislation had several shortcomings and, in 2004, Reg. 1935/2004 EC repealed all directives previously in force (Directives 80/590/EEC and 89/109/EEC), allowing food packaging to benefit from innovation.

The purpose of regulation 1935/2004 EC is to ensure the effective functioning of the internal market with regard to the placing on the market of materials and objects intended to come directly or indirectly in contact with food, while ensuring a high level of protection for human's health and consumer's interests. This regulation applies to materials and objects, including active and intelligent materials and objects, intended to come into contact with food. In addition, it describes the general requirements on the safety of these materials and objects, as well as the authorization process for new active and intelligent materials (Restuccia et al., 2010).

The authorization process for new active and intelligent materials is the responsibility of the European Food Safety Authority (EFSA). The applicant shall submit the application for authorization, which must be accompanied by a technical file and submitted to the competent authority of the Member State. Thereafter, EFSA shall inform the other Member States and the Commission about the application submitted, and shall publish the same and any other supplementary information submitted by the applicant. EFSA has a period of six months from the receipt of a valid request to give an opinion on the evaluation of the application. This evaluation takes into account the substance safety criteria and intended conditions for use of the

Active Packaging with Seaweed Compounds 211

Table 5. Recent Publications Reporting the Use of Seaweed Biomass as a Means to Promote Food Shelf-life, either Through Use in Active Packaging, Packaging Materials, or as Additives.

Target product	Seaweeds	Processing	Results	References
Packaging for general foodstuff	Alginic acid (65–70% guluronic acid, 5–35% mannuronic acid) from non-specified brown algae	3% Alginate solutions with 1 to 8% nanocrystalline cellulose were cast on petri dishes and dried at room temperature. Further treatment with 1% $CaCl_2$ solution.	Polymeric matrix displayed the best mechanical and thermal properties at 5% nanocrystalline cellulose content.	Huq et al., 2012
Fresh-cut pineapple quality retention	Food grade commercial sodium alginate	1.29% sodium alginate solution with glycerol and lemongrass (*Cymbopogon citratus*) extract.	Sensory score was negatively affected by the addition of lemongrass extract, but product firmness, microbial profile and overall quality was preserved for longer with the use of the films.	Azarakhsh et al., 2014
Active package for general foods	Commercial carrageenan	Films produced from a carrageenan, glycerol and grape seed extract solution. Cast on a petri dish and dried at room temperature.	Films enriched with grape seed extract exhibited good antimicrobial activity vs. *L. monocytogenes* and *S. aureus*, and moderate activity vs. *E. coli* and *B. cereus*. Physical and mechanical properties were generally worse upon addition of extract.	Kanmani and Rhim, 2014
Modified atmosphere packaged pork patties – Increase on shelf-life	*Ulva lactuca* and *Ulva rigida*	Acid hydrolysis of seaweeds, followed by neutralization of supernatant, ultrafiltration, and freeze-drying. 1000 mg of extract per kg of meat.	Incorporation of extract had a minimal effect on microbiological counts and discoloration vs. control, but a provided a noticeable reduction on lipid oxidation.	Lorenzo et al., 2014
Production of antimicrobial films for general foodstuff	Commercial κ-carrageenan	1% κ-carrageenan solutions mixed with glycerol, Tween 80, and a range of plant extracts. Casting into smooth surfaced dried at 30°C.	Film permeability was significantly lower upon addition of plant extract. These also provided enhanced antioxidant and antimicrobial activities.	Shojaee-Aliabadi et al., 2014

Table 5 contd. ...

... Table 5 contd.

Target product	Seaweeds	Processing	Results	References
Production of antimicrobial films for general foodstuff	Food grade commercial agar	1.5% agar solution mixed with glycerol, nanocrystalline cellulose, Tween 80 and *Satureja hortensis* extract. Casting into petri dishes and dried in a 40°C oven.	Films yielded high antimicrobial activity against gram-positive bacteria, and moderate activity against gram-negative bacteria. Activity was exclusive to films enriched with the plant extract.	Atef et al., 2015
General purpose films based on gelatine and casein. Supplementation with *A. nodosum* and evaluation its effects	*Ascophyllum nodosum*	Aquous extracts obtained from freeze-dried seaweed. Films obtained by solvent casting, drying on petri dish. Control films had no seaweed extract.	Films displayed antioxidant activity correlated with extract concentration, but not comparisons with standard antioxidants were made. Roughness and general physical properties of the films were negatively impacted by the presence of extract. Indicates direct incorporation of seaweed extracts is faced with severe limitations concerning film quality.	Kadam et al., 2015
Evaluation of compound migration kinetics into food stimulants	*Fucus spiralis*	20% ethanolic extracts were obtained by solid-liquid extraction. Polylactic acid films were produced by extrusion, of which 8% was focus extract.	*Fucus spiralis* extracts by themselves demonstrated antimicrobial activity. No comparison with standard antibiotics. Diffusion coefficients for sorbic acid were effectively the same in the presence of *Fucus extract*, meaning diffusion is not compromised.	Rodriguez-Martinez et al., 2016
Biodegradable films for plasticulture applications	*Kappaphycus alvarezii*	Dried seaweed was immersed in water, along with glycerol and CaCO₃. Films were cast unto a tray and dried in a 40°C oven.	Physical and mechanical properties greatly enhanced by the addition of $CaCO_3$. Evident potential use of these films in both agricultural and food applications.	Abdul Khalil et al., 2018

Pork liver pâté – Increase on shelf-life	*Ascophyllum nodosum*; *Fucus vesiculosus*; *Bifurcaria bifurcata*	Solid-liquid aqueous extractions. 500 mg/kg of the pâté formulation.	No effect on microbiological development. Similar degree of oxidative damage as a 50 mg/kg BHT formulation. Promising use as a natural antioxidant, further study on economic viability required.	Agregán et al., 2018
Production of edible films and testing of its properties	*Sargassum latifolium* as a source of alginate and fucoidan	Two-stage solid-liquid extraction for both polysaccharides. Acid solution wash followed by filtration and precipitation with ethanol for fucoidan. Alkaline treatment of residue followed by filtration and precipitation with ethanol for alginate. Ca^{2+} added in some formulations. Films obtained by solvent casting, drying on petri dish.	Films showed good mechanical and visual properties. Calcium ions greatly influenced film thickness and physical properties. Polyphenol release was highly dependent on film formulation, with simple alginate + fucoidan films having the highest diffusion. Films seem adequate for preservation of low-moisture foods. Moderate antioxidant activity.	Gomaa et al., 2018
Production of edible films for general use	*Sargassum latifolium* as a source of alginate and fucoidan	Two-stage solid-liquid extraction for both polysaccharides. Acid solution wash followed by filtration and precipitation with ethanol for fucoidan. Alkaline treatment of residue followed by filtration and precipitation with ethanol for alginate. Films obtained by solvent casting, drying on petri dish.	Films showed good mechanical and visual properties and some antioxidant activity. No comparison with commercial synthetic antioxidants. No analysis of film shelf-life or long-term stability.	Gomaa et al., 2018

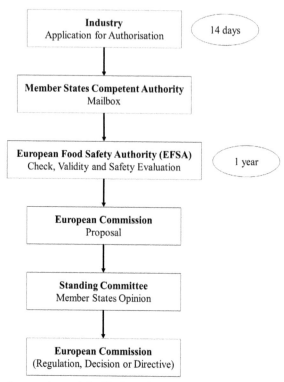

Fig. 8. Authorization procedure as defined by Reg. 1935/2004 EC (adapted from Restuccia et al., 2010).

material or object in which the substance is used. If necessary, EFSA may extend the evaluation period by a maximum of six months, in which case an explanation should be given to the applicant, the Commission, and the Member States (Fig. 8) (Restuccia et al., 2010). The implementation of Reg. 1935/2004 EC allowed for the first time the opportunity for the use of active and intelligent packaging in Europe. Subsequently, the general requirements set out in this regulation were assimilated by Reg. 450/2009 EC.

Regulation 450/2009 EC lays down the specific requirements for the marketing of active and intelligent materials and objects intended to come into contact with food. Only those substances included in the community list of permitted substances may be used in components of active and intelligent materials and objects. The implementation of this regulation represented a partial response to the lack of penetration of active and intelligent packaging in the European market compared to the Japan, USA and Australia markets, which have more flexible regulatory guidelines. More recently, plastic materials and objects intended to come in contact with food have been regulated by Regulation 10/2011 EC. In short, Reg. 1935/2004 EC and Reg. 450/2009 EC provide a new basis for the general requirements and specific safety and marketing issues related to active and intelligent packaging. In addition, both ensure a high level of safety and transparency for consumers.

7. Conclusion

This brief approach to the research surrounding the use of seaweed bioactives in active packaging has revealed a young and fertile field of research. A lot of the potential of seaweed bioactives are yet to be properly exploited by most industries, including the major producers of food packaging. Most academic research is only scratching the surface of how to best use these resources, and there are a lot of limitations regarding the effective extraction of compounds from algal biomass, the evaluation of its potential and consequently, its effective implementation on larger scale. It is expected that once mastery of the field of algal bioactives is achieved, the field of packaging will benefit from the surge of resources and active ingredients that are part of the marine environment.

It should also be noted that while there are certainly limitations and drawbacks associated with the use seaweeds as raw materials, particularly their high water content and relatively low activity per weight, they still show great potential as a source of structural polymers for the packaging industry. This is of particular importance to an industry, which is in dire need of changing its well-rooted practices and supplies, now that it faces a world in which sustainability and green processes are of the utmost importance.

Additionally, the funnelling of research towards innovation within the definition of active packaging can limit its ability to actually answer the industry's problems effectively. Several authors have shared this concern in recent reviews, with Holman et al. (2018b) stating that the categorization of packaging innovations within the active/intelligent/smart packaging definitions can 'limit their application as a troubleshooting tool for industry, retailers and other stakeholders' and keep them from being a 'cost-effective barrier against hazardous contaminants that simultaneously contributes to the longevity of product quality and enhancement of consumer appeal.'

References

Abdul Khalil, H.P.S., C.K. Saurabh, Y.Y. Tye, T.K. Lai, A.M. Easa, E. Rosamah, M.R.N. Fazita, M.I. Syakir, A.S. Adnan, H.M. Fizree, N.A.S. Aprilia and A. Banerjee (2017). Seaweed-based sustainable films and composites for food and pharmaceutical applications: A review, *Renew. Sustain. Energy Rev.*, 77: 353–362.

Abdul Khalil, H.P.S., E.W.N. Chong, F.A.T. Owolabi, M. Asniza, Y.Y. Tye, H.A. Tajarudin, M.T. Paridah and S. Rizal (2018). Microbial-induced-$CaCO_3$ filled seaweed-based film for green plasticulture application, *J. Clean. Prod.*, 199: 150–163.

Agregán, R., D. Franco, J. Carballo, I. Tomasevic, F.J. Barba, B. Gómez, V. Muchenje and J.M. Lorenzo (2018). Shelf-life study of healthy pork liver pâté with added seaweed extracts from *Ascophyllum nodosum, Fucus vesiculosus* and *Bifurcaria bifurcata, Food Res. Int.*, 112: 400–411.

Ahn, G.N., K.N. Kim, S.H. Cha, C.B. Song, J. Lee, M.S. Heo, I.K. Yeo, N.H. Lee, Y.H. Jee, J.S. Kim, M.S. Heu and Y.J. Jeon (2007). Antioxidant activities of phlorotannins purified from *Ecklonia cava* on free radical scavenging using ESR and H_2O_2-mediated DNA damage, *Eur. Food Res. Technol.*, 226: 71–79.

Anaëlle, T., E. Serrano Leon, V. Laurent, I. Elena, J.A. Mendiola, C. Stéphane, K. Nelly, L.B. Stéphane, M. Luc and S.P. Valérie (2013). Green improved processes to extract bioactive phenolic compounds from brown macroalgae using *Sargassum muticum* as model, *Talanta*, 104: 44–52.

Atef, M., M. Rezaei and R. Behrooz (2015). Characterization of physical, mechanical and antibacterial properties of agar-cellulose bionanocomposite films incorporated with savory essential oil, *Food Hydrocoll.*, 45: 150–157.

Azarakhsh, N., A. Osman, H.M. Ghazali, C.P. Tan and N. Mohd Adzahan (2014). Lemongrass essential oil incorporated into alginate-based edible coating for shelf-life extension and quality retention of fresh-cut pineapple, *Postharvest Biol. Technol.*, 88: 1–7.

Azmir, J., I.S.M. Zaidul, M.M. Rahman, K.M. Sharif, A. Mohamed, F. Sahena, M.H.A. Jahurul, K. Ghafoor, N.A.N. Norulaini and A.K.M. Omar (2013). Techniques for extraction of bioactive compounds from plant materials: A review, *J. Food Eng.*, 117: 426–436.

Boisvert, C., L. Beaulieu, C. Bonnet and É. Pelletier (2015). Assessment of the antioxidant and antibacterial activities of three species of edible seaweeds, *J. Food Biochem.*, 39: 377–387.

Bolumar, T., M.L. Andersen and V. Orlien (2011). Antioxidant active packaging for chicken meat processed by high pressure treatment, *Food Chem.*, 129: 1406–1412.

Boulho, R., C. Marty, Y. Freile-Pelegrín, D. Robledo, N. Bourgougnon and G. Bedoux (2017). Antiherpetic (HSV-1) activity of carrageenans from the red seaweed *Solieria chordalis* (Rhodophyta, Gigartinales) extracted by microwave-assisted extraction (MAE), *J. Appl. Phycol.*, 29: 2219–2228.

Charoensiddhi, S., C. Franco, P. Su and W. Zhang (2015). Improved antioxidant activities of brown seaweed *Ecklonia radiata* extracts prepared by microwave-assisted enzymatic extraction, *J. Appl. Phycol.*, 27: 2049–2058.

Chemat, F., V. Tomao and M. Virot (2010). Ultrasound-assisted Extraction in Food Analysis, *Handbook of Food Analysis Instruments*, CRC Press.

Chen, J. and A.L. Brody (2013). Use of active packaging structures to control the microbial quality of a ready-to-eat meat product, *Food Control*, 30: 306–310.

Costa, L.S., G.P. Fidelis, S.L. Cordeiro, R.M. Oliveira, D.A. Sabry, R.B.G. Câmara, L.T.D.B. Nobre, M.S.S.P. Costa, J. Almeida-Lima, E.H.C. Farias, E.L. Leite and H.A.O. Rocha (2010). Biological activities of sulfated polysaccharides from tropical seaweeds, *Biomed. Pharmacother.*, 64: 21–28.

Dang, T.T., Q. Van Vuong, M.J. Schreider, M.C. Bowyer, I.A. Van Altena and C.J. Scarlett (2017). Optimisation of ultrasound-assisted extraction conditions for phenolic content and antioxidant activities of the alga *Hormosira banksii* using response surface methodology, *J. Appl. Phycol.*, 29: 3161–3173.

Davis, T.A., B. Volesky and A. Mucci (2003). A review of the biochemistry of heavy metal biosorption by brown algae, *Water Research*, 37: 4311–4330.

Dehghani, S., S.V. Hosseini and J.M. Regenstein (2018). Edible films and coatings in seafood preservation: A review, *Food Chem.*, 240: 505–513.

Devi, K.P., N. Suganthy, P. Kesika and S.K. Pandian (2008). Bioprotective properties of seaweeds: *In vitro* evaluation of antioxidant activity and antimicrobial activity against food borne bacteria in relation to polyphenolic content, *BMC Complement. Altern. Med.*, 8: 38.

Draget, K.I., O. Smidsrød and G. Skjåk-Bræk (2005). Alginates from Algae. *In*: Biopolymers Online, A. Steinbüchel (Ed.).

Fang, Z., Y. Zhao, R.D. Warner and S.K. Johnson (2017). Active and intelligent packaging in meat industry, *Trends Food Sci. Technol.*, 61: 60–71.

Flórez-Fernández, N., M. López-García, M.J. González-Muñoz, J.M.L. Vilariño and H. Domínguez (2017). Ultrasound-assisted extraction of fucoidan from *Sargassum muticum*, *J. Appl. Phycol.*, 29: 1553–1561.

Ghaani, M., C.A. Cozzolino, G. Castelli and S. Farris (2016). An overview of the intelligent packaging technologies in the food sector, *Trends Food Sci. Technol.*, 51: 1–11.

Glicksman, M. (1987). Utilization of seaweed hydrocolloids in the food industry, *Hydrobiologia*, 151–152: 31–47.

Gomaa, M., M.A. Fawzy, A.F. Hifney and K.M. Abdel-Gawad (2018a). Use of the brown seaweed *Sargassum latifolium* in the design of alginate-fucoidan-based films with natural antioxidant properties and kinetic modeling of moisture sorption and polyphenolic release, *Food Hydrocoll.*, 82: 64–72.

Gomaa, M., A.F. Hifney, M.A. Fawzy and K.M. Abdel-Gawad (2018b). Use of seaweed and filamentous fungus derived polysaccharides in the development of alginate-chitosan edible films containing fucoidan: Study of moisture sorption, polyphenol release and antioxidant properties, *Food Hydrocoll.*, 82: 239–247.

Górka, B. and P.P. Wieczorek (2017). Simultaneous determination of nine phytohormones in seaweed and algae extracts by HPLC-PDA, *J. Chromatogr. B*, 1057: 32–39.

Gupta, S. and N. Abu-Ghannam (2011). Bioactive potential and possible health effects of edible brown seaweeds, *Trends Food Sci. Technol.*, 22: 315–326.

Heffernan, N., T.J. Smyth, R.J. FitzGerald, A. Soler-Vila and N. Brunton (2014). Antioxidant activity and phenolic content of pressurised liquid and solid-liquid extracts from four Irish origin macroalgae, *Int. J. Food Sci. Technol.*, 49: 1765–1772.

Heo, S.-J., P.-J. Park, E.-J. Park, S.-K. Kim and Y.-J. Jeon (2005). Antioxidant activity of enzymatic extracts from a brown seaweed *Ecklonia cava* by electron spin resonance spectrometry and comet assay, *Eur. Food Res. Technol.*, 221: 41–47.

Heo, S.J., S.C. Ko, S.M. Kang, H.S. Kang, J.P. Kim, S.H. Kim, K.W. Lee, M.G. Cho and Y.J. Jeon (2008). Cytoprotective effect of fucoxanthin isolated from brown algae *Sargassum siliquastrum* against H_2O_2-induced cell damage, *Eur. Food Res. Technol.*, 228: 145–151.

Holman, B.W.B., J.P. Kerry and D.L. Hopkins (2018a). Meat packaging solutions to current industry challenges: A review, *Meat Sci.*, 144: 159–168.

Holman, B.W.B., J.P. Kerry and D.L. Hopkins (2018b). A review of patents for the smart packaging of meat and muscle-based food products, *Recent Pat. Food. Nutr. Agric.*, 9: 3–13.

Huq, T., S. Salmieri, A. Khan, R.A. Khan, C. Le Tien, B. Riedl, C. Fraschini, J. Bouchard, J. Uribe-Calderon, M.R. Kamal and M. Lacroix (2012). Nanocrystalline cellulose (NCC) reinforced alginate based biodegradable nanocomposite film, *Carbohydr. Polym.*, 90: 1757–1763.

Kadam, S.U., S.K. Pankaj, B.K. Tiwari, P.J. Cullen and C.P. O'Donnell (2015). Development of biopolymer-based gelatin and casein films incorporating brown seaweed *Ascophyllum nodosum* extract, *Food Packag. Shelf-Life*, 6: 68–74.

Kang, H.S., H.Y. Chung, J.H. Jung, B.W. Son and J.S. Choi (2003). A New phlorotannin from the brown alga *Ecklonia stolonifera*, *Chem. Pharm. Bull.* (Tokyo), 51: 1012–1014.

Kanmani, P. and J.W. Rhim (2014). Development and characterization of carrageenan/grapefruit seed extract composite films for active packaging, *Int. J. Biol. Macromol.*, 68: 258–266.

Kariduraganavar, M.Y., A.A. Kittur and R.R. Kamble 2014. Polymer Synthesis and Processing. Natural and Synthetic Biomedical Polymers. Elsevier, pp. 1–31.

Kim, S.K. (2011). Handbook of Marine Macroalgae: Biotechnology and Applied Phycology, John Wiley & Sons Ltd, Chichester, UK.

Kumar, K.S., K. Ganesan and P.V.S. Rao (2008). Antioxidant potential of solvent extracts of *Kappaphycus alvarezii* (Doty) Doty—An edible seaweed, *Food Chem.*, 107: 289–295.

Lee, S.Y., S.J. Lee, D.S. Choi and S.J. Hur (2015). Current topics in active and intelligent food packaging for preservation of fresh foods, *J. Sci. Food Agric.*, 95: 2799–2810.

Li, Y., Y. Ying, Y. Zhou, Y. Ge, C. Yuan, C. Wu and Y. Hu (2019). A pH-indicating intelligent packaging composed of chitosan-purple potato extractions strength by surface-deacetylated chitin nanofibers, *Int. J. Biol. Macromol.*, 127: 376–384.

Lim, S.N., P.C.K. Cheung, V.E.C. Ooi and P.O. Ang (2002). Evaluation of antioxidative activity of extracts from a brown seaweed, *Sargassum siliquastrum*, *J. Agric. Food Chem.*, 50: 3862–6.

Liu, B., H. Xu, H. Zhao, W. Liu, L. Zhao and Y. Li (2017). Preparation and characterization of intelligent starch/PVA films for simultaneous colorimetric indication and antimicrobial activity for food packaging applications, *Carbohydr. Polym.*, 157: 842–849.

Lorenzo, J.M., J. Sineiro, I.R. Amado and D. Franco (2014). Influence of natural extracts on the shelf life of modified atmosphere-packaged pork patties, *Meat Sci.*, 96: 526–534.

Martins, J.T., A.I. Bourbon, A.C. Pinheiro, B.W.S. Souza, M.A. Cerqueira and A.A. Vicente (2013). Biocomposite films based on κ-carrageenan/locust bean gum blends and clays: physical and antimicrobial properties, *Food Bioprocess Technol.*, 6: 2081–2092.

McMillin, K.W. (2017). Advancements in meat packaging, *Meat Sci.*, 132: 153–162.

Michalak, I., B. Górka, P.P. Wieczorek, E. Rój, J. Lipok, B. Łęska, B. Messyasz, R. Wilk, G. Schroeder, A. Dobrzyńska-Inger and K. Chojnacka (2016). Supercritical fluid extraction of algae enhances levels of biologically active compounds promoting plant growth, *Eur. J. Phycol.*, 51: 243–252.

Mittal, R., H.A. Tavanandi, V.A. Mantri and K.S.M.S. Raghavarao (2017). Ultrasound assisted methods for enhanced extraction of phycobiliproteins from marine macro-algae, *Gelidium pusillum* (Rhodophyta), *Ultrason. Sonochem.*, 38: 92–103.

Moreira, R., J. Sineiro, F. Chenlo, S. Arufe and D. Díaz-Varela (2017). Aqueous extracts of *Ascophyllum nodosum* obtained by ultrasound-assisted extraction: Effects of drying temperature of seaweed on the properties of extracts, *J. Appl. Phycol.*, 29: 3191–3200.

Murray, P.M., S. Moane, C. Collins, T. Beletskaya, O.P. Thomas, A.W.F. Duarte, F.S. Nobre, I.O. Owoyemi, F.C. Pagnocca, L.D. Sette, E. McHugh, E. Causse, P. Pérez-López, G. Feijoo, M.T. Moreira, J. Rubiolo, M. Leirós, L.M. Botana, S. Pinteus, C. Alves, A. Horta, R. Pedrosa, C. Jeffryes, S.N. Agathos, C. Allewaert, A. Verween, W. Vyverman, I. Laptev, S. Sineoky, A. Bisio, R. Manconi, F. Ledda, M. Marchi, R. Pronzato and D.J. Walsh 2013. Sustainable production of biologically active molecules of marine based origin, *N. Biotechnol.*, 30: 839–850.

O'Sullivan, L., B. Murphy, P. McLoughlin, P. Duggan, P.G. Lawlor, H. Hughes and G.E. Gardiner (2010). *Prebiotics from Marine Macroalgae for Human and Animal Health Applications*, Multidisciplinary Digital Publishing Institute (MDPI).

Ospina, M., H.I. Castro-Vargas and F. Parada-Alfonso (2017). Antioxidant capacity of Colombian seaweeds: 1. Extracts obtained from *Gracilaria mammillaris* by means of supercritical fluid extraction, *J. Supercrit. Fluids*, 128: 314–322.

Otero, P., S.E. Quintana, G. Reglero, T. Fornari and M.R. García-Risco (2018). Pressurized liquid extraction (PLE) as an innovative green technology for the effective enrichment of Galician algae extracts with high quality fatty acids and antimicrobial and antioxidant properties, *Mar. Drugs*, 16.

Oussalah, M., S. Caillet, S. Salmieri, L. Saucier and M. Lacroix (2016). Antimicrobial effects of alginate-based films containing essential oils on *Listeria monocytogenes* and *Salmonella typhimurium* present in bologna and ham, *J. Food Prot.*, 70: 901–908.

Pereira, L. 2018. Biological and therapeutic properties of the seaweed polysaccharides, *Int. Biol. Rev.*, 2: 1–50.

Falkenberg, M., E. Nakano, L. Zambotti-Villela, G.A. Zatelli, A.C. Philippus, K.B. Imamura, A.M.A. Velasquez, R.P. Freitas, L.F. Tallarico, P. Colepicolo and M.A.S. Graminha (2019). Bioactive compounds against neglected diseases isolated from macroalgae: A review, *J. Appl. Phycol.*, 31: 797–823.

Popkin, B.M., L.S. Adair and S.W. Ng (2012). Global nutrition transition and the pandemic of obesity in developing countries, *Nutr. Rev.*, 70: 3–21.

Qi, H., Q. Zhang, T. Zhao, R. Chen, H. Zhang, X. Niu and Z. Li (2005). Antioxidant activity of different sulfate content derivatives of polysaccharide extracted from *Ulva pertusa* (Chlorophyta) *in vitro*, *Int. J. Biol. Macromol.*, 37: 195–199.

Qiu, X., S. Chen, G. Liu and Q. Yang (2014). Quality enhancement in the Japanese sea bass (*Lateolabrax japonicus*) fillets stored at 4°C by chitosan coating incorporated with citric acid or licorice extract, *Food Chem.*, 162: 156–160.

Realini, C.E. and B. Marcos (2014). Active and intelligent packaging systems for a modern society, *Meat Sci.*, 98: 404–419.

Restuccia, D., U.G. Spizzirri, O.I. Parisi, G. Cirillo, M. Curcio, F. Iemma, F. Puoci, G. Vinci and N. Picci (2010). New EU regulation aspects and global market of active and intelligent packaging for food industry applications, *Food Control*, 21: 1425–1435.

Rhim, J.-W. (2011). Effect of clay contents on mechanical and water vapor barrier properties of agar-based nanocomposite films, *Carbohydr. Polym.*, 86: 691–699.

Rocha de Souza, M.C., C.T. Marques, C.M. Guerra Dore, F.R. Ferreira da Silva, H.A. Oliveira Rocha and E.L. Leite (2007). Antioxidant activities of sulfated polysaccharides from brown and red seaweeds, *J. Appl. Phycol.*, 19: 153–160.

Rodriguez-Jasso, R.M., S.I. Mussatto, L. Pastrana, C.N. Aguilar and J.A. Teixeira (2011). Microwave-assisted extraction of sulfated polysaccharides (fucoidan) from brown seaweed, *Carbohydr. Polym.*, 86: 1137–1144.

Rodríguez-Martínez, A.V., R. Sendón, M.J. Abad, M.V. González-Rodríguez, J. Barros-Velázquez, S.P. Aubourg, P. Paseiro-Losada and A. Rodríguez-Bernaldo de Quiros (2016). Migration kinetics of sorbic acid from polylactic acid and seaweed-based films into food simulants, *LWT - Food Sci. Technol.*, 65: 630–636.

Rojas-Graü, M.A., R.M. Raybaudi-Massilia, R.C. Soliva-Fortuny, R.J. Avena-Bustillos, T.H. McHugh and O. Martín-Belloso (2007). Apple puree-alginate edible coating as carrier of antimicrobial agents to prolong shelf-life of fresh-cut apples, *Postharvest Biol. Technol.*, 45: 254–264.

Roohinejad, S., M. Koubaa, F.J. Barba, S. Saljoughian, M. Amid and R. Greiner (2017). Application of seaweeds to develop new food products with enhanced shelf-life, quality and health-related beneficial properties, *Food Res. Int.*, 99: 1066–1083.

Salmieri, S. and M. Lacroix (2006). Physicochemical properties of alginate/polycaprolactone-based films containing essential oils, *J. Agric. Food Chem.*, 54: 10205–10214.

Sanches-Silva, A., D. Costa, T.G. Albuquerque, G.G. Buonocore, F. Ramos, M.C. Castilho, A.V. Machado and H.S. Costa (2014). Trends in the use of natural antioxidants in active food packaging: A review, *Food Addit. Contam. - Part A Chem. Anal. Control. Expo. Risk Assess.*, 31: 374–395.

Santoyo, S., M. Plaza, L. Jaime, E. Ibañez, G. Reglero and J. Señorans (2011). Pressurized liquids as an alternative green process to extract antiviral agents from the edible seaweed *Himanthalia elongata*, *J. Appl. Phycol.*, 23: 909–917.

Saravana, P.S., Y.J. Cho, Y.B. Park, H.C. Woo and B.S. Chun (2016). Structural, antioxidant and emulsifying activities of fucoidan from *Saccharina japonica* using pressurized liquid extraction, *Carbohydr. Polym.*, 153: 518–525.

Saravana, P.S., Y.N. Cho, H.C. Woo and B.S. Chun (2018). Green and efficient extraction of polysaccharides from brown seaweed by adding deep eutectic solvent in subcritical water hydrolysis, *J. Clean. Prod.*, 198: 1474–1484.

Schumann, B. and M. Schmid (2018). Packaging concepts for fresh and processed meat – Recent progresses, *Innov. Food Sci. Emerg. Technol.*, 47: 88–100.

Shibata, T., K. Ishimaru, S. Kawaguchi, H. Yoshikawa and Y. Hama (2008). Antioxidant activities of phlorotannins isolated from Japanese Laminariaceae, *J. Appl. Phycol.*, 20: 705–711.

Shojaee-Aliabadi, S., H. Hosseini, M.A. Mohammadifar, A. Mohammadi, M. Ghasemlou, S.M. Hosseini and R. Khaksar (2014). Characterization of κ-carrageenan films incorporated plant essential oils with improved antimicrobial activity, *Carbohydr. Polym.*, 101: 582–591.

Siracusa, V., A.S. Sant'Ana, J. Bacenetti, C. Ingrao, A. Lo Giudice, A. Mousavi Khaneghah and R. Rana (2015). Foamy polystyrene trays for fresh-meat packaging: Life-cycle inventory data collection and environmental impact assessment, *Food Res. Int.*, 76: 418–426.

Tavassoli-Kafrani, E., H. Shekarchizadeh and M. Masoudpour-Behabadi 2016. Development of edible films and coatings from alginates and carrageenans, *Carbohydr. Polym.*, 137: 360–374.

Tierney, M.S., T.J. Smyth, M. Hayes, A. Soler-Vila, A.K. Croft and N. Brunton (2013). Influence of pressurised liquid extraction and solid-liquid extraction methods on the phenolic content and antioxidant activities of Irish macroalgae, *Int. J. Food Sci. Technol.*, 48: 860–869.

Tiwari, B.K. (2015, September 1). *Ultrasound: A Clean, Green Extraction Technology*. Elsevier.

Topuz, O.K., N. Gokoglu, P. Yerlikaya, I. Ucak and B. Gumus (2016). Optimization of antioxidant activity and phenolic compound extraction conditions from red seaweed (*Laurencia obtuse*), *J. Aquat. Food Prod. Technol.*, 25: 414–422.

Vanderroost, M., P. Ragaert, F. Devlieghere and B. De Meulenaer (2014). Intelligent food packaging: The next generation, *Trends Food Sci. Technol.*, 39: 47–62.

Veiga-Santos, P., L.T. Silva, C.O. de Souza, J.R. da Silva, E.C.C. Albuquerque and J.I. Druzian (2018). Coffee-cocoa additives for bio-based antioxidant packaging, *Food Packag. Shelf-Life*, 18: 37–41.

Wang, T., R. Jónsdóttir and G. Ólafsdóttir (2009). Total phenolic compounds, radical scavenging and metal chelation of extracts from Icelandic seaweeds, *Food Chem.*, 116: 240–248.

Wu, Y., C.L. Weller, F. Hamouz, S. Cuppett and M. Schnepf (2001). Moisture loss and lipid oxidation for precooked ground-beef patties packaged in edible starch-alginate-based composite films, *J. Food Sci.*, 66: 486–493.

Yang, J.S., Y.J. Xie and W. He (2011). Research progress on chemical modification of alginate: A review, *Carbohydrate Polymers*, 84: 33–39.

Yuan, Y. and D. Macquarrie (2015). Microwave-assisted extraction of sulfated polysaccharides (fucoidan) from *Ascophyllum nodosum* and its antioxidant activity, *Carbohydr. Polym.*, 129: 101–107.

Yuan, Y.V., D.E. Bone and M.F. Carrington (2005). Antioxidant activity of dulse (*Palmaria palmata*) extract evaluated *in vitro*, *Food Chem.*, 91: 485–494.

Zhang, Q., N. Li, G. Zhou, X. Lu, Z. Xu and Z. Li (2003). *In vivo* antioxidant activity of polysaccharide fraction from *Porphyra haitanesis* (Rhodephyta) in aging mice, *Pharmacol. Res.*, 48: 151–155.

Zhang, R., A.K.L. Yuen, M. Magnusson, J.T. Wright, R. de Nys, A.F. Masters and T. Maschmeyer (2018). A comparative assessment of the activity and structure of phlorotannins from the brown seaweed *Carpophyllum flexuosum*, *Algal Res.*, 29: 130–141.

Zhang, Z., F. Wang, X. Wang, X. Liu, Y. Hou and Q. Zhang (2010). Extraction of the polysaccharides from five algae and their potential antioxidant activity *in vitro*, *Carbohydr. Polym.*, 82: 118–121.

Žlabur, J., S. Voća, M. Brnčić and S. Rimac-Brnčić (2018). New Trends in Food Technology for Green Recovery of Bioactive Compounds From Plant Materials, pp. 1–36. *Role of Materials Science in Food Bioengineering*, Academic Press.

CHAPTER 9

Edible Polymers for Shelf-Life Extension of Perishables

An Insight into Films and Coatings

Sucheta,[1,]* *Nidhi Budhalakoti*[1] and *Kartikey Chaturvedi*[2]

1. Introduction

Food products may be categorized into perishables (fruits or vegetables, dairy products, etc.), semi-perishables (cereal products, tubers and intermediate moisture-baked foods) and non-perishables (cereal grains, pulses and oilseeds) depending upon storage stability of more than 15 days and if required, any additional preservation procedures. The non-perishables can be preserved with efficient packaging systems rather than applying stringent preservation methodologies, whereas fruits or vegetables and dairy products need to be preserved properly with packaging, temperature regulations and innovative preservation procedures. Fruits and vegetables continue to respire (climacteric); however, some of these do not respire (non-climacteric); both undergo loss of moisture and are susceptible to microbial infestation if stored in unfavourable conditions. Respiratory fruits or vegetables (guava, apple, banana, jackfruit, melon, tomato, etc.), continue to produce ethylene and CO_2 after harvest and cannot be packaged in restricted gaseous atmosphere or higher temperature conditions, unlike non-climacteric foods, as they undergo anaerobic fermentation inside, thus, leading to spoilage. Traditionally, refrigeration along transportation and retail chain are used to minimize respiration bursts and enhance the shelf-life which, however, limits the storage capacity and increases the retailing costs due to added energy and transportation (Simson and Straus, 2010). Dairy products, like cheese, vary

[1] Center of Innovative and Applied Bioprocessing, Mohali, India.
[2] National Institute of Food Technology Entrepreneurship and Management, Sonipat, India.
* Corresponding author: suchetakkr@gmail.com

diversely in composition and possess different maturation rates, water content and mechanical stability. Therefore, the packaging requirements are also typical of cheese, aimed at resisting biochemical and microbial actions to prevent degradative changes in color, texture and flavor (Robertson, 2006). Moisture loss, contamination by bacteria, yeasts and molds, development of off-flavors are major issues in commercialization of cheese. Vacuum and modified atmosphere packaging are being currently used commercially to preserve cheese. Recently, coatings have also been studied as an individual packaging material to provide additional protection if used with outer packaging. However, conventional materials like paraffin wax and polyvinyl chloride are under legislative restrictions, which paves way and makes edible films or coatings a promising future alternative for cheese packaging (Costa et al., 2018). Baked foods including buns, bread or cakes are also included in category of perishables affected by staling and mold spoilage. These products lose quality in a series of degradative reactions of staling and easy availability of nutrients by microflora in the bread. In case of fried foods, less fat absorption on frying and less inner moisture enhances the eating quality. This can also be achieved by the use of coating such foods prior to frying (Falguera et al., 2011). All of the above food categories of perishables are nowadays packaged or coated with edible polymers as they restrict the transfer of gases and moisture, thus aiding in prevention of microbial growth. This chapter deals with the basic packaging needs of the above food commodities; evolution of edible polymers as packaging from modified atmosphere packaging; principles underlying formation and effectiveness of edible films and coatings for fruits and vegetables, cheese, baked and fried foods; regulatory aspects and marketing trends of edible films worldwide and biodegradability of such films.

2. Packaging and Food Commodities: Shelf-Life Aspects

Shelf-life preservation of different food commodities depends on the kind of packaging material. A functional packaging system should be able to protect the food product against loss of quality and spoilage. For fruits and vegetables, packaging material must encompass potential of controlling physiological activity and restricting entry of microorganisms. For retailing and short-distances, thermoformed PET trays, plastic crates, cardboard boxes, sacks and gunny bags are used. These limit the storage period of perishables due to lack of prevention in microflora entry as well as non-favorable respiration conditions. Excess exposure to atmospheric conditions in such packages over a period of time leads to higher microbial infestation and higher respiration rates of perishables, which deteriorates the quality and thus leads to spoilage. However, pre-packaging, like molded trays or shrink film wraps, reduces decay of fresh produce as it creates a micro-climate with lower oxygen and high carbon dioxide levels. Dairy cheese packaging should be able to protect moisture loss, prevent chemical changes, microbial growth and allow controlled fermentation. The package should allow the release of carbon dioxide evolved during fermentation inside the package. Also, due to high moisture content inside the package, oxygen must be restricted inside the product's environment to prevent the fungal growth in such conditions. Paraffin wax and polyvinyl chloride are conventional coating materials for cheese. However, the risk of migration of coating material into cheese has led to imposition of legislative restrictions over such materials, leading to emergence of biopolymer based edible coatings (Costa et al., 2018). Packaging requirements for cheese varies with its composition, which is determined by gaseous permeability of the package, mechanical stability and antimicrobial efficacy. Bread has a limited shelf-life of less than a week due to its high-water activity (0.96), which makes it susceptible to mold growth. During storage, water absorption also contributes to microbial spoilage as well as degradative chemical changes. Conventional plastic bags (LDPE, PP) used pose a higher risk of spoilage on storage of bread, buns, muffins, etc., due to high relative humidity inside the package. Addition of chemical preservatives decreases consumer preferences for baked foods. Modification of atmosphere around package using MAPs and active packaging suggests prevention of quality deterioration of such foods on storage. Nearly 60–80 per cent of CO_2 has been reported to effectively enhance the shelf-life of bread (Fik et al., 2012). Deep-fat fried foods, like doughnuts and French fries, undergo loss of moisture and simultaneously fat absorption to form a crunchy crust which makes them appetizing and palatable to consumers. In few cases, about one-third of the fat-fried foods face decrease in acceptability due to health concerns by consumers (Falguera et al., 2011). Higher oil uptake through capillaries created by moisture loss also increase the chances of oxidation during storage. Recently edible coating of such foods prior to frying has been suggested to reduce oil uptake and enhance the storage stability of deep-fat fried products (Marquez et al., 2014). An overview of shelf-life aspects of above food commodities is represented in Fig. 1.

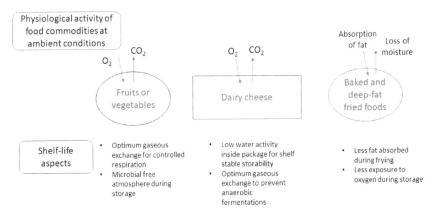

Fig. 1. Shelf-life aspects of food commodities for edible biopolymers-based packaging.

3. Elementary for Advances: Modified Atmosphere Packaging

Edible films or coatings may be defined as "edible polymers applied as a layer of coating over food surface or as package from self-standing films developed, which create or modify atmosphere around the food and allow controlled exchange of gases without loss of aroma volatiles from inside the produce" (Pavlath and Orts, 2009). Edible films and coatings have been originated in the form of wax application on citrus fruits during twelfth and thirteenth centuries. The basic functions of edible films and coatings find resemblance with modified atmosphere packaging of food products (using synthetic polymers, etc.) (Debeaufort et al., 1998). However, MAP emerged later than wax application as per scientific findings and was first defined by Young et al. (1988). Like MAP, films and coatings restrict the gaseous exchange of food products with the surroundings and moisture loss/evapo-transpiration of fresh produce, these also prevent loss of aroma volatiles from the fresh produce. The applicability of MAP is stringently restricted to respiratory behavior of fresh produce and post-processing behavior of baked foods. It is also a costly shelf-life enhancement application in comparison to edible films and coatings. However, the coatings may be applied directly over the surface of fresh produce or product and do not require any stringent atmosphere for storage of food products. In addition, biopolymer films used for packaging possess sufficient permeability so as not to lead to anaerobic conditions inside the package. Therefore, edible films or coatings have actually evolved through the basic mechanism of shelf preservation, i.e., modification of produce atmosphere, but with utilization of economical, biodegradable and edible biopolymers.

4. Difference between Synthetic and Edible Polymers

Synthetic polymers are man-made polymers, such as polyethylene, polypropylene, polyvinyl chloride, elastomers, fibers, etc. They are generally non-biodegradable and used as external packaging material for food products. Natural or edible biopolymers consist of polypeptides, polysaccharides and lipids. Nucleic acids are also a form of biopolymers but these are not used for producing edible packaging. Edible polymers can be extracted from naturally occurring sources, such as vegetables and fruit peels, marine flora and fauna, or can also be synthesised enzymatically through genetic route. Synthetic polymers are generally synthesised chemically and have properties depending upon their mode of production. Polymers can also be produced by valorisation of biomass. Polymers can consist of homopolymer units or copolymeric units. Depending upon their utility and properties required to be incorporated into the food products, their degree of polymerisation and the type of polymeric units can be modified (Ali and Ahmad, 2018).

Generally, edible polymers consist of food-grade materials, whereas synthetic polymers are not food grade. Therefore, they are non-edible. They cannot be used as coating or film-forming materials as the toxic chemicals might leach out into the food matrix during processing.

5. Edible Polymers, Films and Coatings

Biopolymers, such as starch, cellulose, hemicellulose, pectin, waxes, zein, casein, etc., are composed of polymer network comprising –C, –OH, –COOH, –NH2, –SH, etc., functional groups, which can be linked together through hydrogen, disulphide and hydrophobic interactions. This bonding ability assists them in forming gels upon heating and this property contributes towards film-forming ability of such polysaccharides. However, different polysaccharides undergo gelation at different temperatures, some requiring additional cross-linking agents to polymerise. Such mechanisms of some of the polymers and their film- or coating-forming ability is explained below:

5.1 Polysaccharides

Starch. It is composed of a network of linear as well as branched amylose and amylopectin, which are tightly packed in semi-crystalline form inside the granules. When the granules are heated, water enters through amorphous regions in between micelles, leading to a maximum absorption of 30 per cent of the granule. The process is gelatinization in which water forms hydrogen bonds with amylose and amylopectin, being higher in the former one. Upon cooling, after casting of polymer solution, there is a tendency of native hydrogen bond formation between amylose and amylopectin. This process is retrogradation, which aggravates on drying and leads to self-standing film formation (Sucheta et al., 2019a).

Cellulose. It is a widely available polysaccharide source for film formation and is easily isolated from wood, cotton and other plant-based materials. It is composed of anhydrogluco units joined by β–1-4 linkages. Due to the presence of covalent linkages, it is insoluble in water and soluble in a few ionic solvents, like NaOH, ionic liquids, LiCl/N,N-dimethylacetamide, etc. (Cazon et al., 2017). Edible films or coatings from cellulose are usually prepared from cellulose suspensions, after dissolution of cellulose and derivatizing cellulose. Widely used for film formation, cellulose derivatives, i.e., carboxymethyl cellulose, methyl cellulose and hydroxypropylmethyl cellulose are produced after swelling the native cellulose with water followed by treatment with alkalis, like chloroacetic acid, methyl chloride and propylene chloride (Bourtoom, 2008).

Pectin. Native pectin, widely extracted from the citrus group, is a polymer of α–1-4 galacturonic acids, esterified with $-CH_3$ groups at carboxylic group residues (degree of methylation). Higher degree of methylation > 50 per cent of pectin (HMP) forms easily gel at lower temperatures in the presence of acid and sugar, due to increased availability of carboxyl ionic groups, which are involved in hydrogen bonding with adjacent pectin molecules (Espitia et al., 2014). Low molecular weight pectin (LMP) requires presence of certain divalent cations to act as a bridge between the individual pectin molecules. Polymerization in case of LMP occurs through electrostatic interactions ($-Ca^{2+}$ and $-COO^-$) and van der Waals forces (among bridged dimer chains), whereas HMP polymerizes through aggregation of polymer chains.

Plant seed and seaweed gums. Gums are soluble in water and contribute to the viscosity in food systems. They have the significant property of binding to water and therefore are able to form films and coatings owing to intermolecular forces, like ionic, hydrophobic, cross-linking and hydrogen-bonding (Phillips and Williams, 2000). A number of hydrogen bonds and other hydrophilic moieties are present in the gum structure and during film-formation process (heating, casting, drying). The long polymer network is disrupted with the formation of new intermolecular H-bonding forming a film matrix (Janjarasskul and Krochta, 2010). There are many seed gums occurring naturally such as gum arabic, locust bean gum, tragacanth gum, almond gum, psyllium gum, basil seed gum, fenugreek gum, etc. However, the major scientific findings on applicability of gums for films and coatings focused on locust bean gum, guar gum and gum arabic. These are galactomannans, comprising galactose and mannose sugars joined together by β-(1-4)-linkage, and mannose is the backbone sugar of the polymer network. The different functionality of these gums resides in variable mannose/galactose ratio whereas the solubility in water is majorly due to galactose residues in the polymer chain. Locust bean gum is less soluble in water than guar gum due to lower mannose/galactose ratio (3:5), galactose substitution being lesser. It is able to form a film but of low strength and elasticity; therefore, needs substitution with other gums, like carrageenan and xanthan. It is mostly used for its synergistic action over functionality of edible films from other different polysaccharides (Barak and Mudgil, 2014). It is 70–85 per cent soluble in water when heated to a temperature of 80°C for 30 min. (Dakia et al., 2008). Due to hydrophilicity, these also support incorporation of certain additives and other bioactives in films for improved functionality. Like locust gum, guar gum is also made up of backbone consisting D-mannopyranose units associated with D-galactopyranose at specific D-mannose units. Presence of extensive branching makes it easier to hydrate guar gum in comparison to locust bean gum. Also, higher galactose residues in guar gum increase solubility of guar gum in comparison to locust bean gum (Saha et al., 2017). The synergistic effect of guar gum with other gums, like xanthan, agar, carrageenan, etc., have also been well explored. Gum arabic is an excellent emulsifier and is a calcium, potassium or magnesium salt of arabic acid. The main chain is composed of β-D-galactopyranosyl residues linked in β-1-3 linkages with a branched chain of two-five units of β-D-galactopyranosyl units linked to the main chain by β-1-6 linkages. It possesses high solubility in cold water. Due to low viscosity, it shows poor rheological properties, but forms films at lower concentrations due to higher hydrophilicity than other gums. Agar is a heterogenous polysaccharide with main components as D-galactopyranose and 3,6-anhydro-L-galactopyranose in α-(1-4) and β-(1-3) linkages. Agar is insoluble in cold water and soluble in hot water. It forms clear, flexible, homogenous and mechanically stable films. However, fluctuations in temperature and relative humidity affect agar crystallinity and may lead to the formation of micro-fractures and polymer embrittlement (Khalil et al., 2017). Upon dissolution in water and followed by heating, the initial coil-helix transition of polymer chain leads to aggregation of helices. Final gel formation is the overall result of conformational transition, molecular cross-linking and phase-separation processes. Carrageenan are a category of seaweed gums, extracted from red seaweed. These are made up of 3,6-anhydro-D-galactose units bonded with α-1-3 and β-1-4 linkages. The anhydro-galactose units exist in sulphated and non-sulphated forms. Kappa carrageenan is made up of only one sulphate group and is therefore less hydrophilic, less soluble in water films but in the presence of cations, it forms a gel of highest strength. Iota carrageenan possesses two sulphate ester groups, although in the presence of potassium ion, it is insoluble in water but sodium makes it readily soluble. The gelling mechanism of kappa and Iota carrageenan is similar to the agar. Dissolution in water (in the presence of cation) results in the formation of double helices which further bind other segments of the molecule and intertwingled to form a gel (Rudolph, 2000).

5.2 Lipids

Waxes, cocoa butter and resins are possibly used as edible coatings while fatty acid esters, glycerol esters, essential oils, oil and fats are used with other biopolymers (proteins or polysaccharides) as composite edible films. Due to their poor mechanical strength, they can not form self-standing edible films or coatings. The lipids, including waxes, resins, fatty acids, are generally solids and soft-solids at room temperature. They exhibit varied phase transition temperatures and can be shaped by casting at higher temperatures (Han, 2014). Lipids differ in film formation owing to variation in chain length, fatty alcohols and alkanes present in their structure. These days, edible coatings from waxes, like beeswax, carnauba wax, candelilla wax, etc., are used commercially for protection of damage to fresh fruits. Fatty acids and monoglycerides are mainly used as emulsifiers and dispersing agents in film formation. Pure lipids are used as edible films in combination with hydrocolloids, such as starches, cellulose, proteins in three ways: by adding lipids to film-forming solution to form an emulsion, by applying in a layer above the formed film or adding biopolymers to continuous phases of lipids. The lipids enhance the water-barrier properties of films; however, due to high oxygen solubility, oxygen permeability of films increases when incorporated with lipids. The increased water vapor permeability (WVP) barrier of lipid-incorporated films and coatings is due to increased molecular aggregations among lipid molecules on drying. Also, crystallization of fatty acids in the film-forming polymer matrix contributes to enhanced solids in dried film, leading to elevated permeability barrier. Shorter-chain fatty acids aggregate more due to higher molecular mobility, thus, resulting in a higher barrier as compared to long-chain fatty acids. Apart from chain length, structural integrity also plays a significant role in determining the barrier efficiency of lipid-based edible films or coatings. The hydrophobicity of chemical groups in a lipid molecule decides the affinity of water. In a study, beeswax was most hydrophobic followed by stearyl alcohol and stearic acid being the least one. This was due to less hydrophilicity of hydroxyl group of stearoyl alcohol than carboxyl group of stearic acid (Kester and Fennema, 1989). The optimum concentration of lipid also determines the effectiveness in enhancing the barrier properties added to films. Higher soybean oil concentration increases the WVP of starch-based films and coatings while the lower concentration of 2g/L of coating solution results in lowest WVP (Garcia et al., 2000). A few examples of utilization of lipids in biopolymer-based films or coatings to enhance WVP barrier properties are shown in Table 1.

5.3 Proteins

Globular and fibrous proteins are associated with hydrogen, ionic, covalent and hydrophobic bonds which determine their film-forming properties. For optimum film-forming abilities, proteins can be subjected to different modification treatments, like heat denaturation, pressure, irradiation, enzymatic, cross-linking and acid or alkali hydrolysis (Chiralt et al., 2018). The mechanisms of milk, meat, cereal and pulse proteins to form films are explained below:

Milk proteins: Casein proteins are composed of four subunits which differ in aggregation behavior and affect film-forming abilities of casein. Alpha s1 is ca2+ sensitive and precipitates in the presence of calcium ions even in low concentrations; alpha s2 is also ca2+ sensitive but is more hydrophilic than other casein subunits; beta caseins are also sensitive to calcium but aggregates at a temperature of below 4°C and kappa caseins are not calcium sensitive but is strongly amphipathic (Dangaran et al., 2009). Whey proteins, majorly beta-lactoglobulin and alpha-lactalbumin,

Table 1. Few Examples Showing the Effect of Addition of Lipids on Functionality of Edible Films or Coatings.

Kind of lipid	Concentration	Biopolymer-based edible films or coatings	Contributed functionality	References
Acrylated epoxidized soybean oil	Films soaked in solution	High amylose corn starch with amylose 80%	Improvement in moisture permeability	Ge et al., 2019
Beeswax	10, 20, 30 and 40%	Canadian yellow field pea starch (35–40% amylose)	Decreased WVP, increased oxygen permeability, reduced tensile strength, increased elastic modulus	Han et al., 2006
Flaxseed oil	1–10%	Soy-protein isolate (5% w/w)	Increased tensile elongation, swelling properties decreased	Hopkins et al., 2015
Almond and walnut oils	0.5 and 1.0%	Whey-protein isolate (8%)	Increased film opacity, increased film opacity, almond oil had higher effect	Galus and Kadzinska, 2016
Rapeseed oil	0, 1, 2 and 3%	Soy-protein isolate (10%)	Opacity increased, decreased WVP, weakened tensile strength	Galus, 2018
Coconut oil	0–1.5 mL	Chitosan (1%)	Increased film thickness, decreased transparency, reduced moisture sorption	Binsi et al., 2013
Oleic acid and stearic acid	0–2 g/100 g	Soy-protein isolate (4%)	Reduced WVP, higher contact angle	Wang et al., 2014
Clove essential oil	50 g/100 g SPI	Soy-protein isolate 5% w/v with added micro fibrillated cellulose	Oxygen permeability increased, WVP decreased, plasticization of protein	Matias Ortiz et al., 2018
Beeswax, shortening, liquid paraffin	0, 10 and 20%	Agar-maltodextrin (10 and 20%)	Improved hydrophobicity and tensile strength	Zhang et al., 2019
Savory essential oil	0.5, 1.0 and 1.5%	Agar (1.5%)	Decreased water solubility, WVP not affected, increased elongation	Atef et al., 2015

consist of negative charges evenly distributed throughout the chain. Hydrophobic R groups are buried within the protein molecule. Gelation is a prerequisite for film formation for all groups of proteins. Partial unfolding of proteins due to the action of subjected treatments (chemical, high hydrostatic pressure, heating, partial enzymatic hydrolysis) leads to unwinding of polymer network and afterwards to gel formation. The denaturation of proteins exposes the thiol group of cysteine–121, which becomes highly available for intermolecular disulphide bond among the polymer chains. This forms gel with strong cohesiveness and the films formed are stiffer. Whey proteins also form gel when pH is close to isoelectric point; it possesses no charge in this condition and resists electrostatic repulsions followed by agglomeration (Ramos et al., 2012).

Plant proteins: Gluten is a component of wheat, composed of gliadin and glutenin and is able to form edible films or coatings due to its cohesiveness and elasticity. It is insoluble in water but is soluble in ethanol solutions. Therefore, films are formed by dissolution of gluten in aqueous ethanol, then casting, followed by drying to evaporate ethanol. Drying also stimulates the newer disulphide bonds, which provide strength to edible films from gluten. Hydrogen and hydrophobic bonds also play a minor role in the formation of gluten-based edible films (Krull et al., 1971; Hassan et al., 2018). Zein, a corn protein is widely used in the formation of edible films or coatings due to its strong water-barrier properties. It is rich in hydrophobic amino acids (leucine and alanine) which make it insoluble in water and soluble in 60–95 per cent ethanol (Chen et al., 2014). Hydrogen and disulphide bonds contribute to film stability. The films made from zein are tough and oil resistant. Soy protein is a majorly globulin protein with 37 per cent beta-conglycinin and 31 per cent glycinin of total proteins. Glycinin (11 S protein) is responsible for film formation from soy proteins or its isolates, due to the presence of acidic and basic polypeptide units (A&B) which link together through disulphide bonds. Inherent hydrophilicity and stiffness of soy-protein films makes it essential to add lipids, waxes and plasticizer (Galus, 2018). Also, the beany flavor, brittleness and poor mechanical properties necessitate modification of soy proteins by physical, chemical or enzymatic treatments prior to edible-film formation.

Meat proteins: Collagen is a most widely-used animal protein for edible films and coatings. It exists as tropo-collagen, a triple helix structure composed of three alpha chains. Glycine-proline-hydroxyproline is a major repeating unit of collagen protein. Similar to plant proteins, collagen also undergoes gelation process if subjected to extreme conditions of heat, acid or alkaline conditions. The partially-denatured collagen reforms triple helix structure and is referred to as gelatin (Dangaran et al., 2009). Fish gelatin is considered a good source for bio-packaging applications. However, films from fish gelatin are brittle due to weaker intermolecular bonds across the polymer network formed upon cooling the film. Blending with other biopolymers improves the WVP and mechanical properties of collagen-based edible films or coatings.

6. Nanomaterials in Edible Films and Coatings

Nanomaterials of dimension less than 100 nm on a nanometre scale when added in film-forming polymer matrix reinforces it and provides additional structural integrity, improves mechanical, color and barrier properties. Metal nanoparticles and those derived from polysaccharides are mostly used to form nanocomposite-based edible films or coatings. One or more polysaccharide-based nanofillers, when added into distinct biopolymer, improve its functionality. These nanofillers based on starch, cellulose and chitin may be produced by acid hydrolysis or mechanical disintegration processes. The nanomaterials tend to aggregate more as a result of extensive probability of hydrogen-bonding. The nano-fillers may be added in polymer matrix by *in situ* polymerization, solution casting, electrospinning or melt intercalation processes (Condes et al., 2017). For cellulose nanocrystals, bacterial cellulose is preferred over plant as it is edible and can be synthesized in pure form. Cellulose microfibrils, considered as a string of nanocrystals, are hydrolyzed under acidic conditions to yield rod-like nanocrystals. Gelatin, when incorporated with cellulose nanocrystals, improves the mechanical properties and reduces the moisture affinity (George and Siddaramaih, 2012).The non-reinforcement of mango puree films with cellulose nanofibers increases the tensile strength (Young's modulus) and improves the water-vapor barrier (Azeredo et al., 2009). Starch suspension treated with diluted sulphuric acid and constantly stirred for seven days yields nanocrystals (NCs). Starch NCs, when added in concentration of 1–5 per cent, disperse uniformly throughout the polymer resulting in a smooth film of high thermal stability (Li et al., 2015).

7. Films and Coatings Formation

Edible films and coatings are defined according to the method of application over food products. Edible films are produced separately and then applied on the surface of food material, whereas coatings are either sprayed, panned or applied by dipping the surface of the food commodities in the film-forming mixture and then left to dry. Edible film matrices are made up of food-grade materials. All materials including carbohydrates, proteins, lipids, plasticizers, solvents, additives, etc., are obtained from edible materials or are food grade. Figure 2 explains the mechanism of edible films and coating formation (adapted from Sucheta et al., 2019b). Various components of edible films have different functions. Polysaccharides provide gas barrier, lipids due to their hydrophobic nature reduce moisture loss in food products, and proteins provide mechanical stability to films and coatings.

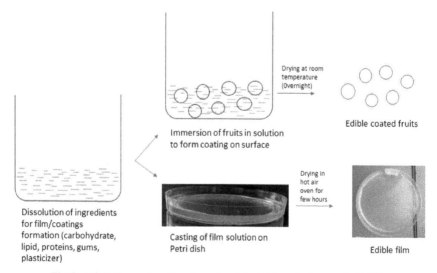

Fig. 2. Edible films and coatings formation (adapted from Sucheta et al., 2019b).

7.1 Mechanisms of Interaction of Polymers as Edible Packaging with Food Commodities

The application of the edible films varies according to their functionality. The major functionalities such as gas barrier properties, water vapour transmittance, thermal and mechanical properties of the edible films determine their use for a particular category of food products. Based on their composition, edible polymers can be of different types: hydrocolloids, lipids and composites. Each type of polymer has different properties. Therefore, the way they interact with the food commodities might also vary. Edible coatings can also be applied to the food commodities prior to osmotic dehydration as there can be loss of essential water-soluble nutrients during the process. The edible film should have selective permeability towards water vapor, so that excess loss of moisture can be prevented. Application of edible films before freeze drying can prevent loss of volatile components, including water-soluble vitamins and other flavoring compounds from food material. Table 2 depicts the effect of different polymers as edible packaging on shelf-life of different food commodities.

Table 2. Effect of Edible Films and Coatings on Shelf-Llife of Different Food Commodities.

Edible biopolymer	Food commodities	Shelf-life enhancement	References
Alginate/CaCl$_2$(EC)	Capsicum	The pomegranate peel extract possessed antimicrobial and antifungal activity	Nair et al., 2018
Pectin/melted bee wax/ sorbitol	Mango	The shelf-life was extended up to 13 days	Moalemiyan et al., 2010
Corn starch/sorbitol	Grapes	Increased sensory attributes after 21 days of storage	Fakhouri et al., 2015
Gelatin/sorbitol or glycerol	Grapes	Increased sensory attributes after 21 days of storage	Fakhouri et al., 2015
Soy protein isolates/ hydroxypropyl methylcellulose/olive oil/potassium sorbate	Pear	Reduced weight loss, overall enhancement in quality and shelf-life	Nandane et al., 2017
Chitosan/glycerol	Cheese	Fungal growth was reduced and organoleptic quality was maintained after 120 days	El-Sisi et al., 2015
Methylcellulose/corn starch/water/glycerol/ cinnamon oil	Cake	Rheological properties of dough, antibacterial and antioxidant, delayed spoilage of cake, specific volume increased and cake retained freshness for longer period	El-Zainy et al., 2014
Cactus mucilage extract/glycerol	Strawberries	Firmness of the coated sample was maintained, shelf-life was prolonged, sensorial quality of the sample maintained up to nine days	Del-Valle et al., 2005

7.1.1 Fruits and Vegetables

Edible films and coatings possess various properties including gas barrier properties, flexibility, transparency, biodegradability, protection against lipid oxidation, water-vapour barrier and other mechanical properties, etc. Compression strength, elastic modulus, burst strength, tensile strength are some other properties of edible films. All these functionalities of edible films and coatings contribute towards shelf-life enhancement of fruits and vegetables. Table 3 shows some effect of different forms of edible films and coatings on major nutrients present in fruits and vegetables.

In case of fruits and vegetables, the major role that edible films play is to prevent moisture loss, thus retaining their firmness and textural quality. There are studies which depict that their sensory quality is also maintained. In general, edible coatings or packaging materials can reduce post-harvest losses by slowing down the rate of physical, chemical and enzymatic changes. Carbon dioxide and oxygen permeability of the edible films depends upon the structural materials used and might vary accordingly. Oz and Eker (2017) conducted a study on edible films-coated processed pomegranate fruits. In this study, pomegranate arils were dipped in chitosan aqueous solutions and distilled water. There was a significant decrease in moisture loss from the coated fruit samples as compared to control. Also, decline in aril anthocyanin content was suppressed after 12 days of storage at 4°C. In a study conducted by Athmaselvi et al. (2012), tomatoes were coated with Aloe-vera gel-based films. After the 20th day, the firmness of the coated tomatoes was comparatively higher than the control samples of tomato. Total soluble solids retained in the coated tomatoes were also found to be higher than the control tomatoes. The retention of total soluble solids is an indicator of reduction in ripening time. Mani et al. (2017) studied the efficacy of aloe-vera gel-based edible coating on *Zizyphus mauritiana* fruits. They found that the coating reduced the weight loss, acid loss, maximized colour retention and also reduced the loss of ascorbic acid from the fruits. Menezes and Athmaselvi (2016) studied the effect of sodium alginate and pectin-based edible films on the shelf-life of sapota fruits. They discovered a significant reduction in the deterioration of physico-chemical quality parameters in coated sapota fruits. The organoleptic property of the fruit was also maintained for 30 days after coating it with polysaccharide-based edible film for 2 min. Li et al. (2017) studied the effect of polysaccharide-based edible coating on strawberries. They found that even after 12 days of coating, there was reduced softening, rot production and total soluble solids were also maintained. The activity of the enzymes responsible for lipid peroxidation and membrane damage, such as peroxidase, catalase, superoxide dismutase, ascorbate peroxidase was also reduced. Ascorbic acid and total phenolic content were also maintained. Polymer coating with anti-browning agent N-acetylcysteine was applied to fresh-cut apples. After 23 days of storage, the firmness and the anti-microbial effect were maintained. Therefore, the overall shelf-life of the apples was enhanced (Rojas-Grau et al., 2007). According to Galus and Kadzinska (2015), addition of lipids to edible films enhances their hydrophobicity, thus preventing moisture loss from fruits and vegetables. Trehalose-based edible films provide gas barrier (CO_2, O_2), thus providing a potential protective effect to fruits and vegetables (Giosafatto et al., 2014).

Table 3. Effect of Edible Coatings on Weight Loss and Major Nutrients in Few Fruits and Vegetables.

Coating material	Food commodity	Gas transmission	Weight loss	Nutrient losses	References
Methyl cellulose, polyethylene glycol, stearic acid, ascorbic acid or citric acid	Apricots	-	9.25 ± 0.05 to $12.8 \pm 0.2\%$	Vitamin C 41.5 ± 0.2 to $73.9 \pm 0.2\%$	Ayranci and Tune, 2004
Methyl cellulose, polyethylene glycol, stearic acid, ascorbic acid or citric acid	Bell peppers	-	2.99 ± 0.05 to $3.47 \pm 0.05\%$	Vitamin C 55.3 ± 0.2 to $70.2 \pm 0.2\%$	Ayranci and Tune, 2004
Zein, starch, polyvinyl acetate, carnauba, carnauba-polysaccharide	Apples	$6–7$ kPa CO_2 and $11–15$ kPa O_2	2.2 kPa to 1.3 kPa	Sucrose equivalence 15.7 to 16.5	Bai et al., 2002
Polyethylene, candelilla, carnauba-shellac, shellac	Apples	5.9 kPa to 11.2 kPa CO_2 and 5.5 kPa to 12.5 kPa O_2	1.7 to 2.9%	-	Bai et al., 2003
Guar gum, glycerol	Tomato	2.8 ml kg^{-1} CO_2	5 to 15%	-	Ruelas-Chacon et al., 2017
Starch, sorbitol, glycerol	Tomato	-	4 to 20%	Ascorbic acid $24–20$ mg/100 g	Nawab et al., 2017
Hydroxypropyl-methylcellulose, ethanolic extract of propolis, water	Grapes	16 mg kg^{-1} h^{-1} to 22 mg kg^{-1} h^{-1} O_2 and 38 mg kg^{-1} h^{-1} to 50 mg kg^{-1} h^{-1} CO_2	2 to 6%	-	Pastor et al., 2011

7.1.2 Dairy Products: Cheese

Edible packaging material creates modified atmospheric conditions, which provide a controlled environment for cheese preservation. There are several factors which are related to cheese preservation and storage, like RH, temperature, oxygen, carbon dioxide permeability and water vapour barrier. Edible coatings can also inhibit or reduce the growth of microorganisms which might cause off flavor and odor properties. Several efforts are now being made to produce novel edible packaging materials.

In a study conducted by Cerqueira et al. (2009) different polysaccharide-based formulations of edible coatings on cheese were studied. Based on their spreading coefficient and opacity towards CO_2 and O_2 it was found that galactomannan-based edible coatings decreased the gas exchange rates in cheese thus, extending its shelf-life. Functional polysaccharides having properties like increased antioxidant, phenolic content, etc. do not only just assist in retaining the quality of the product, but also provide stability in terms of reduced respiration rate. Selective permeability depending upon the structure, source and type of edible film can provide resistance towards mold growth in cheese (Cerqueira et al., 2009). A solution made from *Gleditsiatria canthos* containing galactomannan is considered to be an important structural unit for the construction of edible coating for cheese. They not only provide resistance towards mold growth but also reduce the respiration rate in cheese. Chitosan/whey protein-based edible coatings improve the quality of cheese during ripening time (Yangilar, 2015).

7.1.3 Baked and Fried Foods

Polymer-based coatings are considered to reduce acrylamide formation during frying and also reduce the consumption of oil (Al-Asmar et al., 2018). The fat transfer in fried foods coated with gellan gum-based films is lower than the uncoated foods. Moisture retention in coated samples is higher, however thermal diffusivity is also slightly higher but the increase in temperature is slower than the uncoated samples (William and Mittal, 1999). The moisture loss from deep fat-fried starchy food products coated with hydroxypropyl methyl cellulose or methyl cellulose and corn zein-based edible coating solution is less reported than the uncoated food products (Mallikarjunan et al., 1997). Shelf-life of baked product, mustafakemalpasa (MKP) cheese sweets coated with edible films (whey protein concentrate and corn zein) was prolonged from three to 10 days (Guldas et al., 2010). Bakery products with edible coatings retained crunchiness, delicate and light taste for a longer period of time. Cinnamon oil-based edible coating was used for cake. This led to a decrease in acid value and peroxide value during storage. It also improved the specific volume and firmness of cake and maintained it during the storage period (El-Zainy et al., 2014). Breakfast cereals with raisins are coated with starches to prevent them from moistening. Emulsified edible coatings composed of methyl cellulose, corn starch and soybean oil extended the shelf-life of coated crackers stored at 65 per cent, 75 per cent and 85 per cent relative humidity compared to uncoated ones by reducing moisture uptake (Bravin et al., 2006).

Overall, the edible encapsulating coatings can protect food materials from reaction with environmental contaminants and against loss of volatile compounds. Besides reducing moisture loss, they also provide increased resistance towards mold growth. Wax is the oldest form of edible film used to improve the keeping quality of foods. Increasing demands for biodegradable and environmentally-friendly packaging materials has led to the development of more advanced and improved quality of edible films. These films show different properties and have different mechanisms to protect the food from getting damaged. The exact mechanism of interaction of edible coatings with the food depends not just on the composition of the film, but also on the food commodity it is being coated on. Edible films nowadays are generally constructed with a prior motive to increase microbial resistance, functional property, aroma barrier application, prevent the food like nuts from getting rancid, bioactive compounds enhancement, etc.

7.2 Biochemical Aspects of Edible Packaging

Depending upon the composition of edible films, the biochemical aspects vary according to category of food products coated or packaged with them. The property of the film will vary according to the polymeric compounds used to produce it. The pore size of the film might have a significant effect on the moisture loss, weight loss, total soluble solids and pH of the fruit or vegetable. Also, in some cases, edible coatings or films might protect against structural damage due to freeze drying, spray drying, etc. and other various techniques used during food processing. Similarly, edible coatings or films might also provide protection against nutrient losses, such as lycopene and ascorbic acid in the case of tomatoes. Besides, active edible coatings or films might possess free radical scavenging activity, i.e., oxygen atom with unpaired electron. They can also reduce an instant rise in temperature during frying and formation of acryl-amide molecules, which are considered to be toxic. Utilization of natural sources, such as aloe-vera gels not only provide structural support to the film, but also enrich it with nutrients, such as amino acids and various forms of low-calorie sugars, e.g., mannose. The softening of the fruit tissues which leads to their decay is generally due to the activity of the enzyme polygalacturonase. This enzyme is responsible for the degradation of polygalacturonic acid chain which forms the structural unit of the plant cell wall. There have been studies, which show that edible film/coatings inhibit or delay the synthesis of cell wall-degrading enzyme, thus preventing the softening of the fruit and enhancing the shelf-life (Bhatia et al., 2014). There have been studies that show edible films/coatings made up of lipids, such as mineral oils, can suppress ethylene biosynthesis. Ethylene is a growth hormone secreted by fruits and vegetables. Its synthesis is responsible for maturation. Suppression of this hormone can protect the fruits and vegetables from post-harvest losses and enhanced shelf-life (Davila-Avina et al., 2011). Activity of enzyme pectate lyase is also decreased within fruits and vegetables due to mineral oil coating, which inhibits CO_2 and O_2 availability, thus delaying senescence and enhancing shelf-life. Edible packaging can also prevent oxidative, microbial and hydrolytic rancidity. Hydroxyl peroxide radical formation can be inhibited. Peroxides are responsible for the destruction of nutrients, such as vitamin A and E and also the generation of trans-fatty acid molecules, which are not considered good for health. Besides, antimicrobial properties of edible coatings might protect against rancidity caused by microbes. Microbial rancidity leads to spoilage of food. Anti-oxidative nature of additives added to the edible coatings is also responsible for preventing rancidity. Hydrophobic nature of edible packaging materials can prevent against the uptake of moisture from the environment. Hydrolytic rancidity leads to off odor and spoilage. Further, fat molecules present in food commodity might undergo breakdown into its constituent fatty acids and glycerol molecules. Free fatty acids might undergo oxidation and lead to oxidative rancidity and nutrient losses (Panigrahi et al., 2018). Activity of edible packaging as preservative is also considered to be of importance. Extracts from various natural sources, such as rosemary leaves, can act as powerful preservatives. Citric acid can also be added to enhance the preservative effect. Natural preservatives and plant-based alternatives can enhance shelf-life and play an active role in edible packaging.

8. Current Research Status in Films and Coatings

Mechanical stability of edible films is a major problematic aspect regarding their application in food packaging. The increased complexity of films and coatings has led to an increased understanding of stress-related assessment of these films (Abadias et al., 2018). The advantage of edible films is that any form of active ingredients can be incorporated into their matrix and can be consumed along with the food. The improvement of delivery property (bioactive ingredients) of the edible films and coatings is the major issue to be dealt with in future research (Lopex et al., 2010). Enhancing the functional property of the edible films and coatings might aid in enhancing or maximizing the shelf-life and quality (including the functional properties) of fruits and vegetables (Nayik et al., 2015). Functional elements in edible films and coatings include plasticizers, antioxidants, essential oil pigments, vitamins, anti-microbial agents and other chemical preservatives, etc. (Erkmen and Barazi, 2018). Packaging material from renewable sources and lipids from biological origin could be the major new trend in the production and utilization of edible films. The sensory, microbial, biochemical and physicochemical stability of the edible films is of major importance (Maftoonazad et al., 2013).

Novel probiotic foods have attained huge attention from industries. There are several materials available which can be used for designing edible films. Therefore, their utilization as a probiotic delivery system can be a new area of study. Significant health benefits have been associated with probiotic foods. Modernisation and technological advancement have led to the development of several types of food-packaging materials. Edible films having probiotic health benefits are an integral part of this new trend (Pavli et al., 2018). Edible coatings can also be used as a carrier of anti-microbial properties or as an anti-microbial food additive. This property prevents the food against microbial damage (Valdes et al., 2017).

Delivery of the bioactive compounds at controlled rates and their activity depends upon the type of biopolymers used in the composition mixture for constructing edible films. Edible films having antioxidant property might prevent the development of off-flavor, food oxidation and nutritional losses. Sugarcane bagasse-obtained cellulose nanocrystals reinforced starch-based films were successfully produced by Slavutsky and Bertuzzi (2014). A solution containing 0.25 wt per cent pectin and 5 wt per cent alginate, carrageenan, potato starch (modified or unmodified), gellan gum or cellulose (extracted from soybean chaff or commercial) were used to manufacture composite alginate films (Harper et al., 2014). Proteins extracted from various sources, especially by-products, can be used for the production of edible films. Oil-extracted meal-based protein example canola protein. These proteins might also exhibit some antimicrobial or antioxidant activity (Lee et al., 2015; Chang and Nickerson, 2013). Zein protein is also being extensively studied as an active ingredient for producing edible films. Hydrophobicity is imparted to food coatings by adding lipids. They provide water-vapour barrier to the food (Aydin et al., 2017). Active packaging is a dynamic new technology to preserve foods sensory or safety properties and maintain their quality and shelf-life (Salgado et al., 2015).

9. Regulatory Aspects

Edible films are readily consumable as they do not leave any residue post application. They also have a positive effect on the physical, physiological and biochemical aspects of the food commodity. These films not only reduce respiration rate, but also retain moisture and maintain the firmness of the produce. Edible films might also prevent enzymatic browning of the fruits, such as apples, by deactivating or delaying the action of polyphenolic enzyme, i.e., polyphenol oxidase. Edible films are generally eco-friendly and the commodities used to produce them are generally considered as safe (GRAS). It is also important that the materials used in edible films should be Food and Drug Administration (FDA) approved and used within limits. Any material, including biopolymers, which carry the additives to migrate within, in turn transferring the properties into the food commodity and having a preservative effect, will face regulatory scrutiny (Suput et al., 2015). Toxicity of edible films is of major concern, specially in the case of active packaging materials as they are directly in contact with the food commodity. Interaction of active compounds with the food molecules and the result of that interaction should not produce toxic components. Therefore, their role has to be demonstrated. Maximum and minimum ranges of components used in the composition of edible films should be regulated and may be fixed, although maximum efficiency of edible films or coatings is with incorporation of maximum quantity of active compounds. Edible films and coatings are Food Contact Materials (FCM). Therefore, it becomes extremely necessary to ensure its safety of use and consumption. Excess of anything is generally not considered viable, so, controlled conditions while constructing edible coatings and controlled use of the ingredients is highly recommended. There are certain laws that follow and regulate the use of food packaging materials but their application in terms of edible coatings can only be defined, based on the materials used for producing them.

References

Abadias, C., F.E. Chason, J. Keckes, M. Sebastiani, G.B. Thompson, E. Barthel, G.L. Doll, C.E. Murray, C.H. Stoessel and L. Martinu (2018). Review article: Stress in thin films and coatings: Current status, challenges and prospects, *Journal of Vacuum Science and Technology*, 36: 1–42.

Al-Asmar, A., D. Naviglio, C.V.L. Giosafatto and L. Mariniello (2018). Hydrocolloid-based coatings are effective at reducing acrylamide and oil content of french fries, *Coatings*, 8: 147.

Ali, A. and S. Ahmad (2018). Recent advances in edible polymer based hydrogels as a sustainable alternative to conventional polymers, *Journal of Agricultural and Food Chemistry*, 66: 6940–6967.

Atef, M., M. Rezaei and R. Behrooz (2015). Characterization of physical, mechanical and antibacterial properties of agar-cellulose bionanocomposite films incorporated with savory essential oil, *Food Hydrocolloids*, 45: 150–157.

Athmaselvi, K.A., P. Sumitha and B. Revathy (2013). Development of *Aloe vera* based edible coating for tomato, *International Agrophysics*, 27: 369–375.

Aydin, F., H.I. Kahve and M. Ardic (2017). Lipid-based edible films, *Journal of Scientific and Engineering Research*, 4: 86–92.

Ayranci, E. and S. Tunc (2004). The effect of edible coatings on water and vitamin C loss of apricots (*Armeniaca vulgaris* Lam.) and green peppers (*Capsicum annuum* L.), *Food Chemistry*, 87(3): 339–342.

Azeredo, H.M.C., L.H.C. Mattoso, D. Wood, T.G. Williams, R.J. Avena-Bustillos and T.H. McHugh (2009). Nanocomposite edible films from mango puree reinforced with cellulose nanofibers, *Journal of Food Science*, 74(5): N31–N35.

Bai, J., E.A. Baldwin and R.H. Hagenmaier (2002). Alternatives to shellac coatings provide comparable gloss, internal gas modification, and quality for 'Delicious' Apple Fruit, *Hort. Science*, 37: 559–563.

Bai, J., R.D. Hagenmaier and E.A. Baldwin (2003). Coating selection for 'Delicious' and other apples, *Postharvest Biology and Technology*, 28: 381–390.

Barak, S. and D. Mudgil (2014). Locust bean gum: Processing, properties and food applications— A review, *Int. J. Bio. Macromol.*, 66: 74–80.

Bhatia, S., M.S. Alam, M. Arora and V.K. Sehgal (2014). Polysaccharide-based edible coatings influence the biochemical characteristics and storage behaviour of tomato during ambient storage, *Indian J. Agric. Biochem.*, 27(2): 151–157.

Binsi, P.K., C.N. Ravishankar and T.K.S. Gopal (2013). Development and characterization of an edible composite film based on chitosan and virgin coconut oil with improved moisture sorption properties, *Journal of Food Science*, 78: E526–E534.

Bourtoom, T. (2008). Edible films and coatings: Characteristics and properties, *Int. Food Res. J.*, 15: 237–248.

Cazon, P., G. Velazquez, Ramirez, J.A. and M. Vazquez (2017). Polysaccharide-based films and coatings for food packaging: A review, 68: 136–148.

Cerqueira, M.A., A.M. Lima, B.W.S. Souza, J.A. Teixeira, R.A. Moreira and A.A. Vicente (2009). Functional polysaccharides as edible coatings for cheese, *Journal of Agricultural and Food Chemistry*, 57: 1456–1462.

Chang, C. and M.T. Nickerson (2013). Effect of protein and glycerol concentration on the mechanical, optical, and water vapor barrier properties of canola protein isolate-based edible films, *Food Science and Technology International*, 21: 33–44.

Chen, Y., Y. Ran and J. Liu (2014). Effects of different concentrations of ethanol and isopropanol on physicochemical properties of zein-based films, *Industrial Crops and Products*, 53: 140–147.

Chiralt, A., C. González-Martínez, M. Vargas and L. Atarés (2018). Edible films and coatings from proteins, pp. 477–500. *In: Proteins in Food Processing*, Woodhead Publishing.

Costa, M.J., L.C. Maciel, J.A. Teixeira, A.A. Vicente and M.A. Cerqueira (2018). Use of edible films and coatings in cheese preservation: Opportunities and challenges, *Food Res. Int.*, 107: 84–92.

Dakia, P.A., C. Blecker, C. Robert, B. Wathelet and M. Paquot (2008). Composition and physicochemical properties of locust bean gum extracted from whole seeds by acid or water dehulling pre-treatment, *Food Hydrocolloids*, 22: 807–818.

Dangaran, K., P.M. Tomasula and P. Qi (2009). Structure and Function of Protein-based Edible Films and Coatings. *In:* K. Huber and M. Embuscado (Eds.). *Edible Films and Coatings for Food Applications*, Springer, New York.

Davila-Avina, J.E., J. Villa-Rodriguez, R. Cruz-Valenzuela, M. Rodriguez-Armenta, M. Espino-Diaz, J.F. Alaya-Zavala, G.I. Olivas-Orozco, B. Heredia and G. Gonzalez-Aguilar (2011). *American Journal of Agricultural and Biological Sciences*, 6: 162–171.

Debeaufort, F., J.A. Quezada-Gallo and A. Voilley (1998). Edible films and coatings: Tomorrow's packagings: A review, *Critical Reviews in Food Science*, 38: 299–313.

Del-Valle, V., P. Hernandez-Munoz, Guarda, A. and M.J. Galloto (2005). Development of Cactus-mucilage edible coating (*Opuntia ficus-indica*) and its application to extend strawberry (*Fragaria ananassa*) shelf-life, *Food Chemistry*, 91: 751–756.

El-Sisi, A.S., A.E.S.M. Gapr and K.M. Kamaly (2015). Use of chitosan as an edible coating in RAS cheese, *Biolife*, 3: 1–7.

El-Zainy, A.R.M., H.E.H. Aboul-Anean, L.A. Shelbaya and E.M.M. Ramadan (2014). Effect of edible coating with cinnamon oil on the quality of cake, *Middle East Journal of Applied Sciences*, 4: 1171–1186.

Erkmen, O. and A.O. Barazi (2018). General characteristics of edible films, *Journal of Food Biotechnology Research*, 2: 1–4.

Espitia, P.J.P., W.-X. Du, R. Avena-Bustillos, J. De, N. de, F.F. Soares and T.H. McHugh (2014). Edible films from pectin: Physical-mechanical and antimicrobial properties—A review, *Food Hydrocolloids*, 35: 287–296.

Fakhouri, F.M., S.M. Martelli, T. Caon, J.I. Velasco, L. Lucia Helena and I. Mei (2015). Edible films and coatings based on starch/gelatin: Film properties and effect of coatings on quality of refrigerated red crimson grapes, *Postharvest Biology and Technology*, 109: 57–64.

Falguera, V., J.P. Quintero, A. Jimenez, J.A. Munoz and A. Ibarz (2011). Edible films and coatings: structures, active functions and trends in their use, *Trends Food Sci. Technol.*, 22: 292–303.

Fik, M., K. Surówka, I. Maciejaszek, M. Macura and M. Michalczyk (2012). Quality and shelf-life of calcium-enriched wholemeal bread stored in a modified atmosphere, *Journal of Cereal Science*, 56: 418–424.

Galus, S. (2018). Functional properties of soy protein isolate edible films as affected by rapeseed oil concentration, *Food Hydrocolloids*, 85: 233–241.

Galus, S. and J. Kadzinska (2016). Whey protein edible films modified with almond and walnut oils, *Food Hydrocolloids*, 52: 78–86.

Garcia, M.A., M.N. Martino and N.E. Zaritzky (2000). Lipid addition to improve barrier properties of edible starch-based films and coatings, *Food Chemistry and Toxicology*, 65: 941–947.

Ge, X., L. Yu, Z. Liu, H. Liu, Y. Chen and L. Chen (2019). Developing acrylatede-poxidized soybean oil coating for improving moisture sensitivity and permeability of starch-based film, *Int. J. Bio. Macromol.*, 125: 370–375.

George, J. and Siddaramaiah (2012). High performance edible nanocomposite films containing bacterial cellulose nanocrystals, *Carbohydrate Polymers*, 87: 2031–2037.

Guldas, M., A.A. Bayizit, T.O. Yilsay and L. Yilmaz (2010). Effects of edible film coatings on shelf-life of mustafakemalpasa sweet, a cheese-based dessert, *J. Food Sci. Technol.*, 47: 476–481.

Han, J.H. (2014). Edible films and coatings: A review, pp. 231–255. *In: Innovations in Food Packaging*, Elsevier, USA.

Harper, B.A., S. Barbut, A. Smith and M.F. Marcone (2014). Mechanical and microstructural properties of 'wet' alginate and composite films containing various carbohydrates, *Journal of Food Science*, 80(1): E84–E92.

Hassan, B., S.A.S. Chatha, A.I. Hussain, K.M. Zia and N. Akhtar (2018). Recent advances on polysaccharides, lipids and protein-based films and coatings: A review, *Int. J. Bio. Macromol.*, 109: 1095–1107.

Hopkins, E.J., C. Chang, R.S.H. Lam and M.T. Nickerson (2015). Effects of flaxseed oil concentration on the performance of a soy protein isolate-based emulsion-type film, *Food Research International*, 67: 418–425.

Janjarasskul, T. and J.M. Krochta (2010). Edible packaging materials, *Annual Review of Food Science and Technology*, 1: 415–448.

Kester, J.J. and O. Fennema (1989). Resistance of lipid films to water vapor transmission, *JAOCS*, 66: 1139–1146.

Khalil, H.P.S.A., Y.Y. Tye, C.K. Saurabh, C.P. Leh, T.K. Lai, E.W.N. Chong, M.R.N. Fazita, J.M. Hafiidz, A. Banerjee and M.I. Syakir (2017). Biodegradable polymer films from seaweed polysaccharides: A review on cellulose as reinforcement material, *Materials Research Express*, 5: 244–265.

Kontturi, E., T. Tammelin and O.M. Sterberg (2006). Cellulose model films and the fundamental approach, *Chem. Soc. Rev.*, 35: 1287–1304.

Krull, L.H. and G.E. Inglett (1971). Industrial uses of gluten, *Cereal Sci. Today*, 16: 232–236.

Lee, J.-H., J. Lee and K.B. Song (2015). Development of a chicken feet protein film containing essential oils, *Food Hydrocolloids*, 46: 208–2015.

Li, X., C. Qiu, N. Ji, C. Sun, L. Xiong and Q. Sun (2015). Mechanical, barrier and morphological properties of starch nanocrystals-reinforced pea starch films, *Carbohydrate Polymers*, 121: 155–162.

Li, L., J. Sun, H. Gao, Y. Shen, C. Li, P. Yi, X. He, D. Ling, J. Sheng, J. Li, G. Liu, F. Zheng, M. Xin, Z. Li and Y. Tang (2017). Effects of polysaccharide based edible coatings on quality and antioxidant enzyme system of strawberry during cold storage, *International Journal of Polymer Science*, 2017: 8.

Lopez, O., M.A. Garcia and N.E. Zaritzky (2010). Novel sources of edible films and coatings, *Stewart Postharvest Review*, 6: 1–8.

Maftoonazad, N., F. Badii and M. Shahamirian (2013). Recent Innovations in the area of edible films and coatings, *Recent Pat. Food Nutr. Agric.*, 5: 201–213.

Mallikarjunan, P., M. Chinnan, V.M. Balasubramaniam and R.D. Phillips (1997). Edible coatings for deep-fat frying of starchy products, *LWT – Food Science and Technology*, 30: 709–714.

Mani, A., N. Jain, A.K. Singh and M. Sinha (2017). Effects of aloe vera edible coating on quality and postharvest physiology of ber under ambient storage conditions, *International Journal of Pure and Applied Bioscience*, 5: 43–53.

Marquez, G.R., P. Di Pierro, M. Esposito, L. Mariniello and R. Porta (2014). Application of transglutaminase-cross-linked whey protein/pectin films as water barrier coatings in fried and baked foods, *Food and Bioprocess Technology*, 7: 447–455.

Matias Ortiz, C., P.R. Salgado, A. Dufresne and A.N. Mauri (2018). Microfibrillated cellulose addition improved the physicochemical and bioactive properties of biodegradable films based on soy protein and clove essential oil, *Food Hydrocolloids*, 79: 416–427.

Moalemiyan, M., H.S. Ramaswamy and N. Maftoonazad (2010). Pectin-based edible coating for shelf life extension of Ataulfo mango, *Journal of Food Process Engineering*, 35: 572–600.

Nair, M.S., A. Saxena and C. Kaur (2018). Characterization and antifungal activity of pomegranate peel extract and its use in polysaccharide based edible coatings to extend the shelf-life of capsicum (*Capsicum annuum* L.), *Food and Bioproducts Processing*, 11: 1317–1327.

Nandane, A.S., R.K. Dave and T.V. Raman Rao (2017). Optimisation of edible coating formulations for improving postharvest quality and shelf-life of pear fruit using response surface methodology, *Journal of Food Science and Technology*, 54: 1–8.

Nawab, A., F. Alam and A. Hasnain (2017). Mango kernel starch as a novel edible coating for enhancing shelf-life of tomato (*Solanum lycopersicum*) fruit, *International Journal of Biological Macromolecules*, 103: 581–586.

Nayik, G.A., I. Majid and V. Kumar (2015). Developments in edible films and coatings for the extension of shelf-life of fresh fruits, *American Journal of Nutrition and Food Science*, 2: 16–20.

Nussinovitch, A. (2013). Biopolymer films and composite coatings, *Handbook of Biopolymers and Biodegradable Plastics*, Plastics Design Library, 295–327.

OZ, A.T. and T. Eker (2017). Effect of edible coating on minimally processed pomegranate fruits, *Journal on Processing and Energy in Agriculture*, 21: 197–200.

Panigrahi, J., M. Patel, N. Patel, B. Gheewala and S. Gantait. 2018. Changes in antioxidant and biochemical activities in castor oil-coated *Capsicum annuum* L. during postharvest storage. 3Biotech. 8: 280.

Pastor, C., L. Sanchez-Gonzalez, A. Marcilla, A. Chiralt, M. Chafer and C. Gonzalez-Martinez. 2011. Quality and safety of table grapes coated with hydroxypropyl methylcellulose edible coatings containing propolis extract. *Postharvest Biology and Technology*, 60: 64–70.

Pavlath, A.E. and W. Orts (2009). Edible films and coatings: Why, what, and how? *In: Edible Films and Coatings for Food Applications*, pp. 1–23, Springer, New York, NY.

Pavli, F., C. Tassou, Nychas, G.J.E. and N. Chorianopoulos (2018). Probiotic incorporation in edible films and coatings: bioactive solution for functional foods, *International Journal of Molecular Sciences*, 19: 150.

Phillips, G.O. and P.A. Williams (2000). Introduction to food hydrocolloids, *Handbook of Hydrocolloids*, Boca Raton, Fla: CRC Press.

Ramos, O.L., J.C. Fernandes, S.I. Silva, M.E. Pintado and F.X. Malcata (2012). Edible films and coatings from whey proteins: a review on formulation, and on mechanical and bioactive properties, *Cri. Rev. Food Sci. Nutr.*, 52: 533–552.

Robertson, G.L. (2006). Packaging of dairy products. pp. 524–525. *In:* G.L. Robertson (Ed.). *Food Packaging: Principles and Practice,* Boca Raton: CRC Press.

Rojas-Grau, M.A., R.M. Raybaudi-Massilia, R.C. Soliva-Fortuny, R.J. Avena-Bustillos, T.H. McHugh and O. Martin-Belloso (2009). Apple puree-alginate edible coating as carrier of antimicrobial agents to prolong shelf-life of fresh-cut apples, *Post-Harvest Biology and Technology*, 45: 254–264.

Rooney, M.L. (2005). Introduction to active food packaging technologies, *Plastic Films in Food Packaging*, Plastic Design Library, 127–138.

Rudolph, B. (2000). Seaweed products: red algae of economic significance, pp. 515–529. *In:* R.E. Martin (Ed.). *Marine and Freshwater Products Handbook,* Lancaster, PA: Technomic Pub. Co.

Ruelas-Chacon, X., J.C. Contreras-Esquivel, J. Montanez, A.F. Aguilera-Carbo, L.M. Reyes-Vega, R.D. Peralta-Rodriguez and G. Sanchéz-Brambila (2017). Guar gum as an edible coating for enhancing shelf-life and improving postharvest quality of roma tomato (*Solanum lycopersicum* L.), *Journal of Food Quality*, 9.

Salgado, P.R., C.M. Ortiz, Y.S. Musso, L.D. Giorgio and A.N. Mauri (2015). Edible films and coatings containing bioactives, *Current Opinion in Food Science*, 5: 86–92.

Simson, S.P. and M.C. Straus (2010). *Post-Harvest Technology of Horticultural Crops,* Oxford, Jaipur, India.

Slavutsky, A.M. and M.A. Bertuzzi (2014). Water-barrier properties of starch films reinforced with cellulose nanocrystals obtained from sugarcane bagasse, *Carbohydrate Polymers*, 110: 53–61.

Sucheta, S. Kumar, K. Chaturvedi and S. Kumar (2019a). Evaluation of structural integrity and functionality of commercial pectin-based edible films incorporated with corn flour, beetroot, orange peel, muesli and rice flour, *Food Hydrocolloids*, 91: 127–135.

Sucheta, K. Chaturvedi, N. Sharma and S. Kumar (2019b). Composite edible coatings from commercial pectin, corn flour and beetroot powder minimize post-harvest decay, reduces ripening and improves sensory liking of tomatoes, *International Journal of Biological Macromolecules*, 133: 284–293.

Suput, D.Z., V.L. Lazic, S.Z. Popovic and N.M. Hromis (2015). Edible films and coatings – sources, properties and application, *Food and Feed Research*, 42(1): 11–22.

Valdes, A., M. Ramos, A. Beltran, A. Jimenez and M.C. Garrigos (2017). State-of-the-art of antimicrobial edible coatings for food packaging applications, *Coatings*, 7: 1–23.

Wang, Z., J. Zhou, X.X. Wang, N. Zhang, X.X. Sun and Z.S. Ma (2014). The effects of ultrasonic/microwave assisted protein isolate-based oleic acid/stearic acid blend edible films, *Food Hydrocolloids*, 35: 51–58.

Williams, R. and G.S. Mittal (1999). Low-fat fried foods with edible coatings: Modelling and simulation, *Journal of Food Science*, 64: 317–322.

Yangilar, F. (2015) Evaluation of the effects of chitosan/whey protein (CWP) coating on quality of Göbek Kashar cheese during ripening, *Korean J. Food Sci. An.*, 35: 216–224.

Young, L.L., R.D. Reviere and A.B. Cole (1988). Fresh red meats: A place to apply modified atmospheres, *Food Technol.*, 42: 65–69.

Zhang, R., W. Wang, H. Zhang, Y. Dai, H. Dong and H. Hou (2019). Effects of hydrophobic agents on the physicochemical properties of edible agar/maltodextrin films, *Food Hydrocolloids*, 88: 283–290.

Future Ventilated Packaging Design

Novel Modeling Approaches and Integrated Multi-criteria Performance

Umezuruike Linus Opara,* Alemayehu Ambaw Tsige
and *Tobi Fadiji*

1. Introduction

The demand for fresh fruit and vegetable products has shown consistent growth within the last few decades (Johnson, 2014; Blisard et al., 2004). However, these products continuously and rapidly lose quality after harvest and have limited shelf-life. The aim of cold-chain management is to extend the shelf-life of perishables by ensuring an optimum temperature of a commodity during storage and transport. To do this, chilled air is constantly circulated through the stack by use of air-driving equipment and heat is removed from the produce and other sources, using refrigeration/cooling unit in a pre-cooling, cold-storage, refrigerated-transport and retail-cooling processes. Effective execution of cold-chain management entails an understanding of the dynamics of product temperature changes, the characteristics of the cooling system and the importance of produce packaging (Opara, 2011; Zou et al., 2006a,b).

Packaging allows quick handling and marketing throughout distribution, protecting the commodity by reducing mechanical loads, such as drop, impact, vibration and compression loads. Packaging boxes are normally provided with vent-holes to facilitate the thermal exchange between the produce and the cooling air (Opara, 2011). On the other hand, a vent-hole causes a reduction in the mechanical

Postharvest Technology Research Laboratory, South African Research Chair in Postharvest Technology, Faculty of AgriSciences, Stellenbosch University, Stellenbosch 7600, South Africa.
* Corresponding author: opara@sun.ac.za

strength of the package. The shape, location and size of the vent-holes on the package box are important design parameters to optimize the mechanical and thermal performances (Opara and Fadiji, 2018; Pathare and Opara, 2014; Pathare et al., 2012; Opara, 2011). Hence, package designs should meet the ventilation requirement and yet be strong enough to prevent collapse. Solving this requires understanding and coupling of diverse disciplines: fruit physiological processes of quality loss and their causes produce thermal dynamics, packaging material of construction and strength of material.

Package design and arrangement inside a cooling system have a significant effect on the energy and material used and the carbon footprint of the industry. Particularly, packaging has a strong effect on the rate and uniformity of the produce cooling process. To make the fruit industry more profitable and sustainable, material reduction, reusing, recovering and recycling should be incorporated at the design phase. Packaging technology must balance commodity protection with other issues, including energy and material costs, heightened social and environmental consciousness and strict regulations on pollutants and disposal of municipal solid waste.

Due to the complexity of factors affecting the packaging box design, the approach to analyze the problem experimentally alone would not only necessitate expensive equipment, but would also require substantial amounts of time (Ambaw et al., 2013; Opara and Zou, 2006, 2007). Numerical modeling techniques, such as computational fluid dynamics (CFD) and computational solid dynamics (CSD) offer an effective means of accurately quantifying thermal and structural characteristics of packaged perishables in the postharvest handling line. Thus, integrated multi-criteria performance analysis can be made, using novel modeling approaches and the amount of physical experimentation can be reduced considerably, although, yet not eliminated.

Numerical modeling technique has made impressive progress in the past decade and has evolved into a promising design tool for the analysis and development of packaging boxes. In this chapter, recent advances in the application of mathematical models of fluid dynamics and solid dynamics for optimizing packaging box design for handling horticultural produce are presented. In addition, examples of integrated multi-criteria performance assessment of package performance are discussed, including cooling and energy performance based on CFD studies and mechanical performance of the package and resistance to mechanical damage based on CSD studies.

2. Aspects of Ventilated Packaging in the Fresh Fruit Supply Chain

2.1 Fruits

Thermal properties of fruits dictate the rate of heat exchange with the surrounding environment during cooling/heating processes. Thermal conductivity, thermal diffusivity, specific heat and density of fruit are the main important properties to design and analyzse thermal processes of foods during their storage and processing (Mukama et al., 2019). Additionally, the respiration and transpiration activities of the produce are crucial considerations during the packaging design and operating refrigeration systems. Respiration is an enzymatic process by which fruit and vegetable cells convert sugars and oxygen into carbon dioxide, water and heat. The process of transpiration includes the transport of moisture through the skin of the commodity, the evaporation of this moisture from the commodity surface and the convective mass transport of moisture to the surroundings by which fresh fruits and vegetables lose moisture (Becker and Fricke, 1996; Rico et al., 2007). The colour, shape, size, surface area, volume and mass of fruit are important parameters that determine the market value of fresh produce.

Package design should enable the rapid removal of the respiration heat and the reduction of transpiration process. The understanding and incorporation of the nature of the fruit is vital in designing packaging and operating refrigerated storage facilities. A well-designed package needs to be adapted to the conditions or specific treatments required for the product (El-Ramady et al., 2015).

2.2 Packaging Box

The type and purposes of packaging boxes are numerous as there are products to put in them and depends on the handling requirement. The packaging box can be made in different shape and size from varied materials. There is no uniformity in container size or weight standards for all fruits and vegetables, but individual crops have specific industry packaging standards. If the crops are not packaged accordingly, wholesale customers probably will not buy them. Hence, packaging box design is required to comply with standards in the market for handling and distribution of individual crop (Daniels and Slama, 2010). Depending on the specific nature of the produce, packages may be fitted with a tray, inner bag (liner), clamshells or other accessories to provide cushioning and additional protective properties against gas and moisture migration.

Packaging footprints need to conform to the dimensions of the standard pallet, which measures 40 x 48 inches (101.60 x 121.92 cm). Bulk packages moved by forklifts are handled using wooden bins (Fig. 1a and b) or plastic (Fig. 1c and d).

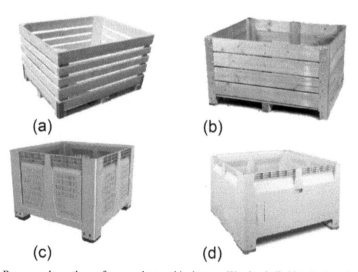

Fig. 1. Boxes used mostly on farms and at packinghouses. Wooden bulk bins (top) and reusable plastic bulk bins (bottom). Bins with a more open structure with good ventilation (left column (a) and (c)) and closed less ventilated (right column (b) and (d)).

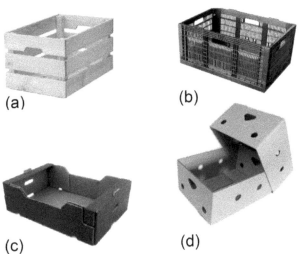

Fig. 2. Boxes commonly used to handleproduce by hand. (a) wooden crats, (b) reusable plastic container, (c) display corrugated fiberboard cartons and (d) telescopic corrugated fiberboard cartons.

These packages weigh as much as 550 kg. Packages of produce commonly handled by hand are usually limited to 25 kg in a wooden, plastic or corrugated fibreboard crates (Fig. 2). Due to their light weight, complete recyclable and biodegradability and as they are more cost-effective than other materials, corrugated cardboards are the predominant packaging mode (Pathare and Opara, 2014). Corrugated cardboard materials are also good in damping mechanical impacts and vibration, which are sources of damage on fruit.

2.3 Palletization

Palletization is stacking of group of packaging boxes on standard sized base structure called pallet (Fig. 3). This makes the loading and unloading of produce into and

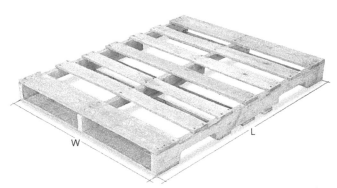

Fig. 3. Representation of a typica pallet structure. The value of width (W) and length (L) specifications depends on standards (see Fig. 4).

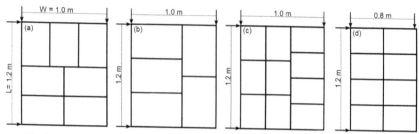

Fig. 4. Schematic of stacking patterns used on the standard ((a) to (c)) and euro (d) pallets.

out of storage warehouse and transport containers more convenient. Palletization is becoming an industry standard for handling large quantities of produce that packaging boxes are designed in such a way that they can be oriented in a pattern (straight or crossed) to fit on the standard (Fig. 4a to c) and euro (Fig. 4d) pallet sizes. Produce should fit optimally inside the box, with little wasted space.

2.4 Precooling

Precooling is the quick removal of the field heat shortly after the harvest of a crop. This can be accomplished by employing one of the different available methods: room cooling, forced-air cooling, vacuum cooling, hydrocooling or spray cooling and package icing. Among these, forced-air cooling (FAC) is the most favoured cooling technique (Ambaw et al., 2017). The tunnel horizontal airflow system is the most common FAC arrangement (Aswaney, 2007; Boyette et al., 1996). The top and back of the tunnel are covered by cloth or plastic sheet. At the front end of the tunnel a fan is mounted to pull chilly air through the stacks and out through the fans. In this arrangement, the chilled air flows horizontally through the stacked produce. Practically, the cooling process continues until 88% of the original temperature difference is removed. This is according to the industrially important characteristics precooling time, called 7/8th cooling time (Boyette et al., 1996), which is the time required to remove 7/8th of the field heat from the crop. The box vent-hole design (area, shape, number and position), box arrangement, the thermophysical properties of the produce, the fruit-stacking pattern within the package, and the ambient conditions determines the rate and uniformity of cooling.

2.5 Cold Storage Room

Cold storage room is a system to keep the quality of agricultural materials beyond their normal shelf-life. In addition, cold storage rooms occasionally used for additional treatments like gassing and fungicide applications (Ambaw et al., 2014; Delele et al., 2012). In the cold storage room, chilled air is constantly circulated through the stack by use of air driving equipment and heat is removed from the produce and other sources by the refrigeration/cooling unit. As in the precooling process, the thermal performance (rate and uniformity of cooling) and energy usage of the storage operation is affected by the design and arrangement of the packaging boxes. For instance, non-uniform stacking of produce can cause non-uniformity flow of air inside the cool store that could cause uneven.

2.6 Transport (shipping) Cooling

Fruits and vegetables for distant market should be kept cool and at the optimum condition during transit. Reefer container is the standard temperature-controlled transportation unit. It is fitted with its own cooling unit as an integral part of the container. The cooling unit is designed and built to take up as little space as possible and, therefore, has very limited refrigeration capacity. Fruit must then be precooled to specified optimum transport temperature prior to loading into reefer containers. Reefers are designed to distribute chilled air from the floor, via specific T-shaped decking (Fig. 5). The air delivery system should be powerful enough to ensure a sufficient and uniform flow of air through the stacked produce inside the shipment.

The incidence and length of temperature breaks that reefer containers may experience within the fresh fruit export supply chain should be analysed as this has an impact on the integrity of the cold chain and the quality of the fresh fruit exported. Knowing the magnitude of the temperature breaks and assessing its consequences increase the capacity and willingness of various stakeholders to address operational challenges.

In addition to the design and operating configuration of the reefer container, the package design (size of boxes and vent-hole design) and arrangement (stacking patterns) significantly influence the cooling rate, cooling uniformity, energy usage and space utilization (Getahun et al., 2017a). Specially, vertical airflow is crucial inside reefer that package design and arrangement should take this into account. Packaging boxes, if not properly designed and appropriately stacked can limit the uniformity and sufficiency of the cold air distribution.

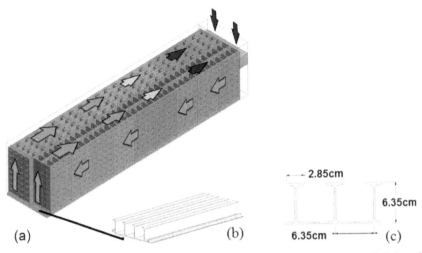

Fig. 5. Schematics showing integrated reefer container (a) arrows indicate airflow, magnified view of the T-shaped floor (b) and dimensions of the T-shaped structure (c).

3. Computational Fluid Dynamic (CFD) Analysis of Ventilated Packaging Box

Application of CFD in the analysis of packaging box aims to understand the airflow through and the accompanying heat exchange processes of stacked produce. The time and space resolutions of airflow and temperature fields delivered by CFD models allows detailed visualization and quantification of airflow and heat transfer processes inside refrigeration systems to perform a detailed and accurate analysis of the thermal performances of the system (Ambaw et al., 2017, 2016; Defraeye et al., 2014; Delele et al., 2013).

4. Modelling Geometries of Postharvest System for Numerical Analysis

The geometry of the postharvest system to be analysed should first be digitized to a format appropriate for the CFD software. Correctly generated geometry is a compulsory to get accurate and realistic CFD model. Three-dimensional (3D) image generation and acquisition methods are rapidly improving and becoming more affordable that it is now possible to develop detailed and accurate geometry of stack of produce with non-spherical shape like pears (Gruyters et al., 2018). Incorporating packaging parts liketrays, bags, polylines in the CFD modelling was impossible until recently, now it is possible to incorporate these accessories and study their influence on the thermal performances of produce cooling processes in the postharvest (Ambaw et al., 2017). The shape and size of individual fruit, specifics of fruit arranged in a box, vent-holes of boxes, etc., are realistically incorporated in CFD models now.

Fruit may be directly dumped into a packaging box without any arrangement or pattern. In this case, fruit are randomly distributed in space inside the box. Describing the geometry of randomly loaded fruit in a box is accomplished using a discrete element model (DEM) for nearly spherical fruit (Delele et al., 2008) and more recently for non-spherical fruit like pear (Gruyters et al., 2018). The DE method is a numerical technique for solving Newton's equations of motion of an assembly of interacting particles. Forces accounted for are the gravity and contact forces due to collision with other spheres or walls. Contact forces are described by a contact force model (Tijskens et al., 2003). This is a useful tool to obtain the three-dimensional array of spheres of different sizes that can be used to develop the model geometry for the CFD model. When the fruit is stacked orderly inside the packaging box, the geometry can be easily developed manually using graphic tool of the CFD software package.

The geometry of the treatment system (precooling, cold storage, reefer container or display cabinet) determines the boundary and shape of the fluid domain of concern (Fig. 6). Interfaces between the solid (produce and packing materials) and the fluid (air) phases are generated using Boolean operator of the CAD software. A single fluid domain, with the solid domain subtracted out and removed is sufficient and computationally reduced alternative to analyse airflow distributions. However, if the heat, mass and other material exchanges between the produce and the overhead fluid is to be modelled, then, the solid domain representing the stacked produce should be retained.

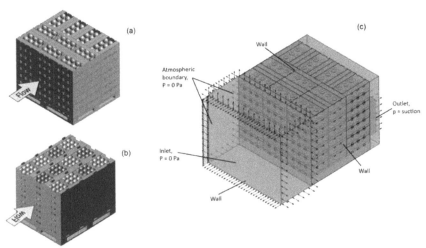

Fig. 6. Schematics showing CFD model geometries. Stack of pomegranate fruit in two different package designs: CT1 stack (a), CT2 stack (b) and illustration of the computational domain and boundary conditions of the model for the airflow distribution (c) (Source: Ambaw et al. 2017).

5. CFD Model Equations

The mathematical equations used in the CFD model are the fundamental laws of conservation of mass, momentum and energy. The continuity (Eqn. 1) and Reynolds-averaged Navier–Stokes equation (or RANS equation) (Eqn. 2) are the basic formulations that govern the motion of incompressible fluid.

$$\nabla \cdot \mathbf{U} = 0 \tag{1}$$

$$\frac{\partial \mathbf{U}}{\partial t} + \nabla \cdot \left(\mathbf{U} \otimes \mathbf{U} \right) - \nabla \cdot \left(\left(\frac{\mu + \mu_t}{\rho_a} \right) \nabla \mathbf{U} \right) - S_U + \frac{1}{\rho_a} \nabla p = 0 \tag{2}$$

where $\mathbf{U}$ is the vector of the velocity (m s^{-1}), ρ_a is the density of air (kg m^{-3}), t is time (s), μ is the dynamic viscosity of air (kg m^{-1}s^{-1}), μ_t is the turbulent eddy viscosity (kg m^{-1}s^{-1}), p is pressure (Pa) and S_U (m s^{-2}) is the momentum source term. Once the velocity *field* is calculated other quantities of interest, such as *temperature* and substance concentrations in air are computed by coupling additional equations (Eqn. 3) and (Eqn. 4) with the RANS equation.

$$\rho_a C_{pa} \left(\frac{\partial T_a}{\partial t} + \mathbf{U} \cdot \nabla T_a \right) - \nabla \cdot \left(\left(k_a + k_t \right) \nabla T_a \right) - Q = 0 \tag{3}$$

C_{pa} (J kg^{-1} K^{-1}) is the heat capacity of air, T_a (K) is the air temperature, k_a (W m^{-1}K^{-1}) is the thermal conductivity of air, k_t (W m^{-1} K^{-1}) is the turbulent thermal conductivity. The turbulent thermal conductivity (k_t) is a function of the turbulent eddy viscosity (μ_t), the heat capacity (C_{pa}) and Prandtl number of the air. The turbulent eddy viscosity is calculated by the turbulence model. Turbulence modelling is a key issue in CFD simulations that a modeller should be able to select the appropriate model among the several different formulations. The selection is performed by comparing the error of prediction, speed and stability of computation of candidate models. The most widely used turbulence models are the k–ε and k–ω eddy viscosity models. Another very popular turbulence model which has been proven very successful in different applications is the Shear Stress Transport k–ω (SST k–ω) model, which combines the best aspects of the k–ε and k–ω models. Most frequently, the SST k–ω turbulence model is used in modelling flow in postharvest cold handling systems (Delele et al., 2008; Defraeye et al., 2013). Modelling additional quantities like gases and moisture distributions needs Eqn. (4) to be coupled to the basic fluid flow equations.

$$\rho_a \frac{\partial G}{\partial t} + \nabla \cdot \left(GU - \left(D_a + D_t \right) \nabla G \right) = m \tag{4}$$

The G in Eqn. 4 is the mass concentration of the transport substance in air with coefficient of diffusivity, D_a while D_t is the turbulent diffusion coefficient and m is the rate of generation of the substance in the system. The turbulent diffusion coefficient, through the empirical turbulent Schmidt number, is a function of the turbulent viscosity ($D_t = \mu_t / Sc_t$).

The temperature and gaseous substance quantities in the internal tissue of the produce may sometimes be required for analysis (Ambaw et al., 2013a). In this case, the air phase and produce phase transport processes must be coupled at the fluid-solid interfaces as given by Eqn. (5) and (6).

$$\left(\rho_a C_{pa} \right) \left(\frac{\partial T_a}{\partial t} + U \cdot \nabla T_a \right) = \nabla \cdot \left(\left(k_a + k_t \right) \nabla T_a \right) + h_{pa} \left(T_p - T_a \right) \tag{5}$$

$$\left(\rho_p C_{pp} \right) \frac{\partial T_p}{\partial t} = \nabla \cdot \left(k_p \nabla T_p \right) + h_{pa} \left(T_a - T_p \right) + Q_r - Q_v \tag{6}$$

where $h_{pa}(T_p - T_a)$ is the heat exchange across the interface between the produce and the cool store atmosphere, Q_r is the respiration heat generation and Q_v is the heat loss due to evaporation of water from the surface of the produce.

Another interesting advance is the coupling of fruit quality models with the CFD model equations. This enables to simulate quality attributes such as taste, firmness, color and flavour during cold storage and refrigerated transportation of fruit. Hence, it is becoming conceivable to simultaneously evaluate the effect of storage conditions like cooling air flow, temperature and humidity on the quality change of stored produce. This will further make possible to identify critical operational requirements during postharvest storage and to improve the decision-making. In this aspect, the work of Wu and Defraeye (2017) can be mentioned. The authors incorporated a generic quality model (Eqn. 7) to the basic CFD model of airflow and heat transfer inside ventilated cartons for different cold chain scenarios and modelled the quality evolution of individual fruit.

$$-\frac{dA}{dt} = \gamma A^n \tag{7}$$

where t is the time (s), γ is the rate constant [s^{-1}], n is the order of the reaction. Experimentally obtained time history data of quality variables such as firmness and peel colour can be modelled using Eqn. (8) to determine the value of γ and n.

γ is normally temperature-dependent, which is often described by an Arrhenius relationship, Eqn. (8).

$$\gamma(T_p) = \gamma_0 e^{\frac{-E_a}{RT_p}} \tag{8}$$

where γ_0 is a constant (d^{-1}), E_a is the activation energy (J mol^{-1}), R is the ideal gas constant (8.314 J mol^{-1} K^{-1}), T is the absolute temperature (K). The constants γ_0 and E_a can be inferred from quality decay data.

6. Discrete and Porous Medium CFD Modelling Approaches

Discrete CFD model includes the detailed geometry of the various parts of the system: individual fruit, packaging box and its vent hole designs, the package arrangement and orientation and the configuration of the cooling system (Fig. 7). A discrete CFD model is used to determine the pressure drop and heat transfer characteristics of a packed produce. Such model is invaluable to assess vent hole design in a more detailed style. The air phase and produce phase transport processes can be incorporated explicitly to perform a realistic and detailed analysis (Ambaw et al., 2013; Berry et al., 2017; Delele et al., 2013; Defraeye et al., 2013). Most of the literature reviewed considers spherical fruits, while only a few consider other particle shapes. In the case of apples and oranges, individual fruit are modelled as a perfect sphere, which simplify the geometry modelling and reduce computational time compared to taking the actual fruit geometries without incurring significant error.

Detailed geometries of the cooling system are only possible for simple systems. However, for large and complex systems like a fully loaded cold storage rooms or reefer container, inclusion of all the geometries would make the model computationally expensive and time consuming (Ambaw et al., 2016; Getahun et al., 2017a). This is avoided by applying the method of porous media approach. In this approach, a packed fruit (Fig. 8a and d) can thus be considered a porous media (Fig. 8b and e). Experimentally obtained pressure drop and heat transfer data for a representative unit

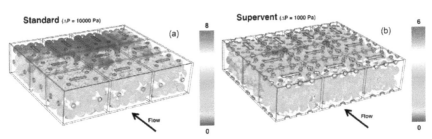

Fig. 7. Contour plot of the convective heat transfer coefficient across the citrus fruit surfaces. Values indicate an adjusted form of the heat transfer rate and emphasis regions with high or low cooling rates (Source: Defraeye et al., 2014).

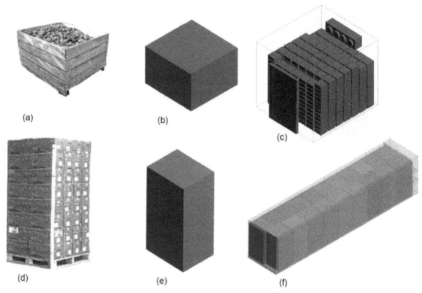

Fig. 8. Schematics showing the porous medium CFD modelling of cold storage system (Top) and reefer container (bottom).

of the full-scale system is used to describe the volume averaged transport equations to model the cold storage (Fig. 8c) and refrigerated container systems (Fig. 8f).

7. CFD Model Validation

The CFD model must first be experimentally validated before using it in any decision-making design steps. The main objective of the validation is to prove the accuracy of the model so that it is used with acceptable levels of uncertainty and error. The level of accuracy required from a CFD analysis depends on the desired use of the results. For qualitative information like the profile and behaviour of a flow field, accuracy requirements are low. On the other hand, absolute quantities like local magnitude of flow velocity, temperature and other transport variables require high level of accuracy. Validation of model predicted absolute quantities require the quantification of the prediction errors. This can be accomplished by comparing the CFD solution with experimental data or against a highly accurate numerical solution.

In the postharvest, CFD models are compared against velocity and temperature measurement data taken from different spatial locations inside the cooling system. Getahun et al. (2017b) measured the local temperature and airflow velocity on sampling points inside a reefer container fully loaded apple fruit. In addition to the property of the cool store atmosphere, time history of fruit core temperature was used to validate the predicted produce cooling rate and temperature distribution during precooling of pomegranate fruit (Fig. 9).

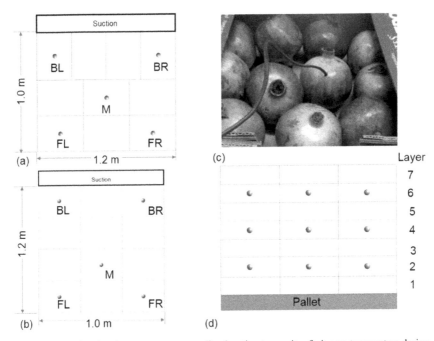

Fig. 9. Schematics showing temperature sampling location to monitor fruit core temperature during pomegranate fruit precooling process (Source: Mukama et al. 2019).

8. Advances in Application of CFD in Ventilated Box Design and Analysis

8.1 CFD Investigation of Airflow and Thermal Processes in the Postharvest

A precooling operation is modelled as a wind tunnel, with the produce to be cooled placed between the inlet and outlet of the tunnel. The air driving fan or blower is modelled at the inlet boundary by specifying temperature, velocity, pressure and turbulent intensity.

In a typical analysis, packaging box designs with a range of vent hole sizes, shapes and positions were evaluated with respect to cooling rate, cooling uniformity and pressure drop (Delele et al., 2013). Clearly, the higher the vent area the better is the cooling rate and cooling uniformity. However, beyond a certain limit, increasing the venthole proportion has no significant benefits on the cooling rate. This limiting value was reported to be 7% (Delele et al., 2013).

In another study, the cooling characteristics of existing package designs and package accessories like liners, were analysed and characterised through a detailed quantification and visualisation of the airflow, pressure drop and temperature distributions across the stack (Ambaw et al., 2017; Mukama et al., 2019). Here, the authors demonstrated that liners had strong effect on cooling rates and delayed cooling time by factors of three compared to stacks with no liners.

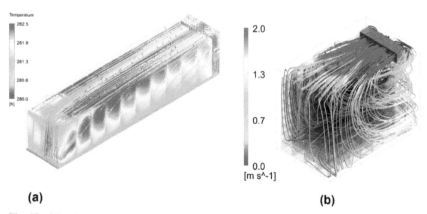

Fig. 10. CFD simulated instantaneous airflow and produce temperature profiles. (a) inside a reefer (Ambaw et al. (unpublished)); (b) inside commercial apple cold storage room (Source: Ambaw et al., 2014).

The modelling approach to study the effect of packaging boxes on the airflow and heat transfers inside larger systems like cold storage room and reefer container is relatively complex (Figs. 10 and 11) compared to modelling a precooling system. These systems are closed systems that the cooling air recirculates inside, extracting heat from the stacked produce and eliminating it to the internally placed refrigeration unit. Inside the cold store the cooling airflow direction is mainly horizontal while in reefer containers airflow is mainly vertical (from bottom to top of the stacked produce). Hence, package designed should take this into account so that enough airflow is attained.

8.2 Packaging Material on Gas Distribution Inside Cold Storage System

Gases like ethylene and 1-methylcyclopropne (1-MCP) are used to modify the ripening rate of stored produce. The challenge here is to uniformly distribute these gases inside the cold store room. This requires incorporating the diffusion, adsorption and reaction kinetics of the gaseous materials into the CFD model. This is accomplished by describing the adsorption and reaction kinetics of the gases as user defined functions and coupling this function with the airflow and heat transfer equations of the CFD model.

Ambaw et al. (2014) quantified the effect of bin and packaging box martial of constructions on the 1-methylcyclopropene (1-MCP) application processes. The developed CFD model simulates the time dependent profiles of 1-MCP concentrations in the air, in fruit and in non-target solid materials (packaging materials). In this work the authors showed that wooden materials taking considerable amount of the 1-MCP from the treatment atmosphere.

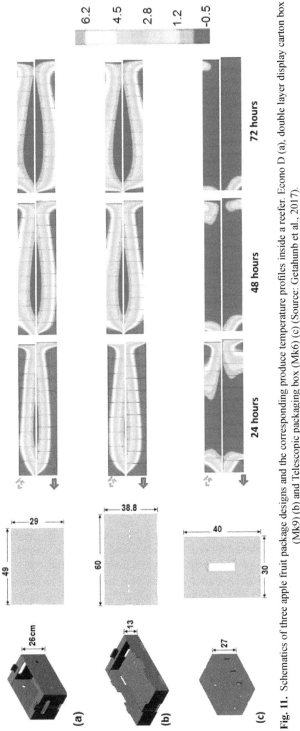

Fig. 11. Schematics of three apple fruit package designs and the corresponding produce temperature profiles inside a reefer. Econo D (a), double layer display carton box (Mk9) (b) and Telescopic packaging box (Mk6) (c) (Source: Getahunb et al., 2017).

8.3 The Impact of Moisture on Packaging Material

Most frequently, high humidity is required to reduce moisture loss and keep produce fresh during storage and transport. On the other hand, high humidity in the storage atmosphere reduces the mechanical strength of cardboard based packaging materials as a result of mechano-sorptive creep. Hence, it is required to know the moisture distributions in the treatment atmosphere and moisture content values of the stacked cartons. This entails the CFD modelling to consider the respiration and transpiration processes of the produce and the heat loss/generation due to evaporation/condensation of water at the surface of the produce and the kinetics of the diffusion-adsorption-desorption of moisture in the cardboard packaging material. Hence, a realistic model of moisture distribution inside fruit storage system is so complex that it is only recently that researchers have started using CFD to address the problem.

Berry (2019) studied the dynamics of moisture transport inside reefer container stacked with apple fruit in corrugated cardboard cartons. Due to the complexity of the system the authors were only able to consider a portion of a fully loaded reefer container. The authors showed relatively low moisture content gradients in the corrugated cardboard cartons under optimal shipping conditions.

8.4 Effect of Packaging Design on the Energy Performance of a Cooling Process

The energy usages of precooling, cold storage and refrigerated transport of fresh produce are influenced by the design and arrangement of the package and the operation of the cooling system (efficiency of individual components, temperature set-point, air driving devices, etc.). Energy savings of 30–40% were achievable by optimising usage of the stores, repairing current equipment and by retrofitting of energy efficient equipment (Foster et al., 2013).

9. Computational Solid Dynamics (CSD) Analysis of Packaging Box

9.1 Structure of Corrugated Paperboard/Cartons

Corrugated paperboard is an important application of paper and paperboard and is considered as the best choice for manufacturing cartons commonly used for handling fresh horticultural produce such as fruit and vegetables (Biancolini, 2005). Corrugated paperboard is a sandwich structure consisting of a flute-shaped or sine wave shaped corrugated medium and two outside paper sheets often referred to as linerboards (Fig. 12). The structure of a corrugated paperboard is an adaptation of the engineering beam principle of two flat, load-bearing panels (linerboards) separated by a structure (corrugated medium). The corrugated medium helps to provide the shear stiffness while the linerboards provide the bending stiffness (Fadiji et al., 2018a).

Depending on the study, corrugated paperboard is referred to as a sandwich structure (Nordstrand, 1995), as a single structure (Beldie et al., 2001), or as a monolithic material (Aboura et al., 2004). Nonetheless, despite the varying approach, the mechanical behaviour of corrugated paperboard is a function of the component

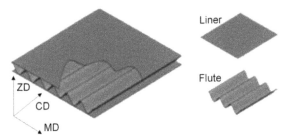

Fig. 12. Basic geometry of a corrugated paperboard (Source: Fadiji et al., 2018a). MD represents machine direction while CD and ZD represent cross direction and thickness direction, respectively.

(linerboard and corrugated medium) properties (Fadiji et al., 2018a). As indicated in Fig. 12, corrugated paperboard is characterised by three directions: MD, the machine direction; CD, the cross direction and ZD, the thickness direction. MD and CD correspond to the direction of manufacturing of the corrugated paperboard and transverse direction, respectively. These directions are often referred to as the in-plane directions. The third direction, ZD is referred to as the out-of-plane direction which corresponds to the thickness direction of the board. The mechanical response of corrugated paperboard is dominated by the stress components in the in-plane direction, i.e., the MD and CD directions while the response in the thickness direction (ZD) is insignificant. This is due to the paper dimension which is smaller in the thickness direction compared to the other two directions (Phongphinittana and Jearanaisilawong, 2013; Jiménez-Caballero et al., 2009).

The packaging requirement influences the number of layers in a corrugated paperboard, although, approximately 80% of corrugated paperboard is manufactured as single-wall board (Niskanen, 2012). However, for more demanding packaging solutions, the double- or triple-wall corrugated paperboard are utilised. The single-wall board consists of three layers, that is, two linerboards and a corrugated medium while the double- and triple-wall board consist of five and seven layers, respectively (Twede and Selke, 2005). Several types of adhesives such as water-soluble, starch based are used for binding the linerboards and the corrugated medium. However, in applications where there is high interaction with moisture that weakens the board structure and causes damage, there is an increasing potential in coating the paper with wax or combining the board with water resistant adhesive.

Commonly, corrugated paperboard is manufactured with different flute standard profiles depending on the end-use (Fadiji et al., 2018a). These are designated by letters A, B, C, E or F (Fadiji et al., 2018a; Hägglund and Carlsson, 2012). These letters are related to the order in which the flutes were invented and not necessarily to the relative sizes. The characteristics of the different flute profiles are shown in Table 1. The most commonly used flutes are C and B flute. Generally, larger flutes offer greater vertical compression strength and cushioning while smaller flutes offer printability advantages as well as structural advantages for retail packaging (Kaushal et al., 2015).

Corrugated paperboard packaging allows for different possible combination of board types, flute types, adhesive types, coatings and treatment. It also allows for varying custom design depending on the end-user's requirements. Corrugated paperboard has been adopted widely for manufacturing paper cartons due to its low

Table 1. Characteristics of different flute profiles (Fadiji et al., 2018a; Budimir et al., 2012).

Flute types	Properties	Flute height (mm)	Take-up factor	Wavelength (mm)	Flute/m length of the corrugated board web
A	First standard board style. The largest flute, seldom used at present.	4.8	1.50–1.55	8.3–10	110
B	Most widely specified profile, difficult to crush, good compactness and high compression strength.	2.4	1.30–1.35	6.1–6.9	150
C	Larger than B flute, higher compression strength but edges can be crushed easily.	3.6	1.40–1.45	7.1–7.3	130
E	Smaller corrugations that B, with excellent flat crush resistance.	1.2	1.15–1.25	3.2–3.6	290
F	Known as the microflutes (very small corrugations), with excellent flat crush resistance and rigidity.	0.5–0.8	1.15–1.25	2.3–2.5	400–550

cost, being lightweight, low weight-to-strength ratio, high stiffness-to-weight ratio and environmentally acceptable attributes (Fadiji et al., 2018a,b, 2016; Giampieri et al., 2011). Other advantages include: can be safely and sustainably manufactured from cellulose-based products; is completely recyclable and biodegradable; and can dampen mechanical impacts and vibration, which are sources of damage on fruit. In addition, the structure of corrugated paperboard gives the cartons the ability to resist buckling and a high stacking strength. The space in the corrugated medium helps to facilitate air movement and serves as thermal insulator, consequently protecting the cartons against fluctuating during transportation, handling and storage. This attribute makes it an ideal choice for packaging, particularly for fresh fruit (Sek et al., 2005). For instance, in the survey by Berry et al. (2015), corrugated paperboard cartons are used mainly for exporting pome fruit (apple and pears) in South Africa. In USA, more than 90% of the packaging utilised in the food industry are manufactured using corrugated paperboard (Little and Holmes, 2000).

Given the competitive nature of the fresh produce export market, corrugated paperboard cartons are a predominant packaging mode, as they are often more cost-effective than other materials. In addition, vent holes are placed on the cartons to enhance rapid and efficient cooling with a minimal amount of internal packaging material (Thompson et al., 2010; De Castro et al., 2005). These are commonly referred to as ventilated corrugated paperboard (VCP) packaging. The design of the vent holes should be such that it can remove the heat build-up due to respiration of the produce inside the carton and provide sufficient and uniform airflow distribution (Pathare et al., 2012). However, a challenge of using corrugated paperboard is ensuring cartons maintain their mechanical strength under cold chain conditions and over extended durations of handling and storage. Factors that can negatively influence carton integrity include high humidity environments, large compression forces (due to pallet stacking) and the configuration of vent holes (Pathare and Opara, 2014). Packaging designers thus need to maintain a strict balance between minimizing cost (materials) and packaging performance (mechanical strength and cooling).

9.2 Types and Combinations of Mechanical Loads on Packaging Boxes

Packages go through various stages in the cold chain during distribution and are thus required to endure different types and combinations of mechanical loads. The level and severity of mechanical damage that occurs in packed produce depends on the energy inputs to the package during transport, storage, handling, and the way in which the energy is dissipated within the package. Experimental measurements and mathematical models are used to analyse and compare the cushioning and damping properties of packaging during design. Research in the structural integrity of packaging carton aimed to increase the understanding of the cushioning, damping properties, impact, compression and vibration strength of ventilated packages. This problem is basically in the field of solid mechanics that studies the behaviour for solid materials, especially their motion and deformation under the action of forces, temperature changes, humidity changes and other external or internal agents. Here, computational solid dynamics (CSD) based on finite element analysis (FEA), usually validated with experimental tests are commonly used to create models to study response of the packages to mechanical loads (Fadiji et al., 2018a,b).

Generally, the loads the package and packed produce are subjected to in supply chain journeys can be categorised as static or dynamic (Singh et al., 2009). During postharvest handling, stacking packages on one another results in an internal pressure of the package within a stack, and in combination with the pressure from the environment lead to static loads. The dynamic loads experienced by the package come from shock and vibration during transportation (Guo et al., 2011; Singh et al., 2009). The static and dynamic loads include drops, impacts, compression and vibration. These loads can cause damage to the package and packed produce either singly or in combination (Opara and Fadiji, 2018; Pathare and Opara, 2014).

9.2.1 Compression Load

During postharvest handling, external forces/loads applied to the sides, faces or corners of the package results in compression damage (Frank, 2014). Compression could either be static or dynamic compression (Pathare and Opara, 2014). Static compression is referred to the resistance of the package to the applied force when stacked vertically for a stipulated time during which the force applied is not changing. Dynamic compression is often as a result of a moving force being pressed on the package and this form of compression may result from vibration and shocks in transportation by load amplification when the packages vibrate at the critical resonant frequencies (Fadiji et al., 2018a).

Inadequate packaging performance may result from: over packing the packages, too high stacking of the packages, collapse of stacked packages during transportation, material handling equipment, shocks and vibration during transportation all generate compression forces that subsequently cause damage to the package and packed produce (Kitthawee et al., 2011). The design of the package influences the effect of compression load, for example for horticultural fresh produce, packages are usually provided with ventilation openings and its structural integrity is a function of the size, location and shape of the ventilation openings (Fadiji et al., 2019, 2016; Pathare et al., 2012).

Appropriate and good packaging offers vital protection against compression loads. The use of strong packages able to withstand multiple stacking can reduce this damage. The packaging should also be shallow enough so as to prevent the bottom layer of the produce from being damaged due to the weight of the top layers.

9.2.2 Impact Load

This occurs during handling, storage and transportation as a result of impacts from forklifts, racks, throwing or dropping of the packages, sudden stopping and accelerating of the vehicle, and shock during transport. Impact load can result in bursting of the package and bruising or crushing of the packed produce. This may occur usually at each stage of handling and is difficult to eliminate (Opara and Pathare, 2014; Gołacki et al., 2009). Depending on the produce, some level of shock protection to prevent damage is required during transportation and handling. Rigid packages with proper cushioning can reduce the damage caused by impact load.

It is paramount to estimate the potential height a package may experience, its resistance to shock damage due to free fall and the fragility of the produce, as this is necessary because the package can fall onto the floor causing damage during transportation and storage (Fadiji et al., 2016a). To measure the resistance of the package to impact load, the drop testing method is used. In the test, the product of the weight of the packed produce and the drop height determines the potential energy of the package (Fadiji et al., 2018a; Poustis, 2005). The drop height is defined as the vertical distance the package under influence of gravity from the point of release onto the impact surface. Drop testing is performed for several reasons: (a) to design impact-tolerant, or rugged, portable products, (b) to replicate the abuse that might occur during manufacturing, shipping and installation and (c) for accelerated life testing (Pathare and Opara, 2014; Goyal and Buratynski, 2000).

9.2.3 Vibration Load

Packages experience vibration load during transportation in transport vehicles such as trucks (especially with bad shock absorbers), planes or ships and also on nearly everything that moves such as conveyors and forklifts. Weak packages with inadequate cushioning, bad or rough roads and transmission vibration also result in vibration damage. During transportation, fruit incur vibration damage when the fruit rub against each other or with the package (Thompson et al., 2008; Acican et al., 2007; Berardinelli et al., 2005; Kader, 2002). Collapse of packages and damage to the produce are the effects of vibration load. Filling the products in the package tightly can reduce vibration of the produce within the package and thus reduce the damage, but it also ensures that fruit does not rub against each other or are forced together. The use of cellular trays, cushioning pads and individual fruit wraps can prevent fruit from rubbing against one another. Proper cushioning can absorb and reduce the adverse effects of vibration on the products.

During transportation, vibration damage occurs when the acceleration experienced by the package and produce is more than the acceleration of gravity (9.8 m s^{-2}). In the study by Fadiji et al. (2016b), peak acceleration increases from the bottom of the carton to the top, hence the top of a stacked package experienced more acceleration impact consequently resulting in more damage to the packed produce. Several factors such as the road roughness, travelling speed, number and load of axles, truck suspension affect vibration during transportation of packed produce (Vursavuş and Özgüven, 2004).

9.3 Basic Concept of Finite Element Analysis (FEA)

Finite element analysis (FEA) is a powerful and prevalent numerical technique that has been developed into an indispensable modern tool for the modelling and simulation of various engineering processes, particularly in food packaging industries (Srirekha and Bashetty, 2010). Similar to other numerical techniques like computational fluid dynamics (CFD), FEA in recent times has become a powerful and prevalent tool in many industries, with the solutions representing a rich tapestry of mathematical physics, numerical methods, user interfaces and futuristic visualisation techniques (Norton and Sun, 2006; Xia and Sun, 2002). FEA is often used as an alternative to various time consuming and expensive experimental tests (Delele et al., 2010)

Generally, the concept of FEA involves piecewise polynomial interpolation. Over the entire structure, the field quantity becomes interpolated in a piecewise form and a set of simultaneous algebraic equation is generated at the nodes which is associated with the elements Eqn. (10).

$$[K]_e\{U\}_e = \{F\}_e \tag{10}$$

where $[K]_e$ is the elementary stiffness matrix which is dependent and determined by the geometry, element and material properties, $\{U\}_e$ is the elementary displacement vector which defines the nodes motion under loading and $\{F\}_e$ is the elementary force vector which defines the applied force on the element. The functions of all the elements are then assembled to form the global matrix equation, i.e., the governing algebraic equations that define and represent the entire structure under study (Cook et al., 2002).

In performing FEA, great accuracy is taken with regards to the actual geometry, boundary conditions and material properties applied. There are three stages involved in FEA and these include pre-processing, analysis (solver), and post-processing (Srirekha and Bashetty, 2010).

9.3.1 Pre-processing

This stage involves modelling of the structure for the analysis using the computer aided design (CAD) program that either comes with the software or provided by another software supplier. The structure is divided into a number of discrete sub-regions known as "elements", connected at discrete points known as "nodes". The structure represented by nodes and element is called the "mesh". The elements not only represent subdivisions of the structure, but also the mechanical properties and behaviour of the structure. In the structure, complex regions such as curves, or regions of interest requires a higher number of elements so as to accurately represent the shape of the geometry, whereas regions with simple geometry can be represented by fewer elements. This stage of the process is very important because the reliability of the solution depends on the size of these elements (Cook et al., 2002). In addition, the constraints, loads, boundary condition and the material properties of the structure are defined in this stage.

9.3.2 Analysis

In this stage, the parameters such as the geometry, constraints, load, and mechanical properties of the structure are used as input to the finite element code to generate matrix equations for each element. Here, a system of linear and nonlinear algebraic equations is solved until convergence is achieved. Nodal results such as displacement values at different nodes, temperature values at different nodes in a heat transfer problem or velocity values in a fluid dynamics analysis are also obtained at this stage (Fadiji et al., 2018b; Cook et al., 2002).

9.3.3 Post-processing

This is the final stage of the process where the data generated from the analysis stage are evaluated and used to create visual representations (either 2D or 3D) such as the structural deflections, stress components or any other animations that enhance the understanding of the problem being analysed. Furthermore, post-processing can also give information on the instantaneous value of all variables at certain positions in the structure. More importantly, the post-processing task is essential for a comprehensive evaluation of the simulation from the point of view of accuracy, authenticity, and satisfaction (Zhao et al., 2016).

10. Application of Computational Solid Dynamics in Box Design and Analysis

10.1 Fruit Impact Damage

Properties such as friction and rolling resistance coefficients, and rupture energy, rupture force, compressibility, and loading slope of fruits are necessary for the proper design and fabrication of postharvest handling equipment including packaging box (Li et al., 2017; Dintwa et al., 2011). Agricultural produce is susceptible to mechanical damage during harvesting, packaging, handling and transportation, which may result in substantial quality reduction (Li and Thomas, 2014). It is of utmost importance to have adequate knowledge of the mechanical characteristics of horticultural produce (fruit and vegetables) to enhance the design and development of farm machinery and packaging.

FEA has been proven as a useful tool in understanding the mechanical damage to fruit (Li et al., 2017), and this will limit reliance on the experimental tests which in terms of cost and time are inefficient (Delele et al., 2010; Salarikia et al., 2017). Chen and De Baerdemaeker, 1993 used FEA to investigate the dynamic behaviour of the pineapple. The spherical resonant frequency was used as a measure for quality evaluation since it is insensitive to skin quality and mainly reflects flesh firmness in terms of modulus of elasticity. The study was able to relate spherical resonant frequency to the refractometric sugar content and the surface chlorophyll content. In another study by Dewulf et al. (1999), the authors evaluated the variation of the dynamic characteristics of a pear due to changes in material characteristics using FEA.

FEA has been used by several authors to study the influence of impact/drop load on different fruit such as tomato (Kabas et al., 2008), apple (Celik et al., 2011), pear (Salarikia et al., 2017; Yousefi et al., 2016), and cantaloupe (Seyedabadi et al., 2015). Kabas et al., 2008 estimated the deformation of tomato under drop load by simplifying the tomato into a spherical solid and assuming a single material using FEA. The deformation behaviour of apple under drop was investigated by Celik et al. (2011). The simulation results were validated with visual investigations using a 3D scanner and high-speed camera, with good agreement. In the study by Ahmadi et al. (2016), FEA was used to study the dynamic impact to apple by applying impact test to collision of the fruit with a flat rigid plate and the collision of an apple with another apple. Results showed that displacement, velocity, acceleration, stress and hydrostatic pressure in the collision of apple to apple were less than that of apple to rigid plate at the velocity of 1 m s^{-1}. Salarikia et al. (2017) used FEA to assess the stress and strain distribution patterns within pear generated by collision of the fruit with a flat surface made of different materials. Simulations were done at two different drop orientations and four different impact surfaces. The largest and smallest stresses, strains and contact forces were developed in collision with the steel and rubber surfaces, respectively in both drop orientations.

Sadrnia et al. (2008) studied the internal bruising of watermelon compressed in both longitudinal and transverse directions using nonlinear FEA. The applied force used the model in compressing the watermelon was 10% of the breaking force. Results showed that the failure stress, failure strain, and modulus of elasticity of red flesh are much lower than those of the rind. In addition, maximum stresses in the models showed that breaking force in the longitudinal compression is less than the breaking force in transverse compression. A finite element model was developed by Dintwa et al. (2011) to understand the deformation behaviour of tomato cells under compression. The model assumed the cell to be a thin liquid sphere permeable wall and used a linear elastic material behaviour for the cell wall. The model successfully reproduced the force–deformation behaviour of a single cell in compression, including its deformed shape.

The results of the finite element model developed by Li et al. (2013) in predicting the internal mechanical damage in tomato compression concluded that the internal structural properties of tomato has an apparent effect on the mechanical damage behaviour of the tissues. The model considered a multiscale structure of the tomato, that is, the exocarp, mesocarp and locular gel tissues of tomatoes. Li et al. (2017) developed a finite element model of a whole tomato fruit under compression to 10% deformation. The vertical compression resulted in a main local deformation response in the contact area and a minor equatorial structural response. In addition, when a vertical force from 0 to 6.5 N was applied, the internal damage volume increased from 0 to 6672 mm^3. The model was validated with experimental tests and was in good agreement. A recent study by Ashtiani et al. (2019) used FEA to simulate the mechanical behaviour of grapefruit under compression loading between two parallel plates. Fruit were compressed in both longitudinal and transverse directions. Results showed that maximum stress was concentrated at the point of contact between the plate and the peel; however, the pulp was more damaged than the peel. An illustration of the changes in von Mises stress distributions as a function of various deformations in the grapefruit model compressed in the longitudinal and transverse directions was presented by the authors (Ashtiani et al., 2019).

Kim et al. (2004) used FEA to analyse the vibration characteristics of the pears in corrugated paperboard container in transit. It was observed that as elastic modulus of the cushion materials of corrugated paperboard pad and internal packaging (tray cup) decreased, the first frequencies of upper and lower pears increased and the peak acceleration decreased. Simulated and measured frequencies of the pears have a relatively good agreement. The variation of the vibration characteristics of 'Huanghua' pear was investigated using FEA by Song et al. (2006). Results showed that with an increase in the Young's modulus of the pear, the eigenfrequency increased. In addition, the eigenfrequency decreased with an increase in both the pear density and volume.

10.2 Corrugated Paperboard Packaging

Finite element model simulation has been a useful tool to perform a more practical and detailed analysis of an individual corrugated paperboard package and it components (paper and paperboard) as well as stack of packages (Fadiji et al., 2018b). This method, however, due to limitation of computational time and cost, homogenisation procedure proposed by Biancolini (2005) is frequently used. This procedure involves transforming the corrugated board into an equivalent homogenous layered structure (Cheon and Kim, 2015).

Pommier and Poustis (1989) using linear elastic FEA predicted the top to bottom compression resistance of corrugated paperboard package by considering the considered shear bending and bending stiffness of the corrugated paperboard. The model results were in correlation with results with experimental results. From the stiffness of the corrugated paperboard obtained from anticlastic bending and four-point tests, Pommier et al. (1991) developed a finite element model to study the buckling of a flapless corrugated paperboard box. A quarter of the box was modelled using symmetry and shell quadratic interpolation rectangular finite elements. Model results of the vertical compression strength was compared with experimental and analytical results. FEA was used by Biancolini and Brutti (2003) to investigate the buckling of corrugated paperboard package. The model was capable of predicting the incipient buckling load of the package, and was in agreement with experimental investigations.

In the packaging of horticultural produce such as fruit and vegetables where the packages require ventilation to allow for uniform cooling of the produce, consideration must be given to the geometrical factors of the package such as vent shape, size, area and location to enhance the design of optimised packages (Fadiji et al., 2018b; Defraeye et al., 2015). The study by Han and Park (2007) investigated the principal design parameters of ventilation holes and hand holes in the facing of corrugated paperboard packages using FEA. The authors studied with respect to stress distribution and stress level, the various designs of ventilation holes. The finite element model was used to determine the appropriate pattern, location and size of the vent/hand holes. A good agreement was found between the simulation results and the actual experimental results. It was found that the most appropriate pattern and location of the ventilation holes was vertical oblong-shaped and symmetrically positioned within a

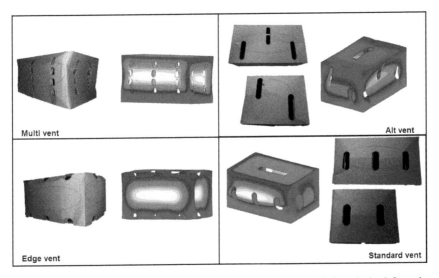

Fig. 13. A typical displacement pattern of different ventilated carton design obtained from the simulation and experimental results (Source: Fadiji et al., 2019).

certain distance to the right and left from the centre of the front and rare facing of the package. A contrast to this conclusion was reported by Jinkarn et al. (2006). According to the authors, circular vent holes have the lowest reduction in compression strength.

Fadiji et al. (2016c) developed a validated FEA model capable of predicting the compression strength of ventilated paperboard packages. The validated model was used to study the effects of vent height, shape, orientation, number and area on the compression strength of the packages. The authors reported that the compression strength of the package was significantly influenced by the package's geometrical designs and environmental conditions. The numerical results were verified with experimental measurements and both techniques were in good agreement. A recent study by Fadiji et al. (2019) used FEA to the role of geometrical configurations on the strength of four different ventilated packages under compression. An illustration of the displacement pattern is shown in Fig. 13. The demonstrated that the functionality of package vent hole design is tied strongly to the properties of the chosen board grade and the strength of the package was adversely influenced by increasing vent area. Maximum stress was also reported to be concentrated at the corners of the packages observed to be concentrated at the corners of the package, irrespective of the design.

Luong et al. (2017) used FEA to simulate the behaviour of corrugated paperboard packages under shocks. Elastoplastic homogenisation approach was used to simplify the package to reduce computational time. The model was able to predict the mechanical behaviour of corrugated cardboard packages under impact dynamics. The authors reported that the drop height of the packed produce is strongly related to the velocity change that produce would experience during transportation and handling. Numerical results were corroborated by experimental results.

11. Conclusion and Future Perspectives

The effects of packaging design on the thermal performance (airflow, cooling rate and cooling uniformity) in cold chain management of fresh produce are, by far, the most frequently studied problems by use of computer models. The computational fluid dynamics (CFD) method is commonly applied to quantify and visualise the effect of vent-holes on the produce cooling rate and cooling uniformity and the accompanying energy requirements. Studies in this aspect are relatively straight forward that most often the available CFD software packages have the required model equations to solve the airflow and heat transfer problems. Faithful accounting of geometries of the packaged produce and temperature and velocity boundary conditions are still challenging to realistically model fully loaded commercial systems like cold storage room and reefer containers. Holistic analysis of the thermal performances of ventilated packages together with the produce quality and system energy usages is interesting advance in this area.

Proper design and implementation of vented-package must consider the effect of mechanical forces on the package and on the biological tissues. Computational solid dynamics (CSD), specifically finite element method (FEM), is normally to analyse the relationship between vent-hole size, proportion, shape and location on the mechanical strength of the box as well as the performances of vented-packages in cushioning and damping of impact, compression and vibration forces. Although FEM is a reliable way to solve static or dynamic problems, there are still challenges due to the complexity of paper material and characterization of temperature and humidity effects on the material properties. Advances are towards incorporation of the stress response of biological tissues into the FEM.

CFD and CSD based package design approach enables researchers to perform abstract conceptual design and analysis of alternatives. This method is creative and novel to perform integrated multi-criteria performance analysis in a more convenient and economical way by reducing the time and cost of experimentation. Nevertheless, CFD/CSD are still requiring experimental validation before implementation and should be approached in a multidisciplinary research. Advances in computer power and computing algorithms offer new prospects toward integrated design and performance analysis of fresh produce packaging. It is now possible to incorporate important geometric and operational details into computer models for a more realistic decision support system. Advances in holistic analysis of the mechanical and thermal performances of ventilated packages together with the produce quality and system energy usages are important for the enhancement of food quality and safety. The novel approaches discussed in this chapter provide the framework for future design of ventilated packaging for optimal performance in the cold chain based on integrated multi-parameter assessment.

Acknowledgement

This work is based on the research supported wholly by the National Research Foundation of South Africa (Grant Numbers: 64813). The opinions, findings and conclusions or recommendations expressed are those of the author(s) alone, and the NRF accepts no liability whatsoever in this regard.

References

Aboura, Z., N. Talbi, S. Allaoui and M.L. Benzeggagh (2004). Elastic behavior of corrugated cardboard: Experiments and modeling, *Composite Structures*, 63: 53–62.

Acıcan, T., K. Alibaş and I.S. Özelkök (2007). Mechanical damage to apples during transport in wooden crates, *Biosystems Engineering*, 96: 239–248.

Ahmadi, E., H. Barikloo and M. Kashfi (2016). Viscoelastic finite element analysis of the dynamic behavior of apple under impact loading with regard to its different layers, *Computers and Electronics in Agriculture*, 121: 1–11.

Ambaw, A., P. Verboven, M.A. Delele, T. Defraeye, E. Tijskens, A. Schenk and B.M. Nicolai (2013a). CFD modelling of the 3D spatial and temporal distribution of 1-methylcyclopropene in a fruit storage container, *Food and Bioprocess Technology*, 6: 2235–2250.

Ambaw, A., M.A. Delele, T. Defraeye, Q.T. Ho, U.L. Opara, B.M. Nicolaï and P. Verboven (2013b). The use of CFD to characterize and design post-harvest storage facilities: Past, present and future, *Computers and Electronics in Agriculture*, 93: 184–194.

Ambaw, A., P. Verboven, M.A. Delele, T. Defraeye, E. Tijskens, A. Schenk, B.E. Verlinden, U.L. Opara and B.M. Nicolai (2014). CFD-based analysis of 1-MCP distribution in commercial cool store rooms: Porous medium model application, *Food and Bioprocess Technology*, 7: 1903–1916.

Ambaw, A., N. Bessemans, W. Gruyters, S.G. Gwanpua, A. Schenk, A. De Roeck, M.A. Delele, P. Verboven and B.M. Nicolai (2016). Analysis of the spatiotemporal temperature fluctuations inside an apple cool store in response to energy use concerns, *International Journal of Refrigeration*, 66: 156–168.

Ambaw, A., M. Mukama and U.L. Opara (2017). Analysis of the effects of package design on the rate and uniformity of cooling of stacked pomegranates: Numerical and experimental studies, *Computers and Electronics in Agriculture*, 136: 13–24.

Ashtiani, S.H.M., H. Sadrnia, H. Mohammadinezhad, M.H. Aghkhani, M. Khojastehpour and M.H. Abbaspour-Fard (2019). FEM-based simulation of the mechanical behaviour of grapefruit under compressive loading, *Scientia Horticulturae*, 245: 39–46.

Aswaney, M. (2007). Forced-air precooling of fruits and vegetables. *Air Conditioning Refrig J.*, 57–62.

Becker, B.R. and B.A. Fricke (1996). Simulation of moisture loss and heat loads in refrigerated storage of fruits and vegetables, *New Developments in Refrigeration for Food safety and Quality*, pp. 210–221.

Beldie, L., G. Sandberg and L. Sandberg (2001). Paperboard packages exposed to static loads–finite element modelling and experiments, *Packaging Technology and Science*, 14: 171–178.

Berardinelli, A., V. Donati, A. Giunchi, A. Guarnieri and L. Ragni (2005). Damage to pears caused by simulated transport, *Journal of Food Engineering*, 66: 219–226.

Berry, T.M., M.A. Delele, H. Griessel and U.L. Opara (2015). Geometric design characterization of ventilated multi-scale packaging used in the South African pome fruit industry, *Agricultural Mechanization in Asia, Africa, and Latin America*, 46: 34–42.

Berry, T.M., T.S. Fadiji, T. Defraeye and U.L. Opara (2017). The role of horticultural carton vent hole design on cooling efficiency and compression strength: A multi-parameter approach, *Postharvest Biology and Technology*, 124: 62–74.

Biancolini, M.E. and C. Brutti (2003). Numerical and experimental investigation of the strength of corrugated board packages, *Packaging Technology and Science*, 16: 47–60.

Biancolini, M.E. (2005). Evaluation of equivalent stiffness properties of corrugated board, *Composite Structures*, 69: 322–328.

Blisard, W.N., H. Stewart and D. Jolliffe (2004). Low-income households' expenditures on fruits and vegetables. Washington, DC: US Department of Agriculture, *Economic Research Service*.

Boyette, M.D. (1996). Forced-air cooling packaged blueberries, *Applied Engineering in Agriculture*, 12: 213–217.

Budimir, I., B. Lajić and S.P. Preprotić (2012). Evaluation of mechanical strength of five layer corrugated cardboard depending on waveform types, *Acta Graphica*, 23: 111–120.

Celik, H.K. and A.E. Rennie and I. Akinci (2011). Deformation behaviour simulation of an apple under drop case by finite element method, *Journal of Food Engineering*, 104: 293–298.

Chen, H. and J. De Baerdemaeker (1993). Modal analysis of the dynamic behaviour of pineapples and its relation to fruit firmness, *Transactions of the ASAE*, 36: 1439–1444.

Cheon, Y.J. and H.G. Kim (2015). An equivalent plate model for corrugated-core sandwich panels, *Journal of Mechanical Science and Technology*, 29: 1217–1223.

Cook, R., D. Malkus, M. Plesha and R. Witt (2002). Concepts and applications of finite element analysis (4th ed.). New York: Wiley.

Daniels, W., A. Diffley, D. Fiser, D. Hollbach, A. Korane, J. Koan and J. Slama (2010). Wholesale Success: A Farmer's Guide to Selling, Postharvest Handling and Packing Produce. (J. Slama, Ed.). Family Farmed.org.

De Castro, L.R., C. Vigneault and L.A.B. Cortez (2005). Effect of container openings and airflow on energy required for forced air cooling of horticultural produce, *Canadian Biosystems Engineering*, 47: 1–9.

Defraeye, T., P. Verboven and B. Nicolai (2013). CFD modelling of flow and scalar exchange of spherical food products: Turbulence and boundary-layer modelling, *Journal of Food Engineering*, 114: 495–504.

Defraeye, T., R. Lambrecht, M.A. Delele, A.A. Tsige, U.L. Opara, P. Cronjé, P. Verboven and B. Nicolai (2014). Forced-convective cooling of citrus fruit: cooling conditions and energy consumption in relation to package design, *Journal of Food Engineering*, 121: 118–127.

Defraeye, T. and P. Cronje, T. Berry, U.L. Opara, A. East, M. Hertog, P. Verboven and B. Nicolai (2015). Towards integrated performance evaluation of future packaging for fresh produce in the cold chain, *Trends in Food Science & Technology*, 44: 201–225.

Defraeye, T., B. Nicolai, W. Kirkman, S. Moore, S. van Niekerk, P. Verboven and P. Cronjé (2016). Integral performance evaluation of the fresh-produce cold chain: A case study for ambient loading of citrus in refrigerated containers, *Postharvest Biology and Technology*, 112: 1–13.

Delele, M.A.E., Y.T. Tijskens, Q.T. Atalay, H. Ho, B.M. Ramon, B.M. Nicolaï and P. Verboven (2008). Combined discrete element and CFD modelling of airflow through random stacking of horticultural products in vented boxes, *Journal of Food Engineering*, 89: 33–41.

Delele, M.A.P. Verboven, Q.T. Ho and B.M. Nicolaï (2010). Advances in mathematical modelling of postharvest refrigeration processes, *Stewart Postharvest Review*, 6: 1–8.

Delele, M.A., B. Vorstermans, P. Creemers, A.A. Tsige, E. Tijskens, A. Schenk, U.L. Opara, B.M. Nicolaï and P. Verboven (2012). Investigating the performance of thermonebulisation fungicide fogging system for loaded fruit storage room using CFD model, *Journal of Food Engineering*, 109: 87–97.

Delele, M.A., M.E.K. Ngcobo, S.T. Getahun, L. Chen, J. Mellmann and U.L. Opara (2013). Studying airflow and heat transfer characteristics of a horticultural produce packaging system using a 3-D CFD model. Part I: Model development and validation, *Postharvest Biology and Technology*, 86: 536–545.

Dewulf, W., P. Jancsok, B. Nicolaï, G. De Roeck and D. Briassoulis (1999). Determining the firmness of a pear using finite element modal analysis, *Journal of Agricultural Engineering Research*, 74: 217–224.

Dintwa, E., P. Jancsók, H.K. Mebatsion, B. Verlinden, P. Verboven, C.X. Wang, C.R. Thomas, E. Tijskens, H. Ramon and B. Nicolai (2011). A finite element model for mechanical deformation of single tomato suspension cells, *Journal of Food Engineering*, 103: 265–272.

Fadiji, T., C. Coetzee, P. Pathare and U.L. Opara (2016a). Susceptibility to impact damage of apples inside ventilated corrugated paperboard packages: Effects of package design, *Postharvest Biology and Technology*, 111: 286–296.

Fadiji, T., C. Coetzee, L. Chen, O. Chukwu and U.L. Opara (2016b). Susceptibility of apples to bruising inside ventilated corrugated paperboard packages during simulated transport damage, *Postharvest Biology and Technology*, 118: 111–119.

Fadiji, T., C. Coetzee and U.L. Opara (2016c). Compression strength of ventilated corrugated paperboard packages: Numerical modelling, experimental validation and effects of vent geometric design, *Biosystems Engineering*, 151: 231–247.

Fadiji, T., T.M. Berry, C.J. Coetzee and U.L. Opara (2018a). Mechanical design and performance testing of corrugated paperboard packaging for the postharvest handling of horticultural produce, *Biosystems Engineering*, 171: 220–244.

Fadiji, T., C.J. Coetzee, T.M. Berry, A. Ambaw and U.L. Opara (2018b). The efficacy of finite element analysis (FEA) as a design tool for food packaging: A review, *Biosystems Engineering*, 174: 20–40.

Fadiji, T., C.J. Coetzee, T.M. Berry and U.L. Opara (2019). Investigating the role of geometrical configurations of ventilated fresh produce packaging to improve the mechanical strength–Experimental and numerical approaches, *Food Packaging and Shelf Life*, 20: 100312.

Ferrua, M.J. and R.P. Singh (2009). Modeling the forced-air cooling process of fresh strawberry packages, Part I: Numerical model, *International Journal of Refrigeration*, 32: 335–348.

Foster, A., C. Zilio, L. Reinholdt, K. Fikiin, M. Scheurs, C. Bond, M. Houska, T. Van Sambeeck and J. Evans (2013). Initiatives to reduce energy use in cold stores.

Frank, B. (2014). Corrugated box compression—A literature survey, *Packaging Technology and Science*, 27: 105–128.

Getahun, S., A. Ambaw, M. Delele, C.J. Meyer and U.L. Opara (2017a). Analysis of airflow and heat transfer inside fruit packed refrigerated shipping container: Part I–Model development and validation, *Journal of Food Engineering*, 203: 58–68.

Getahun, S., A. Ambaw, M. Delele, C.J. Meyer and U.L. Opara (2017b). Analysis of airflow and heat transfer inside fruit packed refrigerated shipping container: Part II–Evaluation of apple packaging design and vertical flow resistance, *Journal of Food Engineering*, 203: 83–94.

Giampieri, A., U. Perego and R. Borsari (2011). A constitutive model for the mechanical response of the folding of creased paperboard, *International Journal of Solids and Structures*, 48: 2275–2287.

Gołacki, K., G. Bobin and Z. Stropek (2009). Bruise resistance of apples (Melrose variety), *Teka Komisji Motoryzacji i Energetyki Rolnictwa-OLPAN*, 9: 40–47.

Goyal, S. and E.K. Buratynski (2000). Methods for realistic drop-testing, *International Journal of Microcircuits and Electronic Packaging*, 23: 45–52.

Gruyters, W., P. Verboven, S. Rogge, S. Vanmaercke, H. Ramon and B. Nicolai (2017). A novel methodology to model the cooling processes of packed horticultural produce using 3D shape models. *In*: AIP Conference Proceedings (Vol. 1896, No. 1, p. 150006). AIP Publishing.

Gruyters, W., P. Verboven, E. Diels, S. Rogge, B. Smeets, H. Ramon, T. Defraeye and B.M. Nicolaï (2018). Modelling cooling of packaged fruit using 3D shape models, *Food and Bioprocess Technology*, 11: 2008–2020.

Han, J. and J.M. Park (2007). Finite element analysis of vent/hand hole designs for corrugated fibreboard boxes, *Packaging Technology and Science*, 20: 39–47.

Hägglund, R. and L. Carlsson (2012). Packaging performance. *In*: K. Niskanen (Ed.). Mechanics of Paper Products. Berlin: De Gruyter.

Jiménez-Caballero, M.A., I. Conde, B. García and E. Liarte (2009). Design of different types of corrugated board packages using finite element tools. *In*: SIMULIA Customer Conference.

Jinkarn, T., P. Boonchu and S. Bao-Ban (2006). Effect of carrying slots on the compressive strength of corrugated board panels, *Kasetsart Journal*, 40: 154–161.

Johnson, R. (2014). The US trade situation for fruit and vegetable products, *Congressional Research Service*, pp. 1–20.

Johnston, W.A. (1994). Freezing and refrigerated storage in fisheries (No. 340), *Food & Agriculture Org.*

Kabas, O., H.K. Celik, A. Ozmerzi and I. Akinci (2008). Drop test simulation of a sample tomato with finite element method, *Journal of the Science of Food and Agriculture*, 88: 1537–1541.

Kader, A.A. (2002). Postharvest technology of horticultural Crops (3rd ed., p. 353), *Davis, California: University of California Department of Agriculture and Natural Resources*.

Kaushal, M.C., V.K. Sirohiya and R.K. Rathore (2015). Corrugated board structure: A review, *International Journal of Application of Engineering and Technology*, 2: 228–234.

Kim, M.S., H.M. Jung and K.B. Kim (2004). Vibration analysis of pears in packaged freight using finite element method, *Journal of Biosystems Engineering*, 29: 501–507.

Kitthawee, U., S. Pathaveerat, T. Srirungruang and D. Slaughter (2011). Mechanical bruising of young coconut, *Biosystems Engineering*, 109: 211–219.

Li, Z., P. Li, H. Yang and J. Liu (2013). Internal mechanical damage prediction in tomato compression using multiscale finite element models, *Journal of Food Engineering*, 116: 639–647.

Li, Z. and C. Thomas (2014). Quantitative evaluation of mechanical damage to fresh fruits, *Trends in Food Science & Technology*, 35: 138–150.

Li, Z., J. Andrews and Y. Wang (2017). Mathematical modelling of mechanical damage to tomato fruits, *Postharvest Biology and Technology*, 126: 50–56.

Little, C.R. and R.J. Holmes (2000). Storage technology for apples and pears (p. 528). Knoxfield: Highway Press Pty Ltd.

Luong, V.D., F. Abbès, B. Abbès, P.M. Duong, J.B. Nolot, D. Erre and Y.Q. Guo (2017). Finite element simulation of the strength of corrugated board boxes under impact dynamics. *In*: Nguyen-Xuan H., P. Phung-Van and T. Rabczuk (eds). Proceedings of the International Conference on Advances in Computational Mechanics.

Mukama, M., A. Ambaw and U.L. Opara (2019). Thermal properties of whole and tissue parts of pomegranate (*Punica granatum*) fruit, *Journal of Food Measurement and Characterization*, 13: 901–910.

Niskanen, K. (Ed.). (2012). Mechanics of paper products, *Walter de Gruyter Incorporated*.

Nordstrand, T.M. 1995. Parametric study of the post-buckling strength of structural core sandwich panels, *Composite Structures*, 30: 441–451.

Norton, T. and D.-W. Sun (2006). Computational fluid dynamics (CFD)—an effective and efficient design and analysis tool for the food industry: A review, *Trends in Food Science & Technology*, 17: 600–620.

Opara, L. and Q. Zou (2006). Novel computational fluid dynamics (CFD) simulation software for thermal design and evaluation of horticultural packaging, *International Journal of Postharvest Technology & Innovation*, 1: 155–169.

Opara, L.U. and Q. Zou (2007). Sensitivity analysis of a CFD modelling system for airflow and heat transfer of fresh food packaging: Inlet air flow velocity and inside-package configurations, *International Journal of Food Engineering*, 3.

Opara, U.L. (2011). From hand-holes to vent-holes: What's next in innovative horticultural packaging? *Inaugural Lecture, Stellenbosch University, South Africa.*

Opara, U.L. and P.B. Pathare (2014). Bruise damage measurement and analysis of fresh horticultural produce—A review, *Postharvest Biology and Technology*, 91: 9–24.

Opara, U.L. and T. Fadiji (2018). Compression damage susceptibility of apple fruit packed inside ventilated corrugated paperboard package, *Scientia Horticulturae*, 227: 154–161.

Pathare, P.B. and U.L. Opara, C. Vigneault, M.A. Deleleand F.A.-J. Al-Said (2012). Design of packaging vents for cooling fresh horticultural produce, *Food and Bioprocess Technology*, 5: 2031–2045.

Pathare, P.B. and U.L. Opara (2014). Structural design of corrugated boxes for horticultural produce: A review, *Biosystems Engineering*, 125: 128–140.

Phongphinittana, E. and P. Jearanaisilawong (2013). A microstructurally-based orthotropic elasto-plastic model for paper and paperboard, *International Journal of Applied Science and Technology*, 5: 19–25.

Pommier, J. and J. Poustis (1989). Box stacking strength prediction: Today McKee, tomorrow? *In*: Proceedings of the 3rd Joint ASCE/ASME Mechanics Conference. CA, USA: San Diego.

Pommier, J.C. and J. Poustis, E. Fourcade and P. Morlier (1991). Determination of the critical load of a corrugated box subjected to vertical compression by finite element methods, pp. 437–447. *In*: Proceedings of the 1991 International Paper Physics Conference, Kona, HI.

Poustis, J. 2005. Corrugated fibreboard packaging. pp. 317–372. *In*: M.J. Kirwan (Ed.). Paper and Paperboard Packaging Technology. Oxford: Wiley-Blackwell.

El-Ramady, H.R. and É. Domokos-Szabolcsy, N.A. Abdalla, H.S. Taha and M. Fári (2015). Postharvest management of fruits and vegetables storage. pp. 65–152. *In*: Sustainable Agriculture Reviews. Springer, Cham.

Rico, D., A.B. Martin-Diana, J.M. Barat and C. Barry-Ryan (2007). Extending and measuring the quality of fresh-cut fruit and vegetables: A review, *Trends in Food Science & Technology,* 18: 373–386.

Sadrnia, H., A. Rajabipour, A. Jafari, A. Javadi, Y. Mostofi, J. Kafashan and J. De Baerdemaeker. 2008. Internal bruising prediction in watermelon compression using nonlinear models, *Journal of Food Engineering,* 86: 272–280.

Salarikia, A., S.H.M. Ashtiani, M.R. Golzarian and H. Mohammadinezhad (2017). Finite element analysis of the dynamic behaviour of pear under impact loading, *Information Processing in Agriculture,* 4: 64–77.

Sek, M., V. Rouillard, H. Tarash and S. Crawford (2005). Enhancement of cushioning performance with paperboard crumple inserts, *Packaging Technology and Science,* 18: 273–278.

Seyedabadi, E., M. Khojastehpour and H. Sadrnia (2015). Predicting cantaloupe bruising using non-linear finite element method, *International Journal of Food Properties,* 18: 2015–2025.

Song, H.Z., J. Wang and Y.H. Li (2006). Studies on vibration characteristics of a pear using finite element method, *Journal of Zhejiang University Science B,* 7: 491–496.

Srirekha, A. and K. Bashetty (2010). Infinite to finite: An overview of finite element analysis, *Indian Journal of Dental Research,* 21: 425–432.

Thompson, J.F., F.G. Mitchell, T.R. Rumsey, R.F. Kasmire and C.H. Crisosto (2008). Commercial cooling of fruits, vegetables, and flowers. *Oakland, California: University of California Department of Agriculture and Natural Resources.*

Thompson, J.F., D.C. Mejia and R.P. Singh (2010). Energy use of commercial forced-air coolers for fruit, *Applied Engineering in Agriculture,* 26: 919.

Tijskens, E., H. Ramon and J. De Baerdemaeker (2003). Discrete element modelling for process simulation in agriculture, *Journal of Sound and Vibration,* 266: 493–514.

Twede, D. and S.E. Selke. 2005. Cartons, crates and corrugated board: Handbook of Paper and Wood Packaging Technology. DEStech Publications, Inc.

Vursavuş, K.K. and F. Özgüven (2004). Determining the effects of vibration parameters and packaging method on mechanical damage in golden delicious apples, *Turkish Journal of Agriculture and Forestry,* 28: 311–320.

Wu, W. and T. Defraeye (2018). Identifying heterogeneities in cooling and quality evolution for a pallet of packed fresh fruit by using virtual cold chains, *Applied Thermal Engineering,* 133: 407–417.

Xia, B. and D-W. Sun (2002). Applications of computational fluid dynamics (CFD) in the food industry: A review, *Computers and Electronics in Agriculture,* 34: 5–24.

Yousefi, S., H. Farsi and K. Kheiralipour (2016). Drop test of pear fruit: Experimental measurement and finite element modelling, *Biosystems Engineering,* 147: 17–25.

Zhao, C.J., J.W. Han, X.T. Yang, J.P. Qian and B.L. Fan (2016). A review of computational fluid dynamics for forced-air cooling process, *Applied Energy,* 168: 314–331.

Zou, Q., L.U. Opara and R. McKibbin (2006a). A CFD Modeling system for airflow and heat Transfer in Ventilated Packaging during Forced-Air Cooling of Fresh Foods: I. Initial Analysis and development of mathematical models, *Journal of Food Engineering,* 77: 1037–1047.

Zou, Q. and L.U. Opara and R. McKibbin (2006b). A CFD modeling system for airflow and heat transfer in ventilated packaging during forced-air cooling of fresh foods: II. Computational solution, software development, and model testing, *Journal of Food Engineering,* 77: 1048–1058.

Index

Milton Keynes UK
Ingram Content Group UK Ltd.
UKHW022038141024
449569UK00014B/653